The Quantum Dice

God does not play dice.

Albert Einstein

God not only plays dice, He also sometimes throws the dice where they cannot be seen.

Stephen Hawking

The Quantum Dice

L I Ponomarev
I V Kurchatov Institute, Russian Scientific Center,
Moscow, Russia

Translated from the Russian by A P Repiev
with illustrations by O Levenok

Institute of Physics Publishing
Bristol and Philadelphia

British Library Cataloguing in Publication Data

A catalogue record for this book is available from the British Library

ISBN 0 7503 0241 0 pbk
ISBN 0 7503 0251 8 hbk

Library of Congress Cataloging-in-Publication Data are available

Published by Institute of Physics Publishing, wholly owned by the Institute of Physics, London

Institute of Physics Publishing, Techno House, Redcliffe Way, Bristol BS1 6NX, England
US Editorial Office: Institute of Physics Publishing, The Public Ledger Building, Suite 1035, Independence Square, Philadelphia, PA 19106, USA

Printed and bound in Great Britain at the University Press, Cambridge

Contents

2. Ideas

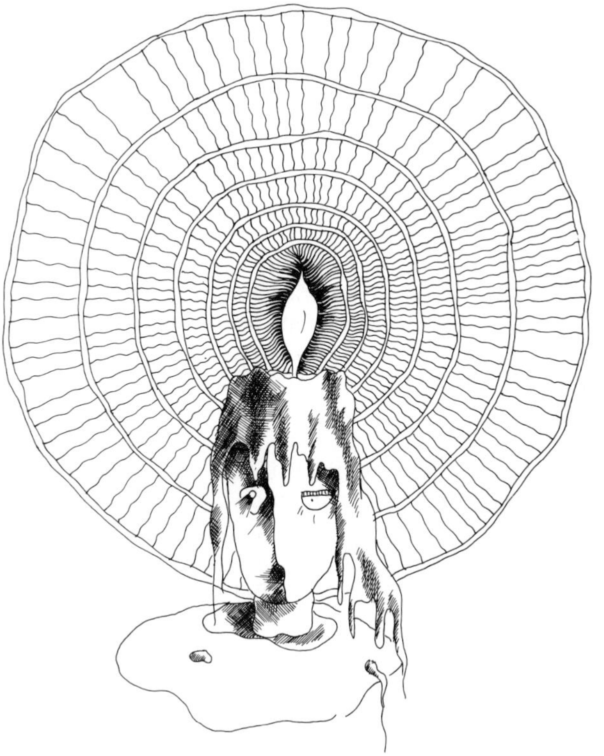

The everyday vocabulary of a peasant is not more than 200 words; he can make do with this number to survive in his day-to-day fight for existence. The words "history", "civilization", "culture", "art", "science" are not necessary for this. Yet only words like these can explain why it was man — the only one of a wealth of biological species — who became the potentate of the planet Earth.

The meaning of the words "quantum physics" is unclear to nearly all; only educated people have heard them; and only a tiny fraction of them understand their real sense.

Admittedly, even without knowing them one can build a house, grow a garden, and bring up children. But nevertheless, in almost every person there is an overmastering thirst for the transcendental, the mystical: we all remember our juvenile longings and silence of adults in the temple. It is this interest in the unknown that has given rise to all religions, art, and science. In the final analysis, it is these abstract ideas that control our existence: religion dictates morals, art shapes ethics, science forms the basis of modern everyday life.

Quantum mechanics is the most important element in the whole system of science. You can hardly gain an insight into any really profound recess of Nature without quantum physics and its theoretical heart — quantum mechanics. Without them you cannot understand the origins of sunlight and stellar light, the laws governing chemical transformations, the composition of the atom, the structure and transmutations of atomic nuclei.

Quantum mechanics was born in a short span of time between the two world wars. Ever since, it has changed the life of civilized peoples more than any previous discovery — yet even educated people know undeservedly little about it. Typical views of quantum mechanics somewhat resemble the pre-scientific attitude of man to lightning and thunder — he did not understand them, and hence unreasonably feared them. For many people the sciences of atoms and quanta will forever be sprinkled with the ashes of an atomic explosion and the atomic physicist will forever be a sort of wicked sorcerer. This is all unfair and to be deplored because such a preconception prevents people from understanding the main thing, that quantum mechanics is not just another one of an infinite variety of contemporary sciences. It is quantum mechanics that has started the current technological era, has caused a drastic revision of the philosophy of knowledge, and has influenced the politics of states. Quantum concepts can only be compared to such revolutions as the Copernican system, Newton's laws, and the discoveries of Faraday and Maxwell. They now belong to human culture, and everyone should have at least a rudimentary understanding of them.

But it is not so much a matter of use or need; quantum mechanics is first of all enthralling and interesting, interesting in the widest sense of the word. It was the compelling logic and the beauty of the teaching of quanta that were the major reasons why I undertook to write this book. Its main emphasis is on the evolution of the ideas and concepts of quantum mechanics.

At first sight they may seem to be unnatural and formidable, just like the ancient chimaeras — sphinx, basiliscus, centaur, mermaid — creatures combining aspects that are incompatible with our everyday experience.

My effort will not be wasted if the reader understands the necessity of that strange world, perceives the inevitability of its quantum chimaeras, accepts them, and at last recognizes that they are natural.

• • •

It is a great privilege and pleasure for me to see my book published in Great Britain, that cradle of modern physics, the native land of Newton and Boyle, Faraday and Maxwell, Dalton and Rutherford, Thomson and Dirac. They will always remain symbols of the beauty of spirit, even in our pragmatic times.

I use this opportunity to thank Alex Repiev who has taught this book to speak English. I must record my deepest obligation to Professor Sydney Dugdale for the outstanding efforts he has taken to adjust my Russian mentality to the British style of thinking.

L.I.P.

1. Origins

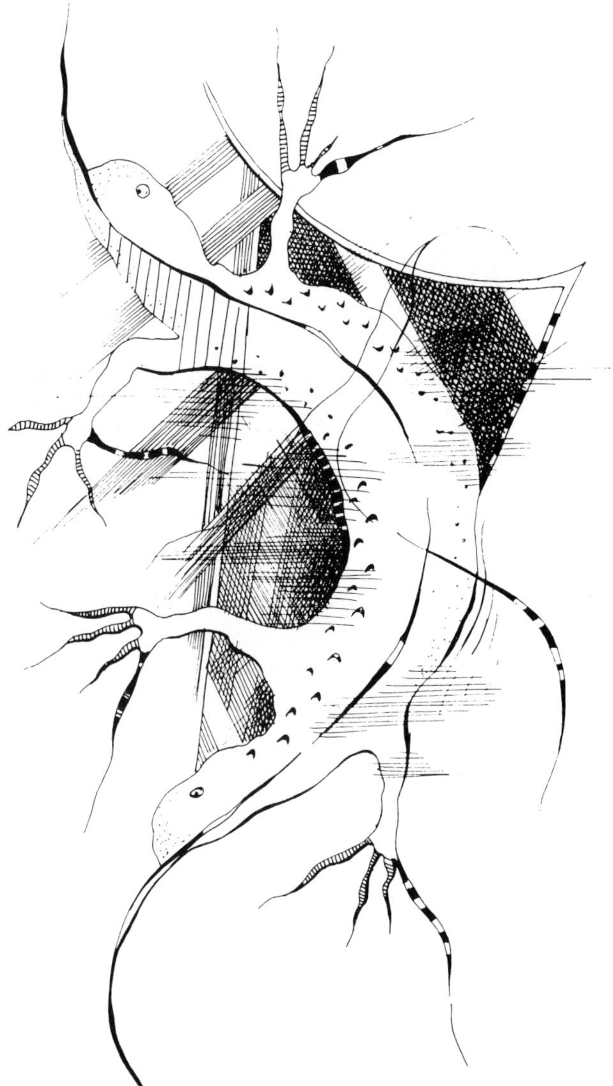

Science! true daughter of Old Time thou art
Who alterest all things with thy peering eyes.
Why prayest thou thus upon the poet's heart,
Vulture, whose wings are dull realities?
How should he love thee? or how deem thee wise,
Who wouldst not leave him in his wandering
To seek for treasure in the jewelled skies,
Albeit he soared with an undaunted wing?

Edgar Allan Poe

Sonnet — *To Science*

Zhig-Guasha, Kabardin-Circassian

Chapter One

Atoms • Waves • Quanta

Quantum mechanics is the science of the structure and properties of quantum objects and phenomena.

Everything is correct in this definition — and yet it is useless unless we explain what the phrases "quantum phenomena" and "quantum object" mean.

The word "quantum" means "how much", "quantity", "share", "portion", and so on. If we place it in front of the words "object" or "phenomenon", we will have something like "quantity phenomenon" or "portion object", i.e., a piece of nonsense, if taken at face value.

Those who have ever learned foreign languages will easily see the reason behind this incongruity: "quantum phenomenon", "quantum object" — just like "quantum physics" for that matter — are idioms, which are not to be taken literally. To grasp their true meaning, you must first get acquainted with the customs and culture of the country whose language you are learning.

Quantum physics is a vast country with a rich and deep culture. If all you know about it is that it has "unravelled the age-old enigma of the mysterious country of the microworld", and also brought about an "upheaval in our entire view of the world", then you know as much about it as tourists know about an unfamiliar country whose culture is foreign to them and whose language they do not understand. Their memory will only retain some highlights — for instance, bright neon signs and posters.

The language of quantum physics is peculiar although in essence it is not different from a conventional language. Like any language it cannot be absorbed overnight. What is needed is patience and method. At first one should memorise some common words and try to construct simple phrases from them, without paying too much attention to grammar. Later on come the ease and fluency which reward the student with the satisfaction and joy of having a firm command of the language.

In order to get used to the language and logic of quantum physics you must start with some seemingly unrelated concepts. To fit them into some sort of consistent picture will take a lot of correlation and speculation. The process of perceiving quantum ideas can be likened to the process of dissolving salt in water: if you drop a pinch of salt into a glass of water it will disappear without trace, but as you add more salt you reach a point where it is sufficient merely to add one more speck for a large regular crystal of salt to start to grow.

This book is a story of the origins, ideas and findings of quantum physics. It will introduce you to its array of images and will illustrate them with some applications. At first you will have to dissolve in your mind some initial concepts, to learn those few, but necessary, words without which you will be unable to construct even a single meaningful "quantum phrase". These words are *atoms*, *waves*, and *quanta*.

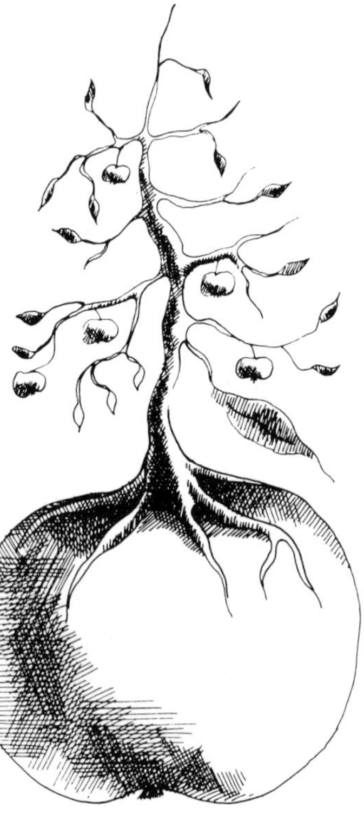

Atoms

There are few people nowadays who do not accept the reality of atoms as they do, say, the motion of the Earth around the Sun. Atoms normally predicate something minute and indivisible. What is the meaning given by today's science to the concept of the atom? What were the origins of this concept, what did the ancients read into it? How did the atomic concept evolve? And why did it have to wait for quantum mechanics to make its appearance to be promoted from just a speculative scheme to a workable theory?

The father of the atomic concept, it is widely believed, was the ancient Greek philosopher Democritus (*c.* 460-370 BC), although history also mentions his teacher Leucippus and, with less certainty, the ancient Hindu philosopher Kanada. Our knowledge of Democritus's life and personality is scarce. We know that he was born in Abdera on the Thracian shore of the Mediterranean; that he was a disciple of Leucippus and studied under some Chaldean scholars and the Persian Magi; that he travelled widely and knew much; that he lived to a great age and was given a public burial by the citizens of his native city, who held him in great esteem. Generations of artists depicted Democritus as a tall man with a short beard wearing a white Grecian tunic and sandals on his bare feet.

Legend has it that one fine morning Democritus sat on a stone by the sea, stared at an apple in his hand, and reasoned: "If I cut this apple in half I will have two halves; if then I cut one half in two I will have two fourths. Now if I keep on dividing the remaining parts in the same way, will I continue to obtain one eighth, one sixteenth, one thirty-second, etc., of an *apple*? Or will I, at some time in the process, reach a point where the divided parts will no longer possess the properties of an apple?" It turned out subsequently that Democritus's doubt (as do almost all disinterested doubts) contained a grain of truth. The philosopher came to the conclusion that a limit to divisibility does exist. He named the ultimate, indivisible particle ατομοσ, the atom (from the Greek for "indivisible"). He published his views in a book entitled *Little World System.* Behold what was written more than 2000 years ago:

"The Universe is made up, in reality, of nothing but atoms and the void; all the rest exists only in the mind. There are countless worlds and each has a beginning and an end in time. And nothing is ever begotten of nothing, nor can anything be destroyed and reduced to nothing. And the atoms are innumerable in size and quantity, moving in all directions in a void, colliding and forming vortices in which all complex substances arise: fire, water, air and earth. The fact is that these, in essence, are but combinations of certain atoms. Atoms are indestructible and unchangeable owing to their hardness."

When Democritus died, Aristotle, the future mentor of Alexander the Great, was fourteen. He was spare, short, and had extremely refined manners. The esteem in which he was held knew no sensible bounds. There were good reasons for this: he commanded the full range of knowledge of that time. Aristotle taught the opposite: an apple can be divided into smaller and smaller pieces indefinitely, at least in principle. (For fairness's sake it should be admitted that the idea of the infinite divisibility of matter for an unsophisticated mind appears as being more natural than the idea that there exists in principle a limit to the divisibility of matter.) Aristotle's views prevailed. Democritus was forgotten for many centuries and his works were ruthlessly destroyed. This is why his writings have come down to us only in fragments and the comments of his contemporaries. Democritus became known in Europe from the poem *De Rerum Natura* ("On the Nature of Things") by Titus Lucretius Carus (*c.* 95-55 BC).

We should not blame the ancients for preferring Aristotle's views to Democritus's; for them both systems were equally reasonable and acceptable. They viewed science not as a means of obtaining some practical applications (which embarrassed them) but rather as a means of reaching by speculation that feeling of harmony of the world that one derives from any consummate philosophy.

It took 2000 years to shake off the erroneous views of the great authority. Physics as a science came into existence in the seventeenth century and quickly replaced ancient natural philosophy.

The new science was based on experiments and mathematics rather than on pure speculation. Instead of merely *observing* nature, people began to *study* it around them, i.e., to stage experiments to check hypotheses and to record the results of these experiments in the form of numbers. Aristotle's idea could not pass such a test, whereas Democritus's hypothesis received support and started atomic theory.

After twenty centuries of oblivion the idea of atoms was resurrected by the French philosopher and sage Pierre Gassendi (1592-1655). In 1647 he published a book devoted to the ideas of atomism. In those days this was fraught with definite risks, for in medieval times scientists were persecuted not only for various hypotheses but also for rigorous facts, if these facts were at variance with universally recognized dogmata. (In Paris, for example, teaching about atoms was prohibited in 1626 under pain of death.)

Nonetheless, the atomistic hypothesis was accepted by all the prominent scientists of the day. Even Newton, with his famous motto *Hypotheses non fingo* ("I frame no hypotheses") accepted and expounded it in his peculiar way at the end of the third volume of his *Opticks.*

It was a compelling hypothesis; yet until it had been verified by experiment it was doomed to remain just a hypothesis.

The first graphic proof that Democritus was right came from the Scottish botanist Robert Brown (1773-1858). In 1827 he was the middle-aged keeper of the department of botany at the British Museum. In his youth he spent four years travelling on expeditions in Australia and brought back about 4000 species of plants. Twenty years later, he was still studying the collections. In the summer of 1827, Brown observed that the finest pollen grains of plants suspended in water move about in an irregular manner due to the action of some unknown forces. He immediately published a paper with a title very typical of that unhurried age: "A Brief Account of Microscopical Observations Made in the Months of June, July and August, 1827, on the Particles Contained in the Pollen of Plants; and on the General Existence of Active Molecules in Organic and Inorganic Bodies."

At first his experiment gave rise to perplexity, which was aggravated by Brown himself, who made an attempt to explain the phenomenon as the result of some "vital force" inherent in all organic molecules. Such a primitive explanation of the Brownian motion could not satisfy scientists, and so they undertook new efforts to study its details. Especially successful were the Belgian Ignace Carbonnelle (1880) and the Frenchman

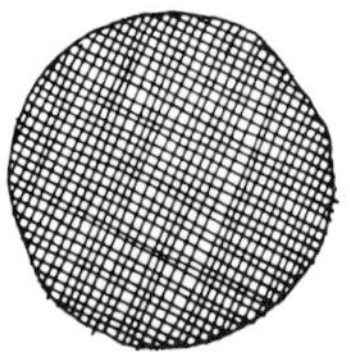

Louis George Gouy (1888). They devised careful experiments and found that Brownian motion did not depend upon such factors as the time of year or day, the addition of salts, or the kind of pollen used, and that "it is observed equally well at night on a subsoil in the country as during the day near a populous street where heavy vehicles pass." It does not depend even upon the type of particles but only on their size and, what is most important, it never ceases." (Nearly twenty centuries before Brown these properties of Brownian motion were pictured by the imagination of Lucretius Carus, who described them in detail in his famous poem.)

It should be noted that this strange motion did not initially attract the attention it deserved. Most physicists had heard nothing about it, and those who had heard of it considered it to be of no interest whatsoever because they assumed it to be similar to the motion of specks of dust in a sunbeam. It was not until forty years later that the idea took shape that the random motions of pollen seen in a microscope were due to chance impulses of tiny invisible particles of the liquid. Following the publication of Gouy's papers almost everybody held this view, and so the atomic hypothesis acquired numerous supporters.

Of course, even before Brown's time many had been quite sure that all bodies were built up of atoms. Certain properties of atoms were obvious to them even without any further investigations. As a matter of fact, all bodies in nature, however different, have mass and size. Evidently, their atoms, too, have mass and size. Precisely these properties formed the basis for the reasoning of John Dalton (1766-1844), an obscure teacher of mathematics and natural philosophy at New College in Manchester and a great scientist who determined the course of development of chemistry for approximately the next hundred years. In 1804, he subjected the available evidence concerning chemical compounds to careful analysis and formulated the notion of the *chemical element*, a substance made up of atoms all of the same species.

One question that immediately arose was: does the diversity of substances signify as great a diversity of atoms (as maintained by Democritus)? The answer turned out to be no. Chemists soon found that elements were not that numerous; about forty were known at that time (now 106). All other matter consists of molecules, which are various combinations of atoms. The atoms of the elements themselves also vary, primarily in mass. The lightest atom is hydrogen. Oxygen atoms are 16 times heavier, iron atoms 56 times heavier, and so on. Thus, for the first time the science of the atom made use of numbers.

As before, however, nothing was known about the absolute dimensions and masses of atoms.

The first successful scientific attempt to evaluate the size and mass of atoms seems to have been made by a physics professor at Vienna University, Joseph Loschmidt (1821-1895). He found in 1865 that the size of all atoms is approximately the same and is equal to 10^{-8} (0.00000001) centimetre, and that a hydrogen atom weighs only 10^{-24} gramme.

This is the first time we have encountered here such minute quantities, and we are ill equipped to comprehend them. In our everyday life the extremes of smallness are "thin as a cobweb" or "light as down". But a cobweb diameter (10^{-3} centimetre) is 100,000 times that of the largest atom, and a down pillow is something quite tangible. In order to fill in the gap somehow between common sense and the smallness of these numbers people normally use comparisons, although as a rule they help little and explain still less. For objects that are so small the very idea of size as a quantity measurable by comparison with a reference becomes pointless. It would be better, therefore, from the very beginning to abandon all efforts to picture such numbers. It is important to understand that, despite their extraordinary smallness, these numbers are not arbitrary; it is exactly such small diameters and masses that must be attributed to atoms for the properties of substances that are made up of these atoms to be as we observe them in nature.

The number of molecules in 1 cubic centimetre of a gas at standard pressure and temperature (atmospheric pressure and the temperature of melting ice) is

$$L = 2.68678 \times 10^{19} \text{ per cubic centimetre}.$$

The number is now known with great precision and is called Loschmidt's number. It is larger than that initially obtained by Loschmidt by a factor of about ten.

Waves

Iron, like any substance, consists of atoms. If you push one end of an iron poker into a furnace it will get hot of course. It is well known now that heat is the energy of atoms in motion and that heating an object increases its energy — you can easily sense this by touching the other end of the poker. But this is not the whole story. As iron is heated its colour changes gradually from a dull cherry-red to a dazzling white. But you cannot even approach a white-hot poker, let alone touch it. This is incomprehensible in terms of the conception of the motion of atoms alone. You did not touch the poker, and the iron atoms did not strike your hand. Why then do you feel the heat?

We are confronted with a new phenomenon and so we must introduce a concept that explains it — *radiation* — which at first sight has nothing to do with the atomic idea.

Light is a form of radiation. We say: to warm oneself in sunlight. Hence radiation can transfer heat. We deal with radiation all the time — when we are sitting at a camp fire, feasting our eyes upon a splendorous sunset, turning the knobs of a radio, or having our chest X rayed. Heat, light, radiowaves and X rays are all forms of the same kind of radiation: electromagnetic radiation. We do not, however, distinguish between them only qualitatively and subjectively, but quantitatively as well. How? Electromagnetic radiation has many features; but, for the moment, we are only interested in one — the wave nature of radiation.

Propagation of waves is such a commonplace phenomenon that it hardly need be discussed here. Yet we will recall some of the principal features of wave motion for the same reason that even large academic encyclopaedias list quite understandable everyday words.

The word "wave" is used widely in physics. Each person visualizes waves differently; some imagine the waves produced by a stone that hit the

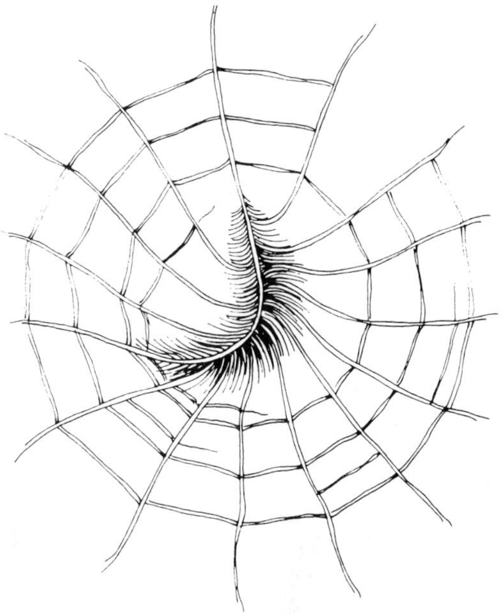

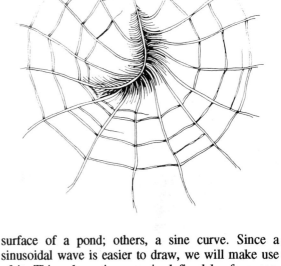

surface of a pond; others, a sine curve. Since a sinusoidal wave is easier to draw, we will make use of it. This schematic wave is defined by four parameters: amplitude A, wavelength λ, frequency ν, and velocity of propagation v.

The amplitude of a wave is its maximum height. The definitions of the wavelength and velocity are clear from the drawing. As for the frequency, we will work out its meaning by observing the motion of a wave in one second. During this time the wave travelling at a velocity of v centimetres per second will cover a distance of v centimetres. By counting the number of wavelengths in this distance we will obtain the frequency of the wave (or radiation) $\nu = v/\lambda$ oscillations per second.

One important property of waves is their capacity for *interference,* i.e., their capacity to cancel out or enhance one another, say in reflection. And it is precisely by this property that you can unfailingly distinguish a wave from a beam of particles.

Another property of a wave that distinguishes it from particles is *diffraction,* i.e., its ability to bend around an obstacle if its size is commensurate with the wavelength. If the obstacle is small enough, the wave may divide into two, due to diffraction, go around the obstacle from both sides and, combining again, cancel itself in exactly the same way as when the direct and reflected waves are added together.

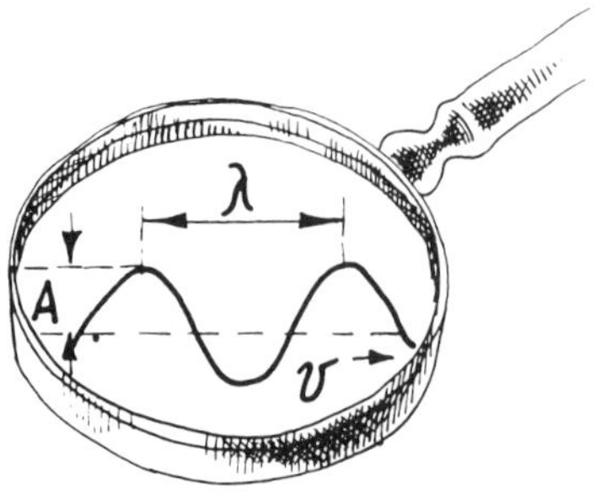

Likewise, by observing the interference and diffraction of X rays and other radiations, it was found that they are all waves, though of different length. It is the wavelength of radiation that is the principal feature by which we quantitatively distinguish between the various types of electromagnetic radiation.

The longest are radiowaves, which range from several kilometres to several centimetres in length; thermal radiation has waves from one to 10^{-2} centimetre; visible light from about 4×10^{-5} to 8×10^{-5} centimetre. The shortest wavelengths are those of X rays: from 10^{-7} to 10^{-8} centimetre. All radiations propagate at the same velocity — that of light $c = 3 \times 10^{10}$ centimetres per second. The frequency of a radiation is readily found from the formula $\nu = c/\lambda$. Clearly, it will be the highest for X rays and the lowest for radiowaves.

We should bear in mind, of course, that no radiation is really the neat sine curve shown in the figure, but it is a *physical process* whose principal characteristics (e.g., periodicity) can, fortunately, be expressed in the language of such simple models. Each kind of radiation has its specific features. Let us concentrate, for the time being, on the radiation that is most important and common to us — solar radiation.

When you bask in the sun, it is highly unlikely that you ponder over the wave composition of the sunshine, although you may have wondered why you get sunburned in the mountains and why you cannot get a tan in the late afternoon. Sir Isaac Newton (1643-1727) lived in England where there is not much sunshine! Yet he wondered what sunlight consisted of. He performed in 1664 an experiment that is now known to every schoolboy. When passing a beam of sunlight through a prism he observed a rainbow of colours thrown on the wall behind the prism, a solar spectrum. A century later another English scientist, Thomas Young (1773-1829), found that corresponding to each colour of the rainbow spectrum is a wavelength of solar radiation: the red colour had the longest wavelengths — 7×10^{-5} centimetre, the green colour had shorter wavelengths — 5×10^{-5} centimetre, and the violet colour yet shorter — 4×10^{-5} centimetre.

The emission spectrum of any body, be it the Sun or a hot steel poker, is said to be specified completely if we, first, know the wavelengths it is composed of and, second, their shares in the total radiation flux. Specifically, the colour of a hot body is determined by the dominant wavelengths in its emission spectrum. As the temperature of the body varies, so does the spectral composition of its emission. As long as the body's temperature is not high, it emits radiation but does not glow, i.e., it only gives off thermal and infrared radiations, which are invisible to the human eye. As the temperature is increased, the body starts to glow — at first a red colour, then orange, yellow, and so on. At 6000°C, for example, a yellow colour becomes predominant (it is this that has enabled the temperature of the solar surface to be measured).

It is customary to describe the spectral composition of radiation using the *spectral function* $u(\nu, T)$, which is a measure of the proportion of radiation with frequency ν in the total radiation flux at a given temperature T of the body. A typical spectral function is shown on the jar in the accompanying figure. It approximates the spectral composition of solar radiation.

Late in the nineteenth century the laws governing thermal radiation began to attract the close attention of scientists. This was largely caused by the needs of metallurgy, notably the invention in 1856 by Heinrich Bessemer (1813-1898) of a new method of steel production, which later came to be known as the Bessemer method. This method relied on the thermal radiation of steel for measurements of its temperature.

To grasp the details of the laws of thermal radiation, the researcher at first had to take into account the fact that at a given temperature the spectrum, and hence the spectral function, of the radiation are sensitive to the material of which the emitting body is made. This can readily be verified by heating in the dark two balls of the same

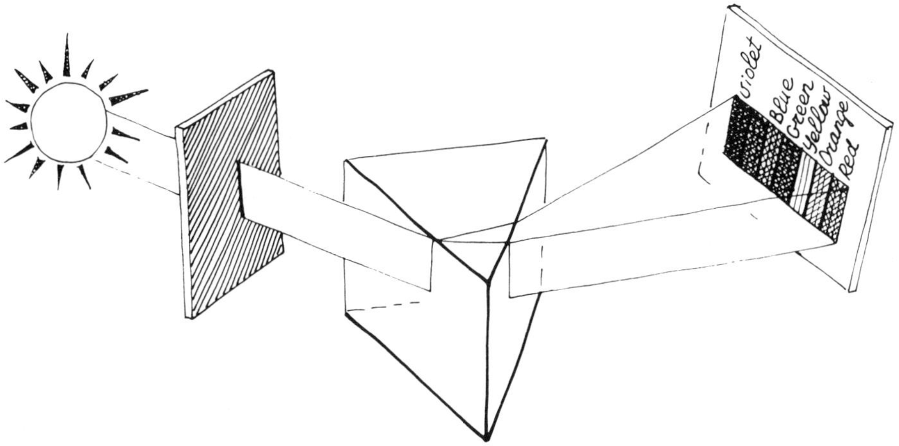

diameter — one made of stone, the other of steel; the former will glow far brighter. Before long it was found that if one heats hollow balls, not solid ones, and observes their radiation through a small hole in the ball's wall, the spectral composition of the radiation will now be independent of the material of the ball. Such a spectrum is called the *spectrum of an absolutely black body*. Incidentally, one example of an absolutely black body is the Sun, and so the function in the figure is the spectrum of its "black body".

The origins of this somewhat unusual name are easily understood. Suppose that instead of heating the ball you do the opposite, namely illuminate it from outside. In that case, you will always see in front of you a black hole, whatever the material of the ball, since nearly all the radiation that has found its way into the void will be repeatedly reflected within it and virtually none of it will escape.

The universal spectral function $u(\nu, T)$, which describes the emission spectrum of an absolutely black body, was introduced into scientific practice by the outstanding German physicist Gustav Robert Kirchhoff (1824-1887) in 1859. Its measurement appeared to be a challenge; it had to await the invention by Samuel Langley (1834-1906) in 1884 of the bolometer, a device to measure radiant energy. That $u(\nu, T)$ is an important function was understood at once, but it was not until forty years later that a theoretical formula for it was deduced. It gave a correct description of the experimental results. Throughout that time, these attempts were never once abandoned: quest for an absolute, it seems, never fails to entice the human mind.

Quanta

In the closing years of the last century, Max Planck (1858-1947), like many before him, was looking for a universal formula for the spectral function of an absolutely black body. He appeared to be more successful than the others. At first he simply guessed it, although not altogether out of the blue. It took Planck two years of arduous speculation to fit the odd pieces of this jigsaw into a coherent picture of the process of thermal radiation.

On 19 October 1900 there was a regular meeting of the German Physical Society, at which the experimentalists Heinrich Rubens (1865-1922) and Ferdinand Kurlbaum (1857-1927) reported on their new, more accurate measurements of the black-body spectrum. During the discussion that followed, the experimenters lamented that none of the then available theories could explain the results they obtained. Planck proposed that they try his formula. That same night Rubens compared his measurements with the predictions of Planck's formula and found that it agreed, to the tiniest de-

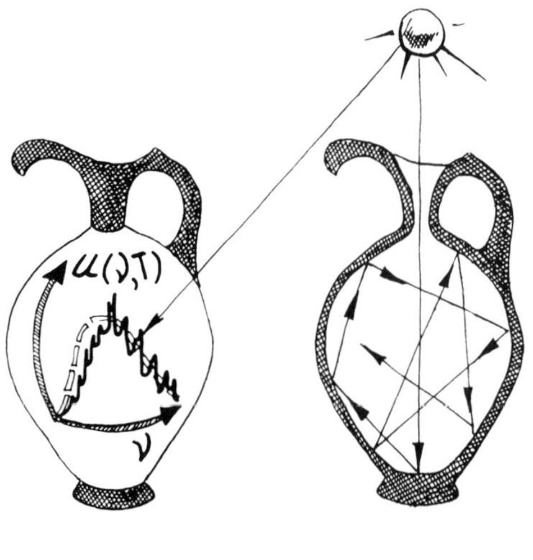

tail, with the observed black-body spectrum. First thing in the morning he congratulated Planck, his colleague and close friend.

Being a theoretician, Planck valued not only the end result of a theory but its intrinsic perfection as well. He did not know yet that he had discovered a new law of nature; he thought it could be derived from previously known laws. And so he sought a theoretical justification for this law of radiation proceeding from the simple principles of the kinetic theory of matter and thermodynamics. Two hectic months ensued and success came. But he had to pay a price for it!

He was forced to assume that radiation E is emitted in discrete portions (or *quanta*), whose size is determined by the formula

$$E = h\nu,$$

where ν is the frequency of the radiation, and h is the universal constant now known as Planck's constant. Only with this assumption could he obtain a correct formula for the spectral function $u(\nu, T)$.

Formally Planck's assumption was clear and simple in the extreme, but fundamentally it contradicted all the previous experience of physics and the intuition instilled by all the previous centuries of scientific endeavour. Recall that we have repeatedly emphasized that radiation is a wave process. If it is so, then radiation must transfer energy continuously, and not in discrete quantities (quanta). Planck, more than anybody else, was aware of this unavoidable contradiction. He was forty two when he derived his famous formula, but he was plagued by the logical imperfections of his own theory almost until his death. This feeling of aggravation was unknown to subsequent generations of physicists; they already knew the end result and had assimilated the new logic. Planck, however, had been brought up in the traditions of classical physics, and belonged heart and soul to its rigid, unhurried world. And now he, of all people, by solving a long-standing riddle of radiation theory, would undermine the beautiful logical consistency of all classical physics. For Planck this was a great shock. He was haunted by the tantalizing question "Have we not paid too dearly for what in essence is a solution to a very special problem?" Twenty years later, in his Nobel Prize Lecture, he recalled that for him to recognize the reality of quanta amounted to the "impairment of all causal relationships". And even in 1933 in his letter to Robert Wood he referred to his hypothesis as an "act of despair".

The feeling of arbitrariness, which an uninitiated man experiences when first exposed to Planck's formula and the history of its discovery, is deceptive. The hypothesis of quanta is not a result of idle speculation; it arose out of painstaking analysis and the distillation of precise experiments. Well, not only analysis. It also took the power of thought, the flight of fantasy, and the audacity to accept the shocking predictions of the theory.

Even before Planck, other physicists such as Rayleigh, Jeans, and Wien had suggested their formulas to describe the black-body spectrum, only to be rejected by the experimentalists, Otto Lummer (1860-1925) and Ernst Pringsheim (1859-1917), after their thorough measurements. They were only satisfied by Planck's formula, which was in stunning agreement with experiment, a fact that did not make it any clearer. Only a quarter of century later a new science, quantum mechanics, would reveal the true meaning of the revolution accomplished, in many respects against his own will, by Max Planck who was guided by the logic of scientific research.

On Friday, 14 December 1900, in the conference hall of the German Physical Society, a new science — the teaching of quanta — was born. Dryly and circumstantially, Max Karl Ernst Ludwig Planck, Professor-in-Ordinary in physics, read before a small audience his paper "On the Theory of the Energy Distribution Law of the

Normal Spectrum". Few people realized that day the grandeur of the moment. The bad weather or the inconsistencies of the theory probably occupied the minds of the learned audience to a greater degree. Recognition came much later, and still later did the scientific community perceive the significance of Planck's constant h to the entire world of atoms. The constant is small,

$$h = 6.626076 \times 10^{-27} \text{ erg second},$$

but it opened the door to the world of quantum phenomena. And invariably, when from the world of customary and classical physics we want to enter the unusual quantum world, we must pass through this narrow doorway.

Quantumalia

Before and After Democritus

I shall conduct the reader over the road that I have myself travelled, rather a rough and winding road, because otherwise I cannot hope that he will take much interest in the result at the end of the journey.

Albert Einstein

Until now the roots and sources of atomistic ideas have remained an enigma for historians of science, although some facts may be considered quite well established. The Indian sage Kanada, a name that means "the devourer of atoms", flourished long before Democritus. In the seventh century BC he founded a philosophical and religious school of thought, to which the notion of the atom was central. According to Kanada, knowledge is obtained through the agency of six positive categories: substance, quality, action, universality, particularity, and inference. Substance, in turn, comes in nine kinds: five of them material (earth, water, air, fire, ether) and four immaterial (time, space, self, mind). The five material kinds of substance are made up of atoms. The tiniest particle available in nature is a speck in a solar beam, which consists of six atoms, connected pairwise "by either will of God or otherwise".

However naive these concrete atomistic conceptions were, we should do justice to the clear-cut formulation of the problem and to the thorough cataloguing of philosophical categories. So, Kanada understood clearly that "we learn about the existence of atoms not through perception but through speculation", and he provided an example of this speculation: if matter were divisible indefinitely, there would be no qualitative difference between a mountain and a mustard grain, because "the infinite is always equal to the infinite".

Was Democritus aware of Kanada's philosophy? Quite probably: he knew much and travelled widely, and perhaps visited India. Was his theory original then? Undoubtedly. Just imagine the difference in epochs in which these thinkers flourished and the difference in traditions, in spiritual atmosphere and in thinking between East and West.

Also mentioned as Democritus's precursors are the Phoenician Moschus of Sidon who flourished in the twelfth century BC, at the time of the Trojan Wars, and Democritus's teacher Leucippus of Miletus. We have no reliable knowledge of Moschus's philosophy but if history throughout the course of more than three millenia has not forgotten the name of a man who was neither a king nor a general, then one is inclined to believe that he may really have left some important

Democritus

Aristotle

legacy. About Leucippus we know virtually nothing — neither his date of birth, nor his works. Aristotle called Democritus a pupil and friend of Leucippus, and always mentioned the teacher in connection with his pupil. History has venerated this tradition.

The theories expounded by Democritus were taken over and enriched by the philosopher Epicurus (341-270 BC), who founded a school of thought, or rather a brotherhood, that persisted for nearly six centuries. Just like Democritus's treatises, none of Epicurus's works has survived; we have learned about his natural philosophy from a poem by Lucretius.

Titus Lucretius Carus

The famous poem *De rerum natura* ("On the Nature of Things") by Lucretius, like many other works of art and literature of antiquity, had been forgotten for many centuries. It was printed in Italy only in 1473, and since then scientists and philosophers have never tired of marvelling at it. The reasons are two-fold: first, this poem is the only consistent exposition of the teachings of the materialists of antiquity, whose works are not extant in the original; second, this is the first of the extant and complete specimens of "novelized science", as it is called in Russian literature.

Lucretius was not only a true poet, but also a thinker with the rare gift of being able to turn abstract concepts of philosophical thinking into dramatic and perceptible images. The following are some excerpts from the poem:

Mark in what scenes thyself must own, perforce,
Still atoms dwell, though viewless still to sense.
And, first th' excited wind torments the deep;
Wrecks the tough bark, and tears the shivering clouds:
Now, with wide whirlwind, prostrating alike
O'er the waste campaign, trees, and bending blade;
. . . .
So vast its fury! — But that fury flows
Alone from viewless atoms, that, combined,
Thus form the fierce tornado, raging wild
O'er heaven, and earth, and ocean's dread domain.
As when a river, down its verdant banks
Soft-gliding, sudden from the mountains round
Swells with the rushing rain — the placid stream
All limit loses, and, with furious force,
In its resistless tide, bears down, at once,
Shrubs, shattered trees, and bridges, weak alike
Before the tumbling torrent: such its power! —
Loud roars the raging flood, and triumphs still
O'er rocks, and mounds, and all that else contend
So roars th' enraged wind: so, like a flood,
Where'er it aims, before its mighty tide,
Sweeps all created things: or round, and round,
In its vast vortex curls their tortured forms. —

J. Dalton

Though viewless, then, the matter thus that acts,
Still there is matter: and, to Reason's ken,
Conspicuous as the visual texture traced
In the wild wave that emulates its strength.
Next, what keen eye e'er followed, in their course,
The light-winged ODOURS? or developed clear
The mystic forms of cold, or heat intense?
Or sound through ether fleeting? — yet, though far
From human sight removed, by all confessed
Alike material; since alike the sense
They touch impulsive; and since nought can touch
But matter; or, in turn, be touched itself.
Thus, too, the garment that along the shore,
Lashed by the main, imbibes the briny dew,
Dries in the sunbeam: but, alike unseen,
Falls the moist ether, or again flies off
Entire, abhorrent of the red-eyed noon.
So fine the attenuated spray that floats
In the pure breeze; so fugitive to sight.
A thousand proofs spring up. The ring that decks
The fair one's finger, by revolving years,
Wastes imperceptibly. The dropping shower
Scoops the rough rock. The plough's attempered share
Decays: and the thick pressure of the crowd,
Incessant passing, wears the stone-paved street.
E'en the gigantic forms of solid brass,
Placed at our portals, from the frequent touch
Of devotees and strangers, now display
The right hand lessened of its proper bulk. —
All lose, we view, by friction, their extent;
But, in what time, what particles they lose,
This envious Nature from our view conceals.
Did no such points exist, extreme and least,
Each smallest atom be, then combined
Of parts all infinite; for every part
Parts still would boast, dividing without end.
And, say, what difference could there, then subsist
'Twixt large and small?

The beginnings of all things are small.

Cicero

. . . .

Those motes minute that, when th' obtrusive sun
Peeps through some crevice in the shuttered shade,
The day-dark hall illuming, float amain
In his bright beam, and wage eternal war.

All are but parts of one stupendous whole.

Alexander Pope

. . . .

From seeds all motion springs; by impulse hence
Through molecules minute of seeds conjoined,
Nearest in power, protruded, though unseen.

Hence urged again, in turn, through things create
Of ampler form, till soon the sense itself
The congregated action marks distinct.
As in the lucid beam's light woof we trace

Still motion visual, though unseen its source.
    Nor small the motive power of primal seeds.

Isaac Newton on Atoms

I. Newton

"All these things being considered, it seems probable to me that God in the beginning formed matter in solid, massy, hard, impenetrable, movable particles, of such sizes and figures, and with such other properties, and in such proportion to space, as most conduced to the end for which He formed them, and that these primitive particles, being solids, are incomparably harder that any porous bodies compounded of them; even so very hard as never to wear or break in pieces; no ordinary power being able to divide what God Himself made one in the first creation!"

"It seems to me, further, that these particles have not only a *vis inertiae*, accompanied with such passive laws of motion as naturally result from that force, but also that they are moved by a certain active principle, such as that of gravity, and that which causes fermentation, and the cohesion of bodies. These principles I consider not as occult properties, supposed to result from the specific forms of things, but as general laws of nature, by which the things themselves are formed; their truth appearing to us by phenomena, though their causes be not yet discovered."

Ba-Chzha, Late Chinese

Chapter Two

Spectra • Ions • Radiant Matter Atoms, Electrons, Waves

In the history of mankind the main concerns are not the names of rulers or the dates of their reign (though it would be hard to imagine history without these). Of equal importance are the birth, rise and decline of civilizations, and the evolution and substance of ideas, which have for ages directed the will of people. We want to understand the causes of the renewal of ideas and of their eventual decline. In exactly the same way, the history of physics is not simply a compendium of facts, but a coherent picture of the origins and evolution of physical ideas, without which science may seem to be a random collection of formulas and concepts. Truths are only productive when they are interrelated and when these relationships can be traced throughout their evolution.

Even savages at the lowest level of development have a history of their own. When history is lost, time gets "out of joint" and people cease being people, just as a person irreversibly degenerates upon losing his memory.

To grasp the completeness and elegance of the concepts of modern physics, it is necessary to trace their sources and their paths of development. Only then will they become congenial and understandable to us, as are the culture, language and history of our native land, which we have imbibed with our mother's milk.

The famous mathematician Felix Klein (1849-1925) once said that the fastest and most reliable way to master any science was to follow it through its whole path of development yourself. This is not the simplest way, but the most fascinating one, and we have chosen it precisely for this reason.

Spectra

A ray of sunlight over the cradle of a baby has always been a symbol of peace. But the ray caries not just caressing heat, it brings to us rich information about the fiery storms and flares on the Sun, and about the elements it is composed of — you need only know how to read this information. If you pass sunlight through a prism, you will see a rainbow spectrum behind it on a screen — a striking phenomenon, although people have got used to it over the 200 years since its discovery. At first glance, there are no sharp boundaries between individual parts of the spectrum: a red shades into an orange, the orange into a yellow, and so on. This is what everyone thought until in 1802 the English physician and chemist William Hyde Wollaston (1766-1828) examined the spectrum more carefully. He built the first spectrograph with a slit and used it to discover several distinct dark lines that crossed the solar spectrum at various places without any apparent order. He did not attach much importance to these lines. He assumed them to be associated either with the quality of the prism, or with the source of light, or with other things of minor importance. He only showed interest in the lines because they separated the coloured bands of the spectrum from one another. Subsequently, as so often in science, these dark lines were named *Fraunhofer lines* after the scientist who investigated them.

Joseph von Fraunhofer (1787-1826) lived a short but amazing life. At eleven years of age, following the death of his parents, he was apprenticed to a master grinder. He had to work such long hours that he had no time left for school. And so up to the age of fourteen he could neither read nor write. One day the ramshackle house of his master collapsed, and as the Fraunhofer boy was being rescued from the wreckage — Oh wonder of wonders! — the Crown Prince rode by in his carriage. He felt pity for the youngster and presented him with a large sum of money. The sum proved sufficient for Joseph to buy himself a grinding machine and even to start schooling himself.

This was during the Napoleonic wars in Europe, a time of great change. Fraunhofer lived in the out-of-the-way town of Benediktbeuern near Munich, ground optical glass and painstakingly investigated the dark lines in the solar spectrum. He discovered 574 lines, named the brightest ones and indicated their exact location in the spectrum. Gradually he found that their positions were strictly invariable. He put this fact to use to control the quality of achromatic lenses — no wonder his telescopes were famous throughout Europe.

Among the brightest lines Fraunhofer distinguished a sharp double line, labelled *D*, which always appeared at the same place in the yellow region. Later he established that in the spectrum of the flame of a spirit lamp at the same place on the spectrograph's screen, there always appeared an identical double line; however, it was not dark but bright yellow. The significance of this observation was not appreciated until much later.

In 1819 Fraunhofer moved to Munich, where he became a professor and a member of the Academy of Sciences; then, in 1823, he was appointed the curator of the physics laboratory. Pondering on the lines in the solar spectrum, he concluded that they are not an optical illusion, but a product of the very nature of sunlight. Curious to unravel the strange nature of these lines, he undertook further studies and found the same lines in the spectra of Venus and Sirius.

Joseph von Fraunhofer died and was buried in Munich in 1826. His gravestone bears the inscription *Approximavit sidera*, which means "He brought the stars closer". But his best monument is his discoveries.

Among them, the most important to us now are his observations of the double *D* line. In 1814, when he published his observations, they passed nearly unnoticed by the scientific community. His ideas were not wasted, however: forty-three years later William Swan (1828-1894) established that the *D* line in the spectrum of a spirit lamp flame appears only when there is some sodium present in the flame. (It enters into the composition of common salt, whose traces can almost always be found in various substances as well as in the spirit lamp flame.) Like many before him, Swan did not realize the significance of his discovery, and so he did not utter the decisive words "The line *belongs* to sodium".

This simple yet important idea came only two years later (1859) to two professors, Gustav Robert Kirchhoff (1824-1887) and Robert Wilhelm Bunsen (1811-1899). At Heidelberg University, in an old laboratory, they carried out a fairly simple experiment. Earlier investigators had passed either only sunlight or only light from a spirit lamp through a prism. Kirchhoff and Bunsen passed both at the same time and discovered a phenomenon that is worth discussing in more detail.

When they directed sunlight alone on to the prism, they saw on the spectroscope screen a solar spectrum with the dark *D* line in its usual place. The line was also there when they put a burning spirit lamp in the path of the sunlight. But when they cut off the sunlight using an opaque screen so that only the lamp illuminated the prism, the bright *D* line of sodium appeared where the dark *D* line had been before. When the two researchers let sunlight in again, the *D* line became dark again.

Then they replaced sunlight with the light emitted by a hot, glowing body. The result was always the same: a dark line appeared in place of the bright yellow one. This suggested that *the flame always absorbed the light it emitted itself.*

To understand why this finding excited the two professors, let us follow their line of reasoning.

The bright yellow *D* line appeared in the spectrum of the flame in the presence of sodium.

At exactly the same place in the solar spectrum, a dark line of unknown origin was observed.

The spectrum of a hot body is continuous, it has no dark lines. If we pass the light emitted by the body through a flame, then the spectrum in the region of the *D* line will not differ in any manner from the solar spectrum: the dark line will appear in exactly the same place. But we are almost sure of the origin of *this* dark line; we can, at least, guess that it belongs to sodium. Consequently, according to the experimental conditions the *D* line of sodium may be either a bright yellow against a dark background or dark against a yellow background. But in both cases the presence of the line suggests that there is some sodium in the flame. Since this spectral line coincides with the dark *D* line in the solar spectrum, there must be sodium in the Sun. This element occurs in the gaseous outer envelope, which is illuminated from inside by the hotter deeper layers of the Sun.

The short article (only two pages) published by Kirchhoff in 1859 contained four discoveries:

Each element has its own line spectrum, i.e., a strictly definite pattern of line.

These lines can be used to analyse the composition of substances not only on the Earth, but also in stars.

The Sun consists of a hot core and a comparatively cooler photosphere made up of incandescent gases.

There is sodium on the Sun.

All these findings and ideas were soon confirmed, including the hypothesis about the structure of the Sun. In 1868 the French Academy of Sciences sent the astronomer César Janssen (1824-1907) to India to observe a total solar eclipse. Janssen's team found that at the moment when the Sun's incandescent core was blotted out by the Moon's shadow and only the corona was visible, all the dark lines in the solar spectrum flared up with bright light.

In 1860 Kirchhoff and Bunsen themselves used their spectroscope to discover two new elements, rubidium and caesium.

From these humble beginnings developed *spectral analysis*, which we now use for example to determine the chemical composition of distant galaxies, to measure the temperature and rotational speed of stars, and so on.

This is an interesting subject in its own right, but for the moment we wish to grasp what Kirchhoff and Bunsen's discoveries have contributed to

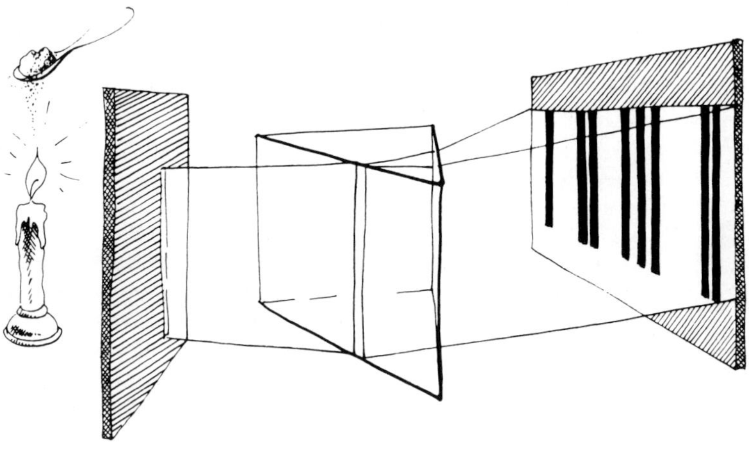

the atomic idea, and how they square with our previous knowledge of atoms.

We now know of two kinds of spectra: continuous (or thermal) and line spectra. The thermal spectrum includes all wavelengths; it is produced by heated bodies, and is independent of the nature of these bodies. The line spectrum is a set of sharp lines; it is produced by heating gases and vapours (when intermolecular interactions are weak). The most important characteristic of this line pattern is that it is unique for each element. Moreover, the line spectra of elements that make up a chemical compound are independent of the nature of this compound. It follows that we must seek an explanation of the spectra in terms of the atomic hypothesis.

That elements are uniquely determined by their respective spectra was soon recognized by everybody; but that the spectrum characterizes a single atom was not realized at first. This realization came later, in 1874, thanks to the work of the famous English astrophysicist Sir Joseph Norman Lockyer (1836-1920). Incidentally, the same ideas had been expressed earlier by Maxwell (1860) and Boltzmann (1866). These insights strongly suggested that since a line spectrum originated inside an atom, *the atom must have a structure!*

Ions

In 1865, when Joseph Loschmidt's papers were published, not much was known about atoms. They were pictured as small solid spheres, about 10^{-8} centimetre in size and weighing from 10^{-24} to 10^{-22} gramme. Each such "sphere" was ascribed an atomic weight, i.e., a number indicating how many times heavier it is than an atom of hydrogen. For example, the atomic weight of oxygen is 16 and that of helium is 4. From this we can infer that 1 gramme of hydrogen, 4 grammes of helium or 16 grammes of oxygen (or, to use the parlance of chemistry, one gramme-atom of any substance) contain the same number of respective atoms. This number $N_A = 6.022 \times 10^{23}$ (called Avogadro's number after the nineteenth-century Italian scientist) is related to Loschmidt's number L by the relationship $N_A = L \times 22{,}414.1$ cubic centimetres, i.e., Avogadro's number is the number of molecules in 22.4 litres of gas at standard temperature and pressure.

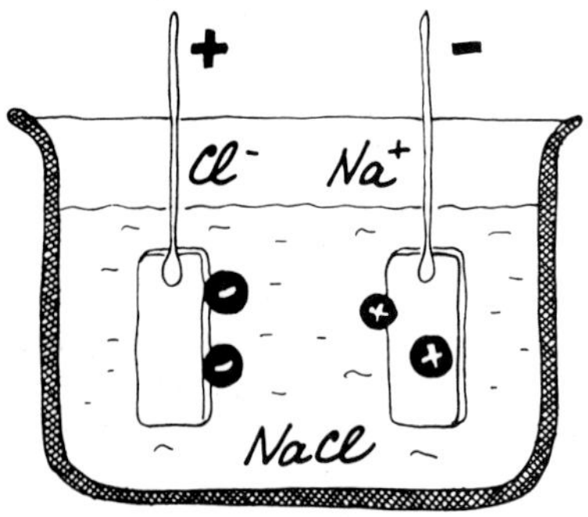

The concept of atoms as solid spheres was sufficient to account for numerous facts in chemistry, the theory of heat, and the structure of matter. By 1870, however, the idea that atoms consist of yet simpler particles had already become well established, and physicists started to look for these particles. First of all they turned to the electrical properties of the atom.

All substances are, as a rule, electrically neutral (if, of course, you do not rub glass with wool, or amber with silk, and so forth). Under some conditions, however, they exhibit electrical properties, e.g., in electrolysis.

If we immerse two electrodes in a molten salt (for instance, common salt, or NaCl) and connect their upper ends to the poles of a battery, some changes will take place in the melt. Pure metallic sodium will be deposited on the cathode (the electrode connected to the negative pole of the battery) and chlorine gas will be liberated at the anode. It follows that the sodium atoms in the melt are positively and the chlorine atoms negatively charged. And so in an electric field they drift in opposite directions.

Michael Faraday (1791-1867) established in 1834 the quantitative laws governing this phenomenon. He found that if the same amount of electricity, namely 96,485.3 coulombs = 2.895×10^{14} esu (electrostatic units), is passed through solutions of substances whose molecules are made up of monovalent atoms, on each electrode one will always find deposited exactly one mole (or gramme-atom) of the substances, i.e., the same number of atoms equal to Avogadro's number $N_A = 6.022 \times 10^{23}$. The deposits will, of course, differ in mass, since so do the atoms. For example, from a solution of NaCl we will obtain 23 grammes of Na and 37.5 grammes of Cl.

Faraday's law of electrolysis is easy to understand by assuming that in the melt of NaCl each atom carries a certain charge, and that the charges on Na^+ and Cl^- are equal in magnitude and opposite in sign. Here Na^+ and Cl^- are called ions. (The name *ion*, or "wanderer", was coined by Faraday. It was suggested to him by William Whewell (1794-1866), the author of the famous *History of the Inductive Sciences, from the Earliest to the Present Times.* He also suggested the names *anion* and *cation* and the now familiar *anode* and *cathode*.)

Knowing Faraday's law, we can readily work out the charge carried by each ion:

$$e = \frac{E}{N} = \frac{2.893 \times 10^{14}}{6.022 \times 10^{23}} = 4.803 \times 10^{-10} \text{ esu}.$$

This value is tiny, but we are becoming accustomed to such small values. What is more astonishing is that no charges smaller than this elementary charge e have ever been found in nature. In 1891 the Irish physicist George Johnstone Stoney (1826-1911) suggested that this minimum possible charge be dubbed an *electron.*

Radiant Matter

In the beginning the word "electron" did not conjure up the notion of a particle. It was merely the name for the minimum amount of charge that can be carried by an ion of any species of atoms. But the latent idea that the electron is a particle had always lurked in the minds of scientists. In fact, we will now take a closer look at the process of electrolysis. An Na^+ ion travelling in a melt under the action of the electric field approaches the cathode. The cathode has a surplus of negative charges; so, when Na^+ reaches the cathode, it receives one negative charge from the cathode and, without any change in mass, sticks to the cathode as a neutral atom of sodium. How is the negative charge transferred from the cathode to Na^+? What is added to the ion, *without any change in its mass*, when it becomes neutral?

It is quite difficult to imagine how all this comes about unless we assume that the elementary charge can exist outside the atom as well. This difficulty was clear, of course, to everybody, but to acknowledge the atomic structure of electricity seems to have been even more difficult, because this would completely disrupt the convenient and customary conception of electricity as a certain fine fluid that readily penetrated all bodies. In his famous *Treatise on Electricity and Magnetism* (1873), Maxwell allowed that in an electrolyte the molecules carry a definite amount of electricity, but he immediately admitted that this tempting hypothesis led to enormous difficulties.

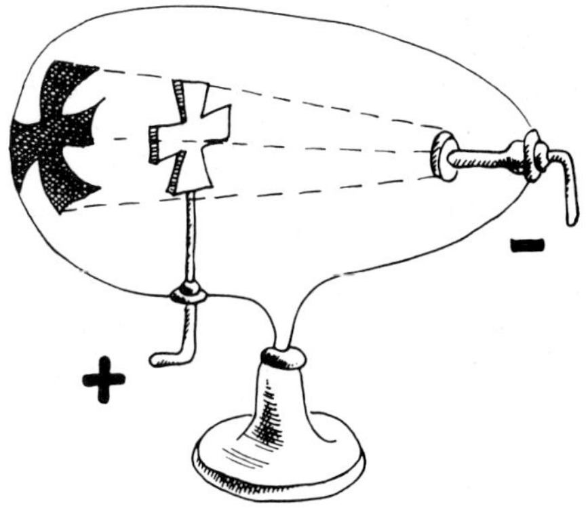

On 16 February 1881 at a meeting of the Chemical Society devoted to the memory of Michael Faraday, Hermann Ludwig Ferdinand von Helmholtz (1821-1894) gave a report "On the Modern Development of Faraday's Conception of Electricity". In his paper, Helmholtz gave the first clear formulation of the idea of the "molecular structure of electricity". "If we accept the hypothesis", he said, "that the elementary substances are composed of atoms, we are led to conclude that electricity, positive and negative, is also divided into definite elementary portions, which behave like atoms of electricity".

Admittedly, this idea as such was not new even then. Way back, in 1749, the great American Benjamin Franklin (1706-1790) made a similar conjecture, but his guess was not supported by the evidence available at the time and so led to no novel consequences. In 1871 the German physicist Wilhelm Eduard Weber (1804-1891) returned to Franklin's ideas, but met no sympathy either. By that time so much was already known about electricity that no hypothesis would be accepted on trust — knowledge presupposes responsibility. Experimental evidence for the idea of electrons was needed. It was sought in the domain of electrical conduction in gases.

Imagine a sealed glass tube filled with some gas (e.g., neon) and with some wires (normally of platinum) fused in at both ends. If we connect the two wires to the opposite poles of a battery, current will flow in the circuit, as in an electrolyte. Probably, it was just this analogy to electrolysis that prompted Faraday in 1838 to make the prototype of such a tube (Faraday's "electric egg"). As we shall see, the analogy is entirely superficial, but gas conduction turned out to be such an interesting phenomenon that many investigators devoted their whole lives to unravelling its mysteries.

In about the middle of the last century, Julius Plücker (1801-1868) abandoned geometry, his earlier occupation, and plunged into the study of conduction in gases. First of all, Plücker established that the conductivity of a gas depended on its concentration in the tube and increased as some of the gas was let out of the tube. In the process each gas begins to glow with its own colour. (When these days you marvel at the riot of colour of neon signs you owe this sight to a German professor of mathematics, because it was Plücker who in 1858 invented these glowing tubes.)

As the tube is further evacuated, a dark space appears near the cathode ("cathode space"). As the pumping proceeds, the dark space spreads and finally fills up the whole tube, and so the glow ceases. But this dark space is not dead; it is pierced by some "rays", although they are invisible to the eye (just as a flying bullet is invisible until it meets some obstacle).

In 1876 a pupil of Plücker, Eugen Goldstein (1850-1931), called this radiation *cathode rays.* Earlier, in 1869, another of his pupils, Johann Wilhelm Hittorf (1824-1914), found that these rays were deflected by a magnetic field. Finally, in 1879, Cromwell Fleetwood Varley (1828-1883) showed that they were negatively charged.

At first these rays were treated in terms of wave concepts (although Varley leaned towards the corpuscular point of view as far back as 1871). This is easy to explain: memories were still fresh of the famous dispute between Newton and Huygens on the nature of light. Therefore, any attempts to account for the above findings using the corpuscular theory were interpreted as a return to the Middle Ages.

Put yourselves in the shoes of those bewildered men of science of the 1870s: you are in possession of a set of fascinating facts, but you see no connection whatsoever between them. On the one hand, the phenomenon of gas conduction closely resembles electrolysis; on the other hand, quite incomprehensible things occur. For example, conductivity increases as the gas density decreases; there are only negative "rays" and no positive ones. Some insight was needed.

Such an insight came from the brilliant experiments performed by William Crookes (1832-1919), an English physicist and chemist. He was an interesting person, who had the rare gift of foreseeing fundamental discoveries.

To begin with, he used a more advanced pump. As he achieved a better vacuum within the tube, he observed another, darker region emerge at the cathode. It also grew in size until it filled the whole tube and the anode then began to glow with a faint greenish light. The day that this happened in 1878 can be regarded as the birthday of the *cathode-ray* tube, the heart of today's television set. This invention alone would have ensured for Crookes the grateful acknowledgement of posterity, but for Crookes himself this was only the beginning. He set out painstakingly to investigate the properties of that "radiant matter" (the term had been introduced by Faraday in 1816). Crookes sensed that he had come across an entirely new phenomenon and proposed that it be called "a fourth state of matter", which is "neither solid nor liquid nor gaseous". He wrote:

"In studying this fourth state of matter we seem at length to have within our grasp and obedient to our control the little indivisible particles which with good warrant are supposed to constitute the physical basis of the universe. We have seen that in some of its properties Radiant Matter is as material as this table, whilst in other properties it almost assumes the character of Radiant Energy. We have actually touched the borderland where Matter and Force seem to merge into one another, the shadowy realm between Known and Unknown which for me has always had peculiar temptations. I venture to think that the greatest scientific problems of the future will find their solution in this Border Land, and even beyond; here, it seems to me, lie Ultimate Realities, subtle, far-reaching, wonderful."

In order that we may appreciate Crookes's audacity, we must recall that in those days scientists divided the whole universe into matter and ether, these two entities being opposite and incompatible: classed with matter were particles, and with ether the medium whose vibrations we perceive as light.

Crookes's "radiant matter" had thus to be a combination of uncombinable properties: those of waves and particles. Half a century later anyone could see that Crookes had been right, but at that time Sir Oliver Lodge, a contemporary and compatriot, wrote that Crookes's supposition would share the fate of those flashes of inspiration that are sometimes permitted writers but are held in derision by the orthodox science of their day.

Irrespective of the meaning that Crookes read into the concept of "radiant matter", his indisputable experiments revealed that the matter had the following amazing properties: it travelled in a straight line; it caused bodies to glow and could even melt them; it was deflected by an electric or magnetic field; it could pass through solids; and its mean free path in air was 7 centimetres, while that of atoms was only 0.002 centimetres. All this led William Crookes to conclude that cathode rays, or radiant matter, are a stream of negatively charged particles that are smaller in size than atoms.

It is easily seen that this hypothesis alone explained all the properties of cathode rays. It explained, in particular, the occurrence of the dark space at the cathode: the size of this region is simply the mean distance covered by the electrons before they collide with atoms of the gas. Clearly, this distance increases as we pump the gas out of the tube. But the main value of the hypothesis lies elsewhere; it became the lodestar in the stormy sea of facts that had been accumulated by that time.

Now physicists knew what was to be done next and what to look for. Above all, it was necessary to isolate that hypothetical "atom of electricity" and to determine its properties, i.e., charge, mass, and size.

This required almost twenty years and the efforts of many outstanding physicists. We cannot describe here all the ingenious and subtle experiments devised by these scientists. We will simply trace how the hypothetical "atom of electricity" gradually acquired real properties until it became, at last, the foundation of physics.

To begin with, in 1895 Jean Perrin (1870-1942) proved once and for all that cathode rays are negatively charged. During the next two years their velocity was found to be about one tenth of the velocity of light, i.e., about 10,000 times that of a rifle bullet or of atoms in thermal motion. Moreover, these and all the other properties of these rays remained the same, regardless of the composition of the gas. This suggested that the cathode particles were invariable components of all atoms.

Finally, J.J. Thomson succeeded in 1897 in measuring the electric charge e and mass m of a single "atom of electricity". The mass turned out to be only about one thousandth of the mass of a hydrogen atom, and the charge equal to the charge of a hydrogen ion measured in electrolytic studies.

This came as a surprise. Judge for yourself. Electrolysis and gaseous conduction were concerns of different sciences — sciences that had taken separate paths and whose concepts had become established over a span of decades. And now, all of a sudden, they appeared to be related to one another. Max von Laue (1879-1960), a Nobel Prize winning pupil of Planck's, said: "Such facts in the history of science are the strongest proof of its truth." Events of this kind are always festive occasions for physicists.

The history of the electron is exemplary of the logical sequence of discoveries in modern physics. If scientists have a large enough body of observational evidence, they put forward their working hypotheses, which are again tested by experiments. The process is crowned by a theory, a condensed explanation of specific phenomena on the basis of a few general principles. The hypothesis of the electron originated from the observations of Faraday, Plücker, and Crookes. Its soundness was tested and retested by the experiments of J.J. Thomson and other physicists. Ultimately, Hendrik Antoon Lorentz (1853-1928) became such a strong believer in the reality of the electron that he, on the basis of the hypothesis, worked out a theory, whose consequences could again be tested. The process is endless, but it is the only way for science to advance.

On 30 April 1897 J.J. Thomson reported his findings at the Royal Society. After forty years of the labours of physicists the first "elementary particle" — the electron — became an entity. This was the most important development in physics since the atom had been accepted as a reality. And

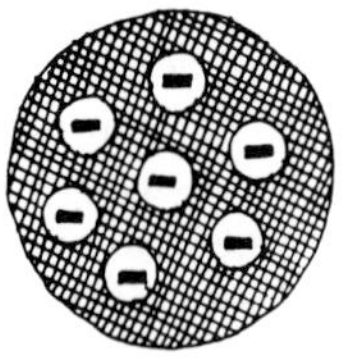

now physicists found out that there existed particles much smaller in size than atoms, that these particles enter into the composition of all atoms, and that not only matter but electricity as well has an atomistic structure. It followed that *in nature there really exists a material carrier of the minimum charge.*

Like the atom, the electron was received with a measure of scepticism. In 1902 Sir Oliver Lodge wrote that the free electron was a purely hypothetical charge. And even as late as 1920 the famous Rö ntgen still questioned the existence of the electron.

The physicists who immediately became converts carefully measured the electron's characteristics, such as its charge e and mass m. Thanks to them, notably to Robert Millikan, who in the years from 1909 to 1940 returned to that problem from time to time, we now know these values with great accuracy:

$$m = 9.109389 \times 10^{-28} \text{ gramme},$$

$$e = 4.803207 \times 10^{-10} \text{ esu}$$

$$= 1.602177 \times 10^{-19} \text{ coulombs}.$$

And what about the size of the electron? This, alas, is something we still know nothing about. We are not even sure whether this question has any clear-cut meaning. We judge the properties of an electron from its interactions with other particles or fields. But to interpret the results of these experiments we only need to know the mass and charge of the electron, not its dimensions. It may well be that electrons really do not have such a property as size. You cannot, say, speak about the thickness of the equator, but you can speak about its length. Another possible hypothesis is that an electron's size varies depending on the experimental conditions. Why not? A comet varies in size depending on its distance from the Sun, even though its mass remains the same. These are not just idle questions, and we will return to them in later chapters.

Atoms, Electrons, Waves

We have just covered a difficult section of the path that investigators had to struggle through at the close of the last century. This was a time when the wealth of new facts and phenomena obscured the simple relationships between them. This was a time when only a firm belief in the harmony of nature could save the scientist from getting lost in the chaos of motley facts and conflicting hypotheses.

A truly great discovery not only answers old questions, but raises new ones. The story of the electron was for physicists a source of inspiration... and new headaches. How are electrons bound in the atom? How many of them are there in the atom? Are they at rest or in motion? And if they move, how are their motions related to atomic radiation? The questions varied in form and nature but before long they all boiled down to the task of finding the number, dimensions and arrangement of the electrons in the atom, and also their influence on the radiation processes.

And nobody gave a thought as to whether or not such questions made any sense at all. The physicists of that time tacitly assumed the electron to be a tiny sphere, 10^{-13} centimetre in diameter, fixed "somehow" within the atom, or flying around there like a gnat in a cathedral. Their prime concern was why the atom emits spectral lines of strictly definite wavelengths and why there are so many lines (e.g., the iron atom has over 3000 lines in the visible range alone). As always when on shaky ground, suitable analogies were sought. For example, a spring with a small weight attached to it vibrates with a frequency dependent on the spring's elasticity. This suggests, according to one school of thought, that the electrons are held in the atom by "some kind" of spring or various types of elasticity. When the atom is excited, the electrons begin to vibrate and emit light with the frequencies of the springs. It followed, according to Lockyer, that the number of electrons in an atom is equal to the number of lines in the spectrum of the element. Furthermore, an atom so structured would most readily absorb light of precisely those frequencies with which it itself emits. But this is just what Kirchhoff and Bunsen had discovered in their famous experiment.

Despite the successes enjoyed by the atomic model with elastically bound electrons, many understood its logical, or rather aesthetic, imperfections. Direct disagreements with experiment were not long in coming. J.J. Thomson was engaged in the studies of X ray scattering by the atoms of various elements; his findings led him to conclude that an atom contains rather few electrons and that their number is about half of the atomic weight of the element. In 1904 J.J. Thomson proposed his atomic model using the hypothesis of William Thomson (Lord Kelvin). The model assumed the atom to be a sphere of diameter 10^{-8} centimetre filled with a uniformly distributed positive charge in which the negative electrons float, being bound quasi-elastically to the sphere. The total charge of the electrons equals that of the sphere so that, on the whole, the atom is neutral, as followed from experiment.

Early in this century most physicists accepted J.J. Thomson's model; only a few proposed models of their own. But be that as it may, all sensed that a new era was dawning in the science of the atom.

Quantumalia

Discovery of Spectral Analysis

The word "spectrum" was introduced by Newton. In classical Latin, which he used for his treatises, "spectrum" means "apparition", "ghost", or "spirit", which is a vivid reflection of the phenomenon — white light passed through a transparent prism flares up into a blaze of colours, a rainbow. For nearly two centuries these silent "ghosts" let scientists watch them, until scientists made them speak in the language of quantum physics.

When spectral analysis came on the scene, it became a subject of keen interest even to the general public, not a frequent occurrence in science in those days. As always in such cases, amateurs were quick to dig up a host of other names to whom the credit for the discovery was allegedly due long before Kirchhoff and Bunsen. Among the candidates for the laurels were the Frenchman Jean Bernard Léon Foucault (1819-1868), who had proposed a similar experiment ten years earlier, the eminent astronomer and chemist Sir John Frederick William Herschel (1792-1871), the inventor of photography on paper William Henry Fox Talbot (1800-1877), and many others. The English contended for many years that spectral analysis had been a brainchild of their famous compatriot George Gabriel Stokes (1819-1903), who had suggested in a conversation with William Thomson (1824-1907) that the *D* line in the solar spectrum is caused by the passage of white solar light through the sodium vapour contained in the gaseous envelope of the Sun. Characteristically, Stokes made no such claims, although he admitted that he had expounded similar ideas to his students, which, however, he considered to be generally known and not of great importance. (Incidentally, at about that time Peter Guthrie Tait (1831-1901) started the custom of publishing scientific reviews. He reproved Stokes and W. Thomson for their devil-may-care attitudes and ignorance of the literature, which had prevented

W. Crookes

J.J. Thomson

them from publishing an obviously novel idea.) Perhaps one should also mention Plücker, who knew that each gas glows with its characteristic colour, but who extracted no conclusions and generalizations from that.

Unlike their numerous predecessors, Kirchhoff and Bunsen were quick to perceive the significance of their discovery. They were the first to understand clearly (and easily to convince others) that spectral lines were a characteristic of the atoms of the source substance, not of the structure of the prism or of the properties of sunlight. Kirchhoff set out to compile an atlas of the Fraunhofer lines of the solar spectrum and to determine the chemical composition of the Sun. He spoiled his eyes while performing this work and was forced to give it up in 1861.

The history and substance of spectral analysis could be the subject of a fascinating narrative, but unfortunately we can hardly afford such a digression here. I will only mention one curious anecdote that took place soon after Kirchhoff and Bunsen's discovery.

On 18 April 1868 on a solar-eclipse expedition to India the French astronomer Janssen already mentioned observed a yellow line of unknown origin in the spectrum of the solar corona. Two months later the English physicist Lockyer found a way to observe the spectrum of the corona without waiting for an eclipse. He observed the same yellow line. He called the unknown element that emitted the line "helium", which means "the solar element". Both astronomers wrote letters describing their findings to the French Academy of Sciences. The two letters arrived together and were read at a meeting of the Academy on 26 October 1868. This coincidence so amazed the Academicians that they decided to have a gold medal issued in commemoration of this event. One side of the medal showed the profiles of Janssen and Lockyer; the other, Apollo in his chariot and the inscription *Analysis of solar prominences.*

On Earth helium was discovered in 1895 by the Scottish chemist William Ramsay (1852-1916).

The Beginnings of Television

The simplest schoolboy is now familiar with truths for which Archimedes would have sacrificed his life.

Ernest Renan

Modern television, which together with the motor car, aeroplane, and telephone is a symbol of our modern civilization, had its humble beginnings in the quiet laboratories of the mid-nineteenth century.

In 1854 Heinrich Geissler (1814-1879) of Bonn, Germany, a glass blower and mechanic, invented the oil pump, which enabled him to obtain a better vacuum in sealed glass tubes, and he learned how to fuse electrodes into them. At about the same time Henrich Daniel Ruhmkorff (1803-1877) of Basel started selling induction devices (so-called "Ruhmkorff coils") that produced sparks several centimetres long. Both inventions were at first used just for amusement — when gas-discharge tubes were connected to Ruhmkorff coils, the former produced a glow so beautiful that it could attract even the idly curious. But scientists, too, were not indifferent to the new phenomenon.

In 1856 Plücker moved from Leipzig to Bonn. He was a mathematician, but for his new assignment he had to lecture on physics.

("Plücker's coordinates" are now known to every mathematician, but at that time his work seems not to have received much recognition.)

Not only did he not believe in spectres, he even was not afraid of them.

Georg Lichtenberg

In Bonn, Plücker became intrigued with experimental physics. He would experiment for hours with Geissler's tubes, improving them. Soon they came to be known as Plücker's tubes — and in another thirty years as Crookes's tubes, Hittorf's tubes, and Lenard's tubes. After J.J. Thomson's work and the discovery of the electron would appear "Broun's tube", a forerunner of the cathode-ray tube, constructed in 1897 by Karl Ferdinand Braun (1850-1918), a German physicist who also invented the crystal detector in 1906 and shared the 1909 Nobel Prize for Physics with Marconi.

William Crookes

Sir William Crookes (1832-1919) was born into a family of a merchant in Regent Street. He was the eldest of 16 children by his father's second marriage; there were also five more children by his first marriage. As he himself was wont to say, hardly anybody in his family knew the meaning of the word "science". He received his his elementary education from his uncle who had a bookshop next door to his father's shop.

At nineteen years of age Crookes graduated from the Royal College of Chemistry in London, which had been founded not long before he entered it. He stayed on at the college as a junior assistant until 1854. At the same time he attended Faraday's lectures at the Royal Institution, which made a great impression on him. In 1861 Crookes discovered the element thallium, and in 1863 he was elected Fellow of the Royal Society, where on 30 November 1878 he delivered a report on the properties of cathode rays.

Life must be lived forward, but can only be understood backward...

Soren Kierkegaard

The rumour persists that Crookes barely missed discovering X rays. The fact is that during his experimentation with cathode rays he had complained to the Ilford Company that the photographic plates they had supplied him were light-damaged. (As we know now, the X rays produced when the electrons collided with the walls of the tube could, of course, have fogged the plates, even through a light-proof casing.) This rumour had not been confirmed. In any case, Crookes never mentioned it in public.

Crookes was a man of amazingly diverse interests: he was an inventor, a stockbroker, the publisher of *Chemical News*, and a pure investigator — all at the same time. He was affable, well-balanced, devoted to his family, and wary of strangers. Crookes was never in anyone's employ; he was totally devoted to science, which however did not prevent him from believing in spiritualism. In 1913 he became president of the Royal Society.

As for his dabbling in spiritualism the story goes as follows. In 1867 in Havana his younger brother Philip died, whom he loved dearly. Cromwell Varley, Crookes's pupil, urged him to communicate with his dead brother. In 1874 Crookes discontinued his spiritualism sessions, although he never recanted.

J.J. Thomson thus characterized Crookes: "He had a singularly independent, original and courageous mind; he looked at things in his own way and was not afraid of expressing views very different from those previously considered orthodox."

Kinetic Theory of Gases

Now let us look at the atom through the eyes of the nineteenth-century researchers. In the nineteenth century other scientists undertook to explain the physical properties of bodies without going into the details of atomic structure. The idea underlying these attempts was extremely simple: *atoms, of which all substances in nature consist, are not at rest, but in continuous motion.*

J. Fraunhofer

It appeared that these views, when properly formulated in the language of numbers, led to many consequences.

Beginning with Newton, who sought to explain the Boyle-Mariotte law, such attempts had been made repeatedly.

Even Francis Bacon (1561-1626) maintained that heat is motion, and Robert Boyle agreed with him. But the credit for the creation of the kinetic theory of matter should perhaps go to Daniel Bernoulli (1700-1782). The Bernoulli family, originally of Dutch stock, has given the world more than 120 famous and distinguished scholars, actors, writers, and statesmen. Against the expressed wish of his father, Daniel began his study of mathematics with his elder brother Nikolaus; he completed his education in Italy. In 1725, he and Nikolaus went to St. Petersburg, where the Petrine reforms attracted many foreigners. Eight months later Nikolaus died, and Daniel, a professor of mathematics, lived seven more years in St. Petersburg, until he could no longer put up with the Russian climate and way of life. It was there that he wrote his *Hydrodynamica;* he published it in 1738 in Basel, five years after his return.

At about that same time and also in St. Petersburg, kindred ideas were being conceived by Mikhail Vasilyevich Lomonosov (1711-1765).

Strange was the fate of the kinetic theory of gases. Bernoulli's *Hydrodynamica* was only noticed 120 years after its publication, in 1859; Lomonosov's work, written in the years 1742-1747, only became known in 1904.

The kinetic theory of gases was re-invented in the nineteenth century only to be "pooh-poohed" again. In 1821 John Herapath (1790-1868), a schoolmaster from Bristol, again proposed the kinetic hypothesis, and again it was ignored. A quarter of a century later, in 1845, John James Waterston (1811-1883), a naval instructor at the East India Company in Bombay, submitted a treatise on the kinetic theory of gases to the Royal Society in London. But one of its reviewers assessed it as "nothing but nonsense, even unfit for reading before the Society". It was only in 1892 that Lord Rayleigh unearthed Waterston's manuscript in the archives and had it published.

Don't ever take a fence down until you know the reason it was put up.

Gilbert Chesterton

Such a unanimous indifference to these works is evidently a result of the general frame of mind of the physicists and, in part, of the philosophical teachings of the day. In the middle of the nineteenth century nearly all philosophers denied the existence of atoms. This is quite strange, because eighteenth-century philosophers considered atoms to be an obvious, and even trivial, fact.

When you don't have any education, you've got to use your brains.

Anonymous

Nonetheless, the ideas of Herapath and Waterston were not lost; they had a decisive influence on the work of James Prescott Joule (1818-1889). In 1851 Joule first estimated the velocity of a gas molecule. It turned out to be unexpectedly high. For hydrogen, for example, it was about 1800 metres per second, twice that of a cannon shell.

The further history of the kinetic theory was even more eventful: it was rediscovered in 1856 by August Karl Krönig (1822-1879) and in 1857 by Rudolf Julius Emanuel Clausius (1822-1888). Then it was refined almost to its present state by James Clerk Maxwell in 1860 and by Boltzmann in 1878. But ten years later it "went out of fashion" again. Boltzmann's works aroused "more astonishment than recognition" and he was called "the last foothold of atomism". He himself admitted sadly: "I am the last to deny the possibility of constructing any other picture of the world except the atomic one". This new wave of disbelief left its trace in textbooks and scientific papers. For example, in the popular textbook (1885) by Tait one comes across the following passage: "The solid atom is still alive (in the form of an improbable, but not yet disproved, hypothesis)... Incomparably more plausible is the theory that matter is continuous, that is, it does not consist of particles with spaces between them." And even as recently as 1898 one of the journals wrote that "...the kinetic theory is just as wrong as the mechanical theory of gravitation".

The avalanche of discoveries at the turn of the century, however, completely swept away those lingering doubts, and from then on the kinetic theory of matter became one of the principal sciences dealing with the structure of matter. It has been instrumental in accounting for the heat capacity and thermal conduction of solids, the elasticity and viscosity of gases, and many other phenomena.

Mikhail Vasilyevich Lomonosov

M. Lomonosov

The first Russian scientist Mikhail Vasilyevich Lomonosov was born on 8 November 1711 in the distant northern village of Denisovka near the town of Kholmogory in Northern Russia. In the winter of 1731, when he was twenty years of age, he walked with a train of loaded sledges to Moscow, where he began his schooling. He died on 4 April 1765, a member of the Russian Academy of Sciences and a member of the academies of Stockholm and Bologna.

Everything about Lomonosov is amazing — his powerful constitution, his wide interests, the power of his creative genius. He was the first to give scientific lectures in Russian. To be able to do this, he had to work out Russian scientific terminology — which he did. He published the first Russian textbook on mineralogy and laid the foundations for modern Russian versification; he supervised the mapping of the territory of Russia and wrote a work entitled "On the Propagation and Preservation of the Russian People". He made mosaic screens out of bits of stained glass that he himself had produced; he organized an expedition to search for a sea passage to India along the northern coast of Russia; he designed navigational instruments and established the first chemical laboratory in Russia.

He was instrumental in founding in 1755 in Moscow the first Russian university, which was later named after him.

In the natural sciences some of Lomonosov's views were well ahead of his time. He was a consistent advocate of atomism and relentless opponent of the phlogiston theory. Forty years before Lavoisier he used scales routinely in his chemical investigations. He conducted his famous experiment with the heating of metals in sealed retorts seventeen years before Lavoisier, and he discovered Venus's atmosphere thirty years before Sir William Herschel.

Lomonosov endured many hardships. Russia at that time was a feudal backward country and scientific pursuits were not considered an honourable endeavour. Lomonosov was forced to look for powerful patrons and to waste his time and efforts on a multitude of things having nothing to do with science. But, to his students, he preached: "Nothing can be more pleasant and useful for posterity than your physico-chemical trials, which you conduct during your free time away from life's more important matters."

Every fool believes what his teachers tell him, and calls his credulity science or morality as confidently as his father called it divine revelation.

Bernard Shaw

Lomonosov held that heat and cold were caused by "the mutual movements of tiny intangible particles". In 1744 he submitted to the Russian Academy of Sciences a dissertation entitled "Reflections on the Cause of Heat and Cold".

The minutes of a session of the Academy contain the admonition that "... Junior Assistant Lomonosov has begun to compose dissertations too early". The low cultural level of the Academy and the increasing isolation of Russia from Western Europe were the reasons why Lomonosov's works have had no influence on the subsequent development of science. They were forgotten. Only later, when preparations for the 200th anniversary of his birth were in progress, were Lomonosov's scientific works unearthed from the archives and the genius of that Russian encyclopaedist appreciated.

Taurt, Egyptian

Chapter Three

The Planetary Atom • Spectral Series Photons • Victory of Atomistics

Man acquires half of his knowledge of the world around him before he is five years old. During the following ten years he learns almost everything about the world, but then his knowledge (save for special skills) is enriched exceedingly slowly. This may be because by that age he has already fallen into the adult habit of asking "And what is this for?" when he is exposed to some new information.

This habit is a particular hindrance when one first gets acquainted with the elements of quantum physics, because at first neither the essence of atomic phenomena nor their relative significance in the general picture is clear. In this situation one should act as a child learning to speak. First the child hears sounds that he does not understand, then he randomly utters words, and, finally, he notices that the worlds are connected in some logical way. He finds that frequently words have no meaning in themselves, but sometimes they acquire an unexpected meaning when pronounced in a certain order. A long time passes, of course, before he is able to perceive the most subtle shades of meaning and moods conveyed by simple combinations of everyday words. It is really only then that he becomes an adult.

In this chapter we will learn many new facts about atoms, waves, and quanta. It is quite possible that the facts that we select and the certainty with which we interpret them will not appear sufficiently well-founded at first, just as the actions of an adult often seem illogical to a child. This cannot be helped, however. When first exposed to the Alice-in-Wonderland reality of atomic physics, we willy-nilly become like children entering a new world. There is no science without facts. And to absorb the facts of quantum physics, let us become children for a time; they always know more than they comprehend.

The turn of the twentieth century is often referred to as the heroic era of physics. This was a time when each year saw new and stunning discoveries, whose fundamental character is obvious even now, more than half a century later. One such discovery was associated with the Crookes's tube mentioned above.

On 8 November 1895 in his university laboratory in Würzburg, Wilhelm Konrad Röntgen

(1845-1923), while investigating cathode rays, discovered a new kind of radiation originating from the point on the anode struck by the beam of electrons. This radiation was unusual — it passed through the human body and could even pierce the closed door of a steel safe.

This discovery alone would have been sufficient to upset the routine of scientific laboratories all over the world. But an era of discoveries had just begun. Several months later, in 1896, Antoine Henri Becquerel (1852-1908) discovered an even stranger kind of radiation. It was spontaneously emitted by a piece of uranium ore and consisted of electrons, gamma-photons and positively charged particles, which Rutherford named alpha (α) particles. Some substances (for example, zinc sulphide) fluoresced when exposed to a beam of alpha-particles. This enabled Crookes to invent in 1903 the spinthariscope — an instrument that permits the observation of flashes produced by single alpha particles striking a zinc sulphide screen.

These two discoveries are well known now, and we only mention them because without them our history of the atom would be incomplete.

The Planetary Atom

A wide variety of conceptions, some quite wild, of the structure of the atom were current in physics early in the twentieth century. For instance, Karl Louis Ferdinand von Lindermann (1852-1939), Rector of the University of Munich (who proved that π is a transcendental number), contended in 1905 that "the oxygen atom has the shape of a ring, and the sulphur atom, the shape of a clot". Still alive was Lord Kelvin's theory of the "vortex atom", according to which the atom looks like a set of smoke rings blown by a smoker. (Kirchhoff said: "This is an excellent theory because it excludes all others.")

Most physicists, however, sided with J.J. Thomson, who held that the atom is a uniformly distributed positive charge in a sphere about 10^{-8} centimetre in diameter in which are floating about negative electrons about 10^{-13} centimetre across. But "J.J." himself (as he was called by his students) showed no enthusiasm for his own model. Many physicists had an altogether different picture of the atom.

Some discussed this in public. One of them was Stoney, who in 1891 conjectured that the electrons moved around within the atom like the satellites around a planet. Another one was Perrin, who in 1901 discussed the "nuclear-planetary structure of the atom". The Japanese physicist Hantaro Nagaoka (1865-1950) in 1902 published his view that the space inside the atom is huge in comparison with its constituent electric grains, or that, in other words, the atom is in a way a complex astronomical system similar to Saturn's rings. Many agreed with these speculations. Among them were the Englishman Sir Oliver Lodge, the Frenchman Paul Langevin (1872-1946), and the Norwegian Carl Anton Bjerknes (1825-1903), to name but a few.

Others only confided their atomic ideas to their diaries. One example is Pyotr Nikolayevich Lebedev (1866-1912). He wrote in 1887 that the frequency of atomic radiation must be determined by the frequency with which an electron spins in its orbit. The voice of another of Russia's atomists, the revolutionary-scientist Nikolai Alexandrovich Morozov (1854-1946), was silenced by the walls of the Schlüsselburg Fortress.

None of the "planetary atomists" could, however, explain the main feature of the atom, namely the stability of a system consisting of a positive core and electrons revolving around it.

An orbiting electron is constantly accelerating, and so, by the Maxwell-Lorentz theory, it should lose energy by radiation. The radiation should be so intense that within 10^{-11} second the electron would fall into the positive centre of attraction. (This result obtained by the German scientist Schott in 1904 long remained a decisive argument in any dispute over atomic structure.)

Nothing like this ever occurs in nature; the atom is not only stable but it restores its structure after being disrupted, thereby, it might seem, supplying evidence for the Thomson model. For more than 200 years physics had been living by the unwritten law that the final choice between hypotheses lies with experiment. And such an experiment was staged in 1909 by the New Zealand-born physicist Ernest Rutherford (1871-1937) and his "boys".

Imagine a large, noisy man confined for long hours inside a dark room, looking into a microscope and counting the tiny flashes of light called scintillations (from the Latin *scintilla* for a spark) produced on the screen of a spinthariscope when struck by alpha particles. The work is excruciating: the eyes get tired in two minutes. At his side are the experienced investigator Hans Geiger (1882-1945) and a twenty-year-old laboratory technician Ernest Marsden (1889-1970). Their apparatus is not complicated. It is essentially a vial containing radium-C, which emits alpha particles, a diaphragm that singles out a narrow beam and directs it onto a zinc sulphide screen, and a microscope for observing the scintillating of the alpha particles on the screen. There is no predicting the location of the next scintillation on the screen; they all appear in a random manner, although together they form on the screen a sharply defined image of the diaphragm slit.

Now if we place a piece of metal foil in the path of the alpha particles, instead of the clear-cut image we will obtain a blurred band on the screen. This band is just slightly wider than the slit image obtained before: the alpha particles have only been deflected by 2 degrees. A fairly simple calculation shows, however, that even such a small deflection can only be accounted for by assuming that atoms harbour intense electric fields, a staggering 200,000 volts per centimetre. There can be no such fields in the positive sphere of the Thomson atom. As to collisions with electrons, we may ignore them: an alpha particle (flying at 20 kilometres per second) is to an electron as a cannon ball is to a pea. And yet the paths of the alpha particles were deflected. In search of an answer, Rutherford proposed that Marsden check whether there were some alpha particles reflected back from the foil, a wild proposition in terms of the Thomson model. Two years later, after Geiger and Marsden had counted over a million scintillations, they had proved that one alpha particle in 8000 is thrown backwards.

Only then, on 7 March 1911, did Rutherford present his historic report "The Scattering of α and β Particles by Matter and the Structure of the Atom" before the Manchester Literary and Philosophical Society, whose president had once been John Dalton. That day the audience learned that the atom resembles the solar system: it consists of a nucleus and electrons revolving about it at a distance of about 10^{-8} centimetre. The nucleus is a tiny thing, only 10^{-13}– 10^{-12} centimetre across, but it contains virtually all of the mass of the atom. The nucleus is positively charged, the nuclear charge, in terms of electron charges, being about one half of the atomic weight of the element.

The comparison with the solar system is jus-

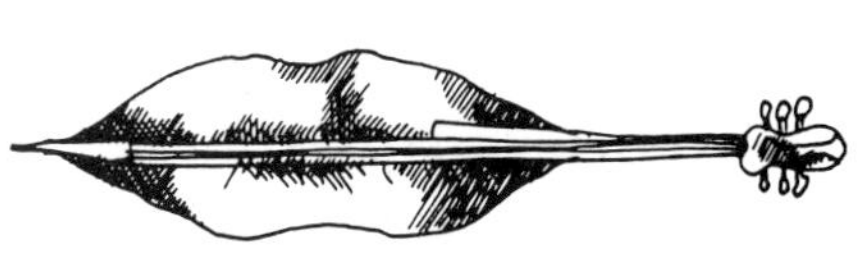

tified: the solar system's diameter (6×10^9 kilometres) is about the same number of times larger than the Sun's diameter (1.4×10^6 kilometres) as an atom (10^{-8} centimetre) is larger than its nucleus (10^{-12} centimetre).

It is difficult for us now to understand why scientists of the calibre of Rutherford should be perplexed. Everything is so obvious: alpha particles are simply deflected by the nuclei of atoms. This picture has remained with us from our school years. But to sketch this picture for the first time took exceptional scientific courage. As mentioned above, over a million scintillations had to be counted; it was also necessary (as Geiger recalled later on in his life) "to overcome enormous difficulties, which now we are even unable to fathom". To begin with, it took ten years (!) to prove that alpha particles are nothing but the nuclei of helium atoms. This was all gradually forgotten; the result is more important and simpler than the path that has led to it.

The physics community reacted to Rutherford's report with reserve. For the next two years Rutherford himself did not put much stock on his own model, although he never put in question the experiments that had led to it. The reason was the same as before. If we are to believe in electrodynamics, this system just cannot exist, because, according to electrodynamic laws, a revolving electron will rapidly and without fail fall into the nucleus. One is thus confronted with the dilemma: electrodynamics or the planetary atom. Physicists silently chose the former — silently, because Rutherford's experiments could neither be forgotten nor disproved. Atomic physics had run into another impasse. It took Niels Bohr to lead science out of it.

Spectral Series

Hypotheses were coming and going, but early on the investigators realized that some important insight into atomic structure could be gained by studying the line spectrum of an atom. Any

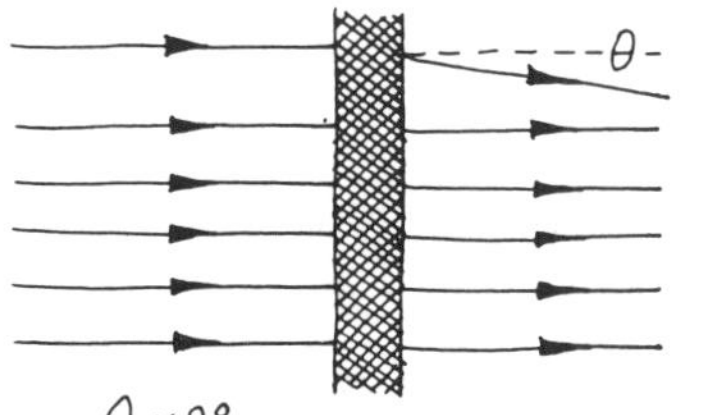

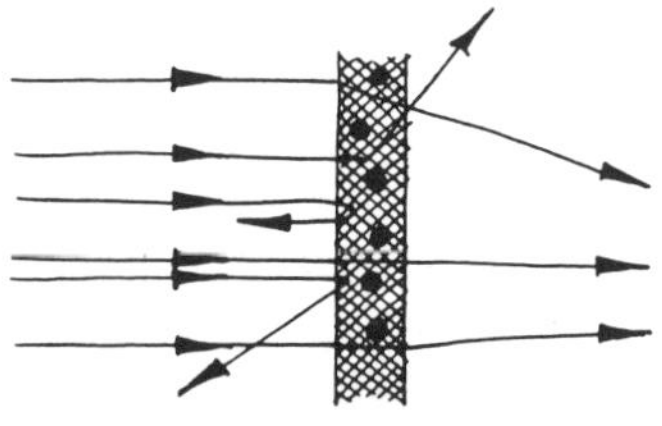

physical research eventually boils down to some sort of measurement. And so it was necessary, above all, to learn to measure wavelengths as accurately as possible, i.e., to examine the structure of line spectra even more painstakingly than Fraunhofer had done.

The first of Kirchhoff's spectrographs was a rather primitive contrivance consisting of two sections of a telescope, a cigar-box and a prism manufactured by Fraunhofer himself. Later this spectrograph was considerably improved, but in time it was replaced by more advanced devices using a diffraction grating, manufactured especially by Henry Augustus Rowland (1848-1901), a representative of the then still young American science community. This device was used over the following decades by Carl David Tolmé Runge (1856-1927), Heinrich Gustav Johannes Kayser (1853-1940), and especially by Friedrich Paschen (1865-1940) and his co-workers at Tübingen, to measure tens of thousands of spectral lines of various elements. The data were carefully listed in long tables. (By 1913 there were already more than 50,000 works on spectral analysis.) In particular, it was found that the famous yellow D line in the spectrum of sodium was actually two very closely spaced lines: D_1 = 5895.9236 angstroms and D_2 = 5889.9504 angstroms (1 angstrom = 10^{-8} centimetre — about the diameter of an atom).

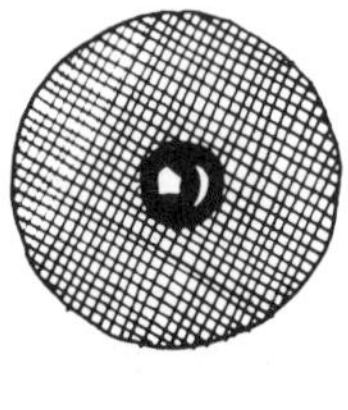

But the ultimate task of any science is not just to pile fact upon fact, but to establish relationships between the phenomena and to find their cause. Everyone felt that those formidable tables contained a huge amount of information about the structure of atoms. But how was it to be extracted? (Probably Egyptologists, before Champolion, had the same feeling when they viewed the hieroglyphics.)

The first step is always difficult and inconspicuous. So, we know very little about Johann Jakob Balmer (1825-1898) who first discovered some pattern in that chaos of numbers. We know that he was born at Lausen in the Swiss canton of Basel on 1 May 1825. He finished secondary school in this town and then went on to study mathematics at the universities of Karlsruhe, Berlin, and Basel. He obtained his Ph.D. in mathematics in 1869 and became a *Privatdozent* at Basel University. He soon left the university, however, to teach physics at a gymnasium for girls.

Balmer was already sixty when he suddenly noticed that the four spectral lines in the visible range of the hydrogen spectrum are located not at random, but form a series that can be described by the formula

$$\lambda = b \cdot \frac{k^2}{k^2 - n^2}$$

where $n = 2$, $k = 3, 4, 5, 6, \ldots$, and $b = 3645.6$ angstroms.

Lines	Measured by Ångström	Calculated by Balmer	k
C	6562.10	6562.08	3
F	4860.74	4860.80	4
G	4340.1	4340.0	5
H	4101.2	4101.3	6

This is a remarkable relation. That it is exact can easily be seen from the table above, which was

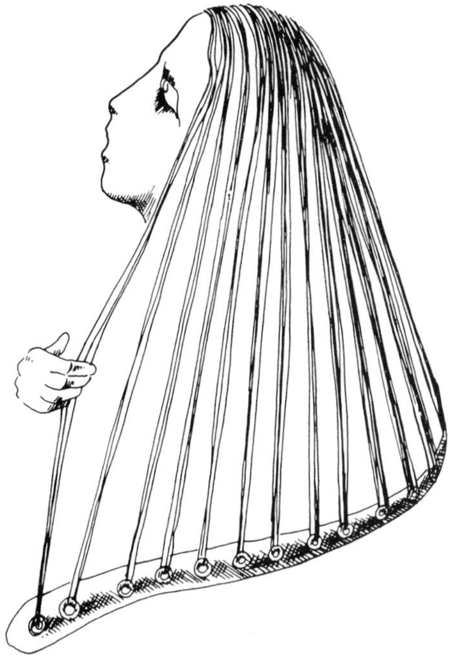

compiled in 1885 by Balmer. The first column lists the names that Fraunhofer gave to the first four spectral lines, the second column lists the same wavelengths in angstroms, which had been measured shortly before with great precision by the Swedish physicist Anders Jonas Ångström (1814-1874). (The unit length 1 angstrom was named after him.) The third column gives the wavelengths calculated by Balmer's formula for the whole numbers k provided in the fourth column. The agreement between measurement and prediction is amazing; these are not chance coincidence, and so Balmer's discovery was not lost in the archives, but rather triggered a series of new studies.

Sometimes Balmer is portrayed as a somewhat eccentric school teacher who, for lack of anything better to do, would divide and multiply various numbers until he by chance came across a simple relationship between them. This is not so, however. He was a well-educated man, wrote papers on various problems of projective geometry and often addressed himself to the most complex problems of the theory of cognition. So, in 1868 he published a paper in which he made an attempt to establish a relationship between scientific research and the philosophical systems of his day. Throughout his life he had been influenced by the Pythagoreans' teachings of harmony and the mystic role played by whole numbers in nature. Like the ancients, Balmer was convinced that the key to the mystery of the unity of all observable phenomena should be sought in combinations of whole numbers. Therefore, when his attention was drawn to the sets of clearly defined spectral lines, he approached the issue with his own yardstick. His expectations were rewarded; the wavelengths of spectral lines did turn out to be related by a simple rational equation.

Balmer's discovery started a new era in the science of the atom. His formula is the essence of the whole theory of the atom. In those days hardly anyone knew this, but perhaps this was in the air. The following year, 1886, Runge noticed that Balmer's formula becomes more elegant if written not in terms of wavelengths λ, but frequency $\nu = c/\lambda$ (here $c = 3 \times 10^{10}$ centimetres per second is the velocity of light):

$$\nu = \frac{4c}{b}\left(\frac{1}{4} - \frac{1}{k^2}\right).$$

And in 1890 the Swedish physicist Johannes Robert Rydberg (1854-1919) proposed the current form of this relation:

$$\nu = cR\left(\frac{1}{n^2} - \frac{1}{k^2}\right),$$

where n and k are integers and R = 109,667.5937 per centimetre has been known ever since as the Rydberg constant for the hydrogen atom. Putting here $n = 2$, one can calculate the entire Balmer's series, which has now been subsequently measured up to $k = 50$.

Another improvement was to write the frequency as the difference of two spectral *terms* T_n and T_k:

$$\nu = \frac{cR}{n^2} - \frac{cR}{k^2} = T_n - T_k.$$

At first this presentation did not seem to offer any noticeable advantages, but in 1908 the young Swiss scientist Walter Ritz (1878-1909) pointed them out. Interested in Rydberg's work, he set forth the so-called *combination principle*: the frequency of any line in the spectrum of any atom can be represented as the difference of two spec-

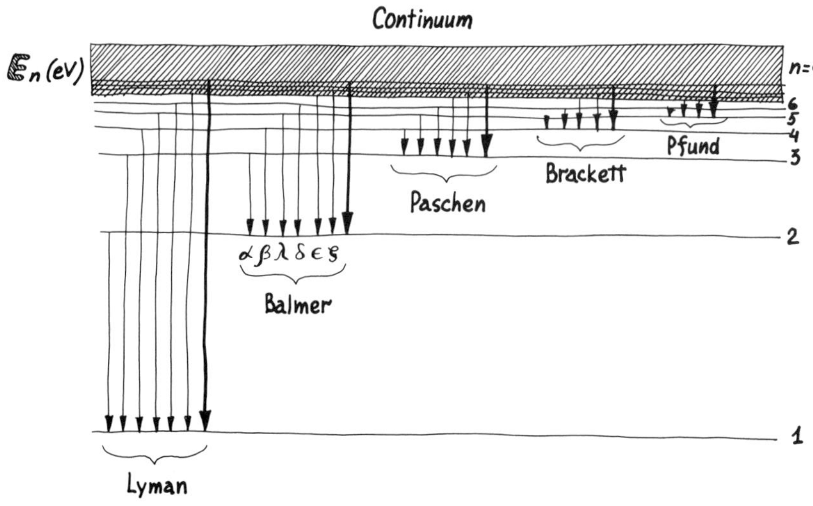

tral terms T_n and T_k:

$$\nu_{nk} = T_n - T_k$$

even if T_n cannot be expressed in a form as simple as that for the hydrogen atom.

At first glance, this is no improvement: instead of a set of frequencies we now have a set of terms. But there is more to it than just that. Justimaginereadingatextwithoutspacesbetween-thewords — you will sense the difference at once, especially if the text is in a foreign language. What is more, there were far fewer numbers to deal with: the frequencies of fifty lines of hydrogen, which were known at the turn of the century, could now be found using a dozen terms.

Suddenly a pattern was discerned in this seemingly chaotic sea of numbers: the disorderly set of lines appeared to consist of series. In an unintelligible text of numbers separate words could now be made out. In the simplest case of the hydrogen atom it even became possible to distinguish the letters of the words. The meanings of words and the origin of letters, however, remained as much a mystery as before: the hieroglyphics had not yet spoken, though they did not seen so puzzling any more. The struggle with the spectra very much resembled blind attempts to decipher a text in an unknown language. This wearisome work dragged on for over a quarter of a century, and the lack of clues baffled some of the greatest scientific minds of the century. A key to the cipher was badly needed.

It was found by Niels Bohr in 1913.

Photons

Radiation originates inside an atom, but, once beyond the atom's confines, it exists on its own. Sometimes it is made up of waves of the same length; such radiation is called monochromatic. The line spectrum of an atom consists of a set of monochromatic radiations peculiar to this atom alone.

Up to this point we have been concerned mainly with only one feature of waves, their frequency. Radiation waves, however, are more complex entities; their properties are characterized not only by the frequency of radiation. A sunbeam is transparent but quite material. It even has mass. So, each second on each square metre of the Earth's surface fall 7.3×10^{-12} gramme of light.

The action of radiation can best be visualized as ocean waves washing ashore. After the work of the Dutch physicist Christiaan Huygens (1629-1695) and the French physicist Augustin Jean Fresnel (1788-1827) this analogy became well established. Each year supplied new proof of this in the form of the phenomena of interference and diffraction of light. In 1873 James Clerk Maxwell (1831-1879) predicted that light incident on the surface of an object should exert a pressure on it (also in agreement with our analogy). Light pressure is very tiny, but, nonetheless, Pyotr Nikolaevich Lebedev (1866-1912) measured it in 1899. It seemed that the wave nature of light had already been proved and that any further experiments to test it would be redundant. Heinrich Rudolf Hertz (1857-1894) wrote in 1889: "Since the days of Young and Fresnel we have known that light is wave motion... To question this fact is no longer possible, to disprove these views is unthinkable for a physicist. For the human race the wave theory is an obvious fact". Hertz experimentally proved the wave nature of electromagnetic radiation and, thereby, the validity of Maxwell's equations.

Fortunately, experiments in physics are performed not only to test theories. In 1887, two years before Hertz wrote the above lines, he himself came across a phenomenon that could only be explained by the corpuscular nature of radiation. The idea behind this phenomenon, which was later named the photoelectric effect, is simple.

If light strikes the surface of a piece of metallic sodium, some electrons are knocked out from the surface. By the end of the nineteenth century most physicists were very much aware that the atom was a complex structure; therefore, this phenomenon came as no surprise to them. Quite soon, physicists realized that the electrons in Hertz's experiments were ejected from sodium atoms by the radiation of the quartz lamp. What they could not comprehend were the laws governing this effect. They were established by Alexander Grigoryevich Stoletov (1839-1896) and Philipp Eduard Anton Lenard (1862-1947) in the closing years of the nineteenth century. These investigators measured the number and velocity of the ejected electrons as functions of the intensity and frequency of the incident radiation.

We already know that radiation originating from different atoms differs not only in wavelength λ (or, what is the same, in frequency ν) but also in intensity. This is easily seen in spectrograms: some lines are much brighter than others. For instance, in the yellow doublet of sodium, the D_2 line is twice as bright as the D_1 line. From previous experience we know that the greater the amplitude, the greater is the action of waves. To see that this is so, we only have to go to the seashore during a storm. Accordingly, increasing the amplitude increases the intensity of radiation. Also, radiation can be made more intense by having more emitting atoms. Therefore, if instead of one quartz lamp we take two, three, or ten of them, the intensity of the radiation will increase two, three, or ten times, respectively. It might seem that the energy of the emitted electrons will grow in the same proportion.

But the energy of the electrons remains the same; only their number increases.

This was the first of the inconsistencies awaiting the researchers at the end of their experiments. The energy, however, was found to depend on the frequency of the incident radiation, and strongly at that. A quartz lamp emits violet and ultraviolet light. If instead we flood our piece of sodium with red light, we will detect no electrons coming from the surface, no matter how intense the incident radiation.

"If radiation is a wave process (and this is a proven fact), then this just cannot happen", contended one group.

"But it does happen!" argued another group.

If large cliff in the sea were to collapse suddenly before your eyes, you would probably look for some external causes of this occurrence. Of course, the surf washes away the shore bit by bit, and rocks do collapse from time to time, but everybody knows how rarely this occurs. But if you see a warship off-shore firing off rounds at the cliff from her large calibre guns, you immediately guess that the unexpected destruction was caused by the shells rather than by the waves, even though the energy of the shells is less than the total energy of sea waves. The energy of waves is uniformly distributed along the seashore; centuries are required for the results of their daily impact to become evident. In comparison with the total work performed by the waves, the energy of a shell is negligible; however, it is concentrated in a small volume and is released in a split second. The shell must be energetic enough to shatter a cliff. A rifle bullet can hardly do this.

This gives an idea of the reasoning followed by Einstein when he came up with his explanation of the photoelectric effect. He was aware of Planck's discovery and doubts, but to him, with his unbiased manner of thinking, the hypothesis of quanta of light did not seem so appalling as it did to Planck. And so Einstein was not only the first to believe in it, but also the first to use it to explain new findings. Einstein stated that light not only is emitted in quanta, as follows from Planck's hypothesis, but it also travels in quanta. (As an aside, the term "quantum" was coined by Einstein; Planck only spoke about "elements of energy".) Therefore, light falling on a metal surface can be likened to shells rather than to sea waves. Besides, each such shell-quantum can only release one electron from an atom.

According to Planck, the quantum has the energy $h\nu$. Einstein reasoned that some of this energy (P) goes into ejecting an electron from the atom, the rest to accelerating it to a velocity v. That is to say, the electron gains the kinetic energy $mv^2/2$. Or

$$h\nu = P + \frac{mv^2}{2}.$$

Once we accept this hypothesis the principle of the effect becomes clear. While shells are small in calibre (red light) they are too weak to knock out an electron from the atom ($h\nu < P$), no matter how many of them are in a salvo. If we go to large calibres (violet light), the energy of a shell becomes sufficient for the task ($h\nu > P$). As before, however, the energy of our shell-quanta will depend only on their size (i.e., on frequency ν), and not on their number.

Sixteen years later, for the classical simplicity of the above equation the Swedish Academy of Sciences awarded Einstein the Nobel Prize for Physics. But back in 1905, when the equation was just published, Einstein was attacked on all fronts. Even Planck joined in the attack. He was fond of Einstein and, urging the Prussian Minister of Education to invite Einstein to teach in Berlin, implored him "not to reproach him [Einstein] too strongly" for his hypothesis concerning the photoelectric effect.

One can understand Planck: he had just introduced the quantum of action h against tradition and his own feelings. Only gradually did he come to accept his own idea. Even as late as 1909 he confessed to Einstein, "I still don't believe very strongly in the reality of light quanta." The mischief had been done, however. "...Planck planted a bee into physicists' bonnets," said Einstein twenty years later. And it gave them no peace, even when they tried to ignore it. At any rate, Planck tried to introduce his quantum of action so as not to do any damage to wave optics — that edifice of exceptional beauty which had been built over two centuries. Therefore, according to Planck, light is only emitted in quanta, but it propagates, as before, as a wave. Only in that way could all the results of wave optics be conserved.

On the other hand, Einstein acted as if no physics had existed before him or, at least, as if he had been a person completely ignorant of the true nature of light.

He displayed one of his most remarkable features — the ability to use logic to perfection, but he had more trust in his intuition and in the facts. For him there were no accidental facts in physics. The photoelectric effect for him was not an annoying exception to the rules of wave optics but one of nature's signals pointing to the existence of profound laws yet to be discovered. It so happened that historically it was the wave properties of light that were studied first. Only when confronted with the photoelectric effect did physicists first run into the corpuscular aspect of light. Reliance on customary patterns of thought was so strong, however, that many just refused to belive in the fact. "Absolutely impossible!" they muttered, like the American farmer who saw a giraffe for the first time in his life.

Einstein, of course, was acquainted with the history of optics just as well as the next man, but his independent mind treated its time-honoured tenets without due respect. What price all of them if they could not explain a single but undeniable finding. For Einstein the unity of nature was an article of faith, and for him one such experiment meant no less than the entire history of optics. And his integrity would not allow him to sweep even a single embarrassing fact under the carpet.

The only truly dangerous thing in science is bad experiments, since experimental results are customarily taken at face value. Any hypothesis, however attractive, is always subjected to an acid test. Even if it turns out to be wrong, the experiment performed to test it may often yield more valuable results than the hypothesis itself, which may well be wrong. Einstein's hypothesis was also tested; it turned out to be true.

In 1911 Robert Andrews Millikan (American, 1868-1953) tested Einstein's equation by using it to calculate the value of Planck's constant h. It coincided with the value obtained by Planck from the theory of thermal radiation. Soon an experiment was made whose idea was similar to the picture of the cliffs collapsing on the seashore. Again the winner was Einstein, not wave optics.

Einstein did not deny, of course, that wave optics existed. Nor did he dispute the experiments proving the wave aspect of light. He simply gave a clear formulation of the contradiction that arose and left it to be resolved by the next generation of physicists. In 1909, addressing a meeting in Salzburg of the Society of German Natural Scientists, he predicted that "the next phase of development in theoretical physics will give us a theory of light that will in a sense be a fusion of wave theory with emission theory". Twenty years later his prediction came true.

Although it elicited a chorus of protest, the idea of light quanta did not wither, but rather struck deep roots. The year was 1913; in that year in Rutherford's laboratory appeared a shy, unhurried Dane, named Niels Bohr.

Victory of Atomistics

With a solemnity worthy of true English tradition and the significance of the event, the centenary of the atomic theory of matter was duly celebrated on 20 May 1904 in Manchester, where John Dalton had spent the most productive years of his scientific career.

The theory won out after a long struggle. Even after Dalton's publications, many regarded atomistics as just a "curious hypothesis, permissible from the standpoint of our cognitive power". When nineteenth century philosophers wholeheartedly rejected the atomic hypothesis, some of their negative attitude rubbed off on physicists. For instance, the German philosopher Arthur Schopenhauer (1788-1860) referred to atoms as nothing but a "fiction of ignorant pharmacists", and the Austrian philosopher and physicist Ernst Mach (1838-1916) called all atomists "a congregation of the faithful" and would interrupt anyone who tried to convert him to this belief with the question: "Have you ever seen an atom yourself?" Only in 1910, when he saw the scintillations of alpha particles on the screen of a spinthariscope did he admit with reserve and dignity, "Now I believe in the existence of atoms". Mach's doubts can be understood: it is difficult for a person to conceive of anything that is indivisible *in principle.* And yet the atomic idea won out at the turn of the century: the mind proved to be capable of understanding even what it was unable to conceive. This came about much sooner than the 300 years predicted by Ludwig Eduard Boltzmann (Austrian, 1844-1906), who was not understood by his contemporaries and took his own life.

Still, the victory came a little too late: Thomson and Rutherford had by then deprived the concept of the "atom" of its previous meaning. It became clear that the atom was not the simplest piece of matter, even though it could not be split up by chemical means. "Unfortunately, the laws of nature become entirely clear only when they are no longer correct", said Einstein. This does not mean, of course, that previously discovered laws all of a sudden become of no significance. In the history of the atom, regardless of later achievements in science, the proof of its existence (even in the old sense of "indivisible") will always be among its most important triumphs.

The final victory of atomistics is also connected with Einstein. In 1905, independently of the Polish physicist Marian Smoluchowski (1872-1917), he gave a mathematical description of Brownian motion. His theory was confirmed experimentally by Perrin, who on the advice of Langevin undertook in 1909 a thorough investigation into Brownian motion. Even before Perrin many physicists had been convinced that the true cause of this motion was the bombarding of the particles by molecules of the liquid — the latter being invisible even in the most powerful microscope. The remarkably elegant experiments of Perrin not only proved this conjecture, but also led to something more important: the mystical motion of particles in a liquid is an exact model of the true motion of invisible molecules, only magnified several thousand times. Therefore, when we watch Brownian particles we obtain a graphic picture of the motion of invisible molecules. (Just as a knowledge of the properties of radiowaves gives us an idea of waves of light or even X rays.)

Now the atomic hypothesis was accepted by all, even by its famous opponent, the German chemist Friedrich Wilhelm Ostwald (1853-1932). In 1909 Rutherford, assisted by his research student Thomas Royds, provided the most convincing proof of the atomic structure of matter.

It had long been observed that minerals containing radioactive thorium, uranium, and radium accumulate helium. It had even been estimated that one gramme of radium in a state of radioactive equilibrium exuded 0.46 cubic millimetre of helium a day, i.e., 5.3×10^{-9} cubic centimetre a second. Now that the nature of alpha particles had been established, there was nothing miraculous about this. But Rutherford and Royds went further. They *counted* the number of alpha particles emitted per second by 1 gramme of radium; it was found to be 13.6×10^{10}. These alpha particles, having captured two electrons each, become helium atoms, each of which occupies a volume of 5.32×10^{-9} cubic centimetres. Hence in one cubic centimetre there are

$$L = \frac{13.6 \times 10^{10}}{5.32 \times 10^{-9}} = 2.56 \times 10^{19} \text{ atoms}.$$

But this is our old acquaintance, Loschmidt's number. As a matter of fact, 1 mole of helium (or of any other monoatomic gas) takes up a volume of 22.4 litres and contains 6.02×10^{23} atoms, i.e., in 1 cubic centimetre there are

$$L = \frac{6.02 \times 10^{23}}{22.4 \times 10^{3}} = 2.69 \times 10^{19} \text{ atoms}.$$

The agreement is convincing.

By 1912 there were more than a dozen ways of determining Avogadro's number. A knowledge of this value was important for explaining such diverse phenomena as Brownian motion and the blue colour of the skies, the viscosity of gases and the black-body spectrum, radioactivity and electrolysis. This number appeared to be enormous, and to demonstrate this Lord Kelvin suggested a thought experiment: toss a glassful of water with atoms labelled in some way or other into the ocean; after the ocean has been stirred properly scoop some water from it on the other side of the ocean — the glass will contain 200 labelled molecules of water (in reality even more — about 1000). Like the global population, Avogadro's number $N_A = 6.022137 \times 10^{23}$ mol^{-1} cannot be fractional. And what is more, we now know it to a considerably greater degree of accuracy than the number of people living on Earth.

"If, in some cataclysm, all of scientific knowledge were to be destroyed, and only one sentence passed on to the next generation of creatures, what statement would contain the most information in the fewest words? I believe it is the atomic hypothesis (or the atomic fact, or whatever you wish to call it) that *all things are made of atoms — little particles that move around in perpetual motion, attracting each other when they are a little distance apart, but repelling upon being squeezed into one another.*

In that one sentence, you will see, there is an enormous amount of information about the world, if just a little imagination and thinking are applied."

So believed Richard Phillips Feynman of the United States (1918–1988), a 1965 Nobel Prize winner. Although he echoes Democritus, nearly word for word, the *concepts* and *ideas* we associate with these words are now entirely different: during twenty five centuries science has learned a great deal about the atom. Acquiring this knowledge was not a simple task — only the results in science are simple.

Quantumalia

The Indivisible Atom

The breath-taking advances in scientific knowledge have overshadowed the old arguments in favour of the existence of atoms — they are now remembered only for their historical value. Some of them, however, may prove of interest to us.

So, those who believed in atoms asked their opponents the simple question: "How can the same amount of matter, if it were not made up of atoms, occupy different volumes as we observe, for instance, in the case of the compression and expansion of gases?" They went on to produce proof that atoms are very small and exist in enormous numbers. For example, one needle-shaped crystal of indigo can colour all of 1 tonne of water. One grain (0.062 gramme) of musk filled a large room with an odour that persisted for 20 years without undergoing any noticeable changes.

You never know what is enough until you know what is more than enough.

William Blake

The advances of the exact sciences undermined people's faith in speculation, in the most convincing way. It was replaced by quantitative estimates.

Newton estimated the thickness of soap films to be 10^{-6} centimetre = 100 angstroms, one fiftieth of the wavelength of light. After him many (including Lord Kelvin) returned to soap bubbles and early in this century it was shown that soap films can have thicknesses as small as twice the molecular diameter.

Ever since Benjamin Franklin poured a spoonful of oil on to the surface of a pond near London, his experiment has been repeated many times in different forms. Specifically, Lord Rayleigh prepared oil films as thin as 16 angstroms and in 1890 Röntgen succeeded in reducing the thickness of the films to 5 angstroms, which is only five times the diameter of the hydrogen atom.

Faraday made gold foil only 10^{-6} centimetre thick, and, by precipitating gold from solution onto glass, he produced gold films 10^{-7} centimetre thick, i.e., one tenth of the average thickness of a soap bubble's wall. These gold films are transparent and their thickness is about ten atomic diameters.

Among other attempts at determining the dimensions of atoms mention should be made of the unjustly forgotten work of Thomas Young (1773– 1829) of England. He studied in 1805 the capillary motion and surface tension of liquids and came to the conclusion that atoms are less than 10^{-8} centimetre across.

M. Planck

The Diffraction Grating

There is no way of knowing what turn the history of the atom would have taken had physicists not invented the diffraction grating. Fraunhofer used it as far back as 1822. Ångström made it the principal tool in his tool-kit and, finally, Rowland refined it almost to its present form. The principle behind the grating is diffraction, i.e., the capacity of waves to bend around an opaque barrier whose size is comparable with their wavelength. Waves of different length diffract differently, which enables them to be separated and measured. This instrument permitted accuracies of measurement in spectroscopy that are astounding even for physics. Early in the century the grating enabled two lines in the visible range to be resolved, if their wavelengths differed by at least 10^{-3} angstrom (now 10^{-4} angstrom).

Just imagine that you were to measure the length of the equator to within 1 metre. The idea is clearly ridiculous, if only because the result would be influenced by each mole-hill along the way. But in spectroscopy such exercises make perfect sense and the history of the atom proved this — despite all the distrust and scoffing that sometimes accompanied such undertakings. One example is the story of the standard metre.

The famous platinum-iridium bar with two transverse parallel scratches, produced by a decision of the French National Convention and kept under a bell glass at the International Bureau of Weights and Measures at Sèvres near Paris, turned out to be inaccurate. This had long been suspected. In 1829 the Frenchman Jacques Babinet (1794-1872) proposed to take as a reference the wavelength of some spectral line "as a quantity absolutely invariable and independent even of cosmic cataclysms". His idea was first realized in 1892 by Albert Michelson (1852-1931). But only in 1958 was the new reference standard legally accepted: 1,650,763.73 wavelengths in a vacuum of the orange-red line of krypton-86.

Just What Hath Rutherford Wrought?

E. Rutherford

In the early twentieth century the concept of the planetary atom was not as rare as many science historians today tend to think. Suffice it to say that as early as 1896 (one year before the discovery of the electron) this idea was used by Lorentz and Larmor to explain the splitting of lines in a magnetic field, an effect discovered by Zeeman. One finds kindred ideas even in the pages of textbooks of that time. By way of illustration we will cite some passages from the third volume of the *Cours d'electricité* published in 1907 by a professor of the University of Paris, Joseph-Solange-Henri Pellat (1850-1900): "...the atom is not an indivisible particle of matter. The emission of light, producing spectral lines that are characteristic for each kind of atom, points to the diversity of atoms. It may be assumed that atoms consist of a large number of corpuscles attracted to some kind of centre, as the planets are attracted to the Sun.

"For the atom to be neutral it is necessary that the positive electric charge, which, as we assumed, is in the centre of the atom, be equal in magnitude to the sum of the negative charge-corpuscles revolving about this centre.

"In a word, all the light, electrical, heat and mechanical phenomena can be explained by assuming the existence of two different kinds of matter: the corpuscle, or the negative electron, and the positive electron, about which we know almost nothing. The central positive charge of the atom consists of a combination of positive electrons whose number varies according to the kind of atom being considered but is quite definite for each kind...

"It is hardly necessary to demonstrate the elegance of this theory, which enables all phenomena known so far to be explained, and to relate so many phenomena and laws together even though they do not seem to have anything in common."

Next year, the famous French physicist and mathematician Jules Henri Poincaré (1854–1912) stated: "the experiments on the conduction of gases... provide us with grounds for regarding the atom as consisting of a positively charged centre, of a mass approximately equal to that of the atom itself, and some electrons, attracted to the centre and revolving around it."

It isn't what people think that is important, but the reason they think what they think.

Eugene Ionesco

Perhaps these quotations will disappoint many: it appears as though Rutherford has not thought up anything new. This common delusion comes from a lack of understanding of the difference between science and natural philosophy. A strict rule in science states: he who proved it, discovered it. And you can prove things in science only by experiments and data. All of the previous statements were pure speculation and amounted to the following: the atom, probably, has such and such a structure. Only Rutherford had the moral right to say: "This must be so. I can prove it with figures. Anybody who wishes can check them if he repeats my experiments."

Mendeleev was wont to say: "Sure, you can say anything, but you go and demonstrate it." Napoleon said the same with a soldier's directness: "It doesn't matter who suggested the idea; what matters is who carried it out." This difference between a vague idea and a scientific proof should always be kept in mind when in the history of science bitter disputes over priority blaze up from time to time. It is reasonable in such cases to consider that originators of a theory are those whose work, by virtue of deep-seated reasons or chance circumstances, has had a decisive influence on the subsequent development of science rather than those who first mentioned it. In purely human terms, there appears to be an element of unfairness about it. But history is not only based on moral considerations. Its task is to establish the true sequence of cause and effect, not to rectify wrongs.

Knowledge is the only instrument of production that is not subject to diminishing returns.

John Clarke

Light Pressure

The hypothesis of light pressure has been known since the days of Johannes Kepler (1571– 1630), who proposed it in 1619 to explain the origin and shape of the tails of comets. Nothing was known then about the magnitude of light pressure and, as so often in such cases, incredible stories were told. For example, in 1696 Nikolas Hartsoeker (a one time mentor of the Russian Czar Peter the Great) related the accounts of travellers who contended that the "current of the Danube River is much slower in the morning when the sun's rays oppose its flow than after midday, when the sun's rays quicken its flow."

In 1746 Leonard Euler again returned to the idea that light waves exert a pressure on exposed bodies, but the idea was rejected by all the authoritative figures of the scientific world.

Until the end of the last century, numerous attempts at detecting the pressure experimentally ended in complete failure. The cause of these failures became entirely clear after the theoretical work of Maxwell and the successful experiments of Lebedev. The pressure turned out to be minuscule. For example, even on a clear sunny day the pressure of sunlight is less than 2 milligrams per square metre, the weight of five poppy seeds.

Janus, Roman

Chapter Four

Pre-Bohr Times • The Bohr Atom
Post-Bohr Times • Formal Model of the Atom

In childhood we all day-dreamed of pirates and treasure ships. In our fervent imagination we pictured the battle and the chase, treasure islands and noble deeds. Almost real to us were those visions of frigates scudding through the blue seas and out of sight, leaving only a foamy wake. During desperate moments, to increase the speed of their sailing ships, buccaneers took a last ditch measure: they would heave their ballast overboard, thus barely escaping pursuit. Often they got away with this trick, but every now and then they were severely punished; the frigate without its ballast became unstable, like an eggshell with a sail. All it took was the first squall to capsize the ship.

At first sight, this chapter may appear so dry and complicated that some readers will consider it to be unnecessary ballast. But this ballast is similar to that which is put into the hold of a frigate: without it the sails of our fantasy are not only useless but even dangerous. Too often, in pursuit of speed and grace, we sacrifice soundness and depth. We generally pay dearly for such carelessness; at some point in our journey our ship of knowledge, made unstable by the lack of hard facts, capsizes and we are back to square one.

There is nothing in this chapter that cannot be comprehended by a thoughtful, patient reader. It will, however, call for systematic logical thinking on his part. His labour and patience will be rewarded later in the book. It may well be that a first reading of the chapter will raise more questions than it will give answers. No harm done. In return you will get a glimpse into the life of quantum mechanics — behind the scenes, as it were. Psychologically, only such insights give you a real sense of the elegance and stability of novel ideas.

Pre-Bohr Times

By the time Niels Bohr arrived at Rutherford's laboratory in Manchester, much was already known about the atom. So much that at times this hampered sifting the wheat from the chaff.

The figure shows only those effects that subsequently provided some important clues to our understanding of atomic structure.

From this observational evidence it was necessary to deduce the architecture of the atom, which is in itself an invisible object. Problems of this type are known as "black-box" problems. We know how our "black box" (an atom) is affected and what comes out of it, i.e., we know what happens and why it happens. But we wish to know more, i.e., the workings of the "black box". This is much more difficult to accomplish than reconstructing a complete stage show from haphazard fragments of music and speech.

Even though we may know all the external manifestations of the internal behaviour of the atom, we will still have to generalize and to synthesize them, using our intuition, to lead us through the gaps in our logical constructs. The figure illustrates the complexity of the problem at hand, which is to give a consistent explanation of all of these extremely diverse experiments. Niels Bohr found such an explanation by assuming that the three physical ideas, *atoms*, *waves*, and *electrons*, are related by the concept of the *quantum*.

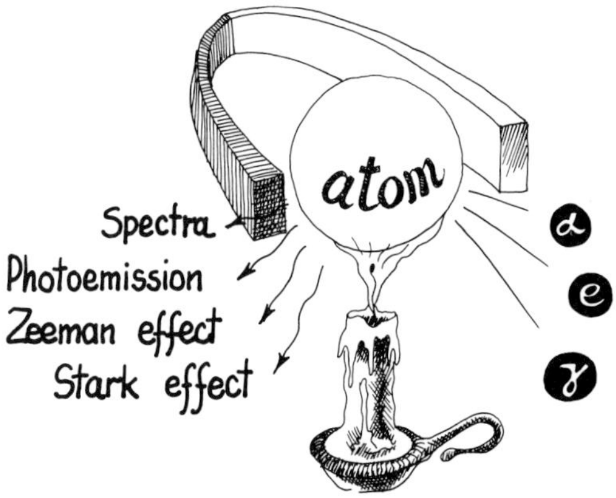

Until that time these ideas had evolved independently. Chemistry and the kinetic theory of matter proved that atoms do exist. Maxwell's electromagnetic theory described the behaviour of light waves. The electrodynamics of Maxwell and Lorentz dealt with the electron.

Even after the work of Einstein and Millikan, nobody in Europe took the quantum of action h seriously, though there were some attempts to make use of it. So in 1910 Arthus Haas (1884-1941) used Planck's relation $E = h\nu$ to determine the boundaries and periods of motion of the electrons in the Thomson atom. John William Nicholson (1881-1955) in 1912 wanted to apply the idea of quanta to the spectra of the Sun and nebulae, and Walther Hermann Nernst (1864-1941) employed it to quantize the rotation of molecules.

The scepticism toward the idea of quanta was best captured by Planck himself in a paper that he presented before the German Chemical Society on 16 December 1911, almost eleven years after he had made his famous report. He said: "The simplest, so to speak, and most naive explanation would be to attribute an atomistic structure to energy itself... This assumption I had made earlier, but I gave it up because I found it too radical..." In a book he wrote a year later Planck said in the same vein: "When you consider the complete experimental confirmation that Maxwell's electrodynamics have received in the most subtle manifestations of interference, when you consider that a rejection of interference would lead to inconceivable difficulties for the whole theory of electrical and magnetic phenomena, you would feel aversion to the destruction of these fundamentals in a single blow. For this reason, in the following we will not deal with the hypothesis of light quanta, the more so since its development is still in the rudimentary stage."

The Bohr Atom

In 1912 Niels Bohr was already working under Rutherford in Manchester. Manchester is separated from Continental Europe by half of England and the English Channel. Perhaps that explains why the attitude toward the quantum hypothesis in Rutherford's laboratory was cautious but without the continental distrust. And perhaps that also explains why, when Planck was still writing his book, Niels Bohr was already firmly convinced that "...the electron structure of Rutherford's atom is controlled by means of the quantum of action". Nevertheless, only after another year of hard reflection did he formulate his famous "Bohr postulates".

What could have been his line of reasoning? When Alexander the Great failed to untie the Gordian knot, he simply cut it with his sword — he was a general and conqueror. Bohr's task was more difficult, but he acted in a similar manner. He reasoned approximately as follows: according to the laws of mechanics the electron in Rutherford's planetary atom must revolve around the nucleus to keep it from falling into it. But, according to the laws of electrodynamics, the revolving electron should radiate energy and, ultimately, it should fall into the nucleus. *We must prohibit it from falling into the nucleus.*

"Just a minute," people objected, "what do you mean by 'prohibit'? Do you mean to say that the electron and the nucleus are bound by electric forces?"

"Yes."

"And that they are described by Maxwell's equations?"

"Yes."

"And that even the mass m and charge e of the electron are determined from electrical measurements?"

"Exactly."

"Then the motion of the electron in the atom must also obey Maxwell's electrodynamics?"

"No."

You must agree that such a way of conducting an argument can anger even the calmest of people. "But, you see, the atom is stable, nonetheless!" Bohr repeated in answer to all the objections. "And we know of no reason for this stability. It just exists." And many years later he reminisced: "The starting point for me was... the stability of matter, which from the point of view of the earlier physics is a genuine miracle."

In his search for supporting evidence for this indisputable fact, Bohr ran across a book by Johannes Stark called *Principles of Atomic Dynamics* in which he first saw the Balmer and Rydberg formulas. "Everything became clear to me at once," recalled Bohr. "After many attempts to apply quantum ideas in a more rigorous form it occurred to me, in the early spring of 1913, that the key to the problem of atomic stability lay in the amazingly simple laws that govern the optical spectrum of elements."

Now he could formulate his famous postulates:

First postulate (on stationary states): *In the atom there are orbits in which an orbiting electron does not radiate.*

Second postulate (on quantum jumps): *Radiation only occurs when an electron jumps from one stationary orbit to another.* The radiation frequency is given by Einstein's formula $h\nu = \Delta E$ for light quanta emitted in transitions between levels with the energy difference $\Delta E = E_i - E_f$, where E_i and E_f are the energies of the initial and final states of the electron, respectively.

To aid our understanding of the postulates we will turn to the obvious analogy between the postulated revolution of an electron about the nucleus and the revolution of a satellite about the Earth. Newton discovered the law of gravitation when pondering over the question: "Why doesn't the Moon fall into the Earth?" This question can now only be found in old jokes, because everybody knows the answer: "Because it is in motion, and at a definite velocity, which is determined by its distance from the Earth." For a satellite not to collapse onto the Earth and also not to fly away into space there must be a definite relationship between the radius r of its orbit and its velocity v along this orbit.

In the hydrogen atom, in which a single electron of mass m and charge e moves around the nucleus, there is a similar relationship, which can be given by the equation

$$\frac{mv^2}{r} = \frac{e^2}{r^2}.$$

This equation is always valid, irrespective of whether the electron radiates or not. It is simply the well-known equality of the centrifugal and attractive forces. If in terms of electrodynamics the electron loses energy by radiation, it will fall into the nucleus, as does a satellite upon re-entry into the atmosphere. But if there exist some special — *stationary* — orbits in which the electron flouts the laws of electrodynamics and does not radiate, there must also be some additional conditions that distinguish these orbits from among all the rest. How these conditions arise is best shown by pursuing our analogy with the satellite.

In addition to the radius r of the orbit and the velocity v along this orbit, circular motion has another characteristic — the angular momentum, *or orbital moment,* l, which is the product of the mass m by v and by r, i.e., $l = mvr$. For a satellite the value of l is arbitrary, depending on the values of r and v. Bohr stated that an atomic electron differs from a satellite in that its orbital moment l cannot be arbitrary — it must be equal to an integral multiple of $\hbar = h/2\pi$ (the symbol $\hbar$ was introduced by one of the founders of quantum mechanics Paul Adrien Maurice Dirac (1902-1984), and it is pronounced as "h bar"). Thus,

$mvr = n\hbar$

where n is an integer 1, 2, 3... This is then the additional condition imposed by Bohr that distinguishes stationary orbits (the only permissible ones in the atom) from the infinite number of conceivable ones. And since this condition is based on the quantum of action h, the whole approach has been called *quantization.*

From these two conditions, using only some algebra, we can express in terms of m, e and h the orbital radius r and velocity v of the electron. We can also find the total energy of the electron in orbit $E = mv^2/2 - e^2/r$, which is the sum of the kinetic energy of the electron and the potential energy of its Coulombian attraction to the nucleus. These depend on the number of the orbit in the following way:

$$r_n = \frac{\hbar^2}{me^2} n^2, \quad v_n = \frac{e^2}{\hbar} \cdot \frac{1}{n}, \quad E_n = -\frac{me^4}{2\hbar^2} \cdot \frac{1}{n^2}.$$

Stationary orbits (and hence energy levels) are numbered by integers $n = 1, 2, 3,...$

It is important and significant that no other, intermediate, energies, besides the set E_n, numbered by the integers n, are possible in the atom. This lack of continuity of virtually all the parameters of the atomic electron — energy, velocity, orbital moment — is the most characteristic feature of quantum theory. This fact is somehow very difficult to grasp.

When an electron jumps from the kth level to the nth level it emits the energy $\Delta E = E_k - E_n$, and the frequency of the emission is given by Einstein's formula

$$\nu = \frac{\Delta E}{h} = \frac{\Delta E}{2\pi\hbar}.$$

This immediately leads to Bohr's famous formula for the frequency of radiation of the hydrogen atom:

$$\nu = \frac{me^4}{4\pi\hbar^3}\left(\frac{1}{n^2} - \frac{1}{k^2}\right).$$

If we observe radiations produced by transition of electrons from various levels k to some fixed level n, we will obtain not just a set of spectral lines,

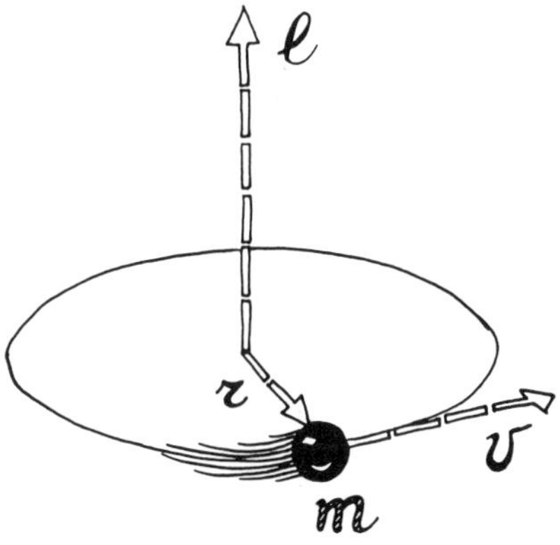

but a *series*. For example, transitions from levels $k = 3, 4, 5, 6, ...$ to the level $n = 2$ yield the Balmer series.

Bohr's formula looks very much like Rydberg's formula for the hydrogen atom, which had been found empirically long before Bohr and was discussed at length in the preceding chapter:

$$\nu = cR\left(\frac{1}{n^2} - \frac{1}{k^2}\right).$$

If Bohr is right, we can, by comparing the two formulas, find Rydberg's constant:

$$R = \frac{me^4}{4\pi c\hbar^3}.$$

This agrees with what was found long ago from spectroscopic measurements.

This was the first triumph of Bohr's theory and it made quite a stir. But this is not the whole story. It followed from Bohr's theory that the radius of the hydrogen atom in its ground (unexcited) state ($n = 1$) is $r_1 = \hbar^2/me^2 = 0.53 \times 10^{-8}$ centimetre = 0.53 angstrom. This implies that the atomic size (10^{-8} centimetre) calculated from this formula is in agreement with that predicted from the kinetic theory of matter.

Finally, Bohr's theory explained how the properties of a line spectrum are related to the internal structure of the atom. This relation had always been sensed intuitively. But only Bohr was able to give it its first quantitative representation. It appeared that the key to such a relationship was Planck's constant h.

This came as a surprise, since the quantum of action came into being as part of the theory of

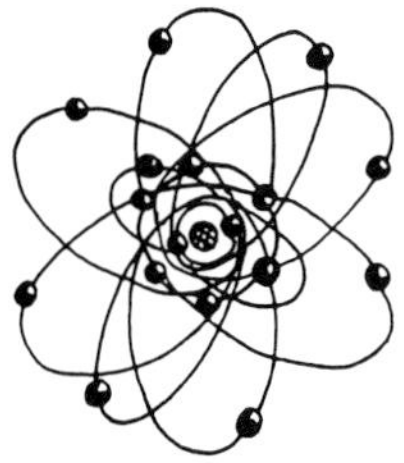

thermal radiation and was in no obvious way connected with atoms or waves emitted by the atoms. And yet it made it possible to estimate the absolute dimensions of the atom and predict the frequency of its radiation. What helped Bohr (as well as many before him) to guess this relationship was his profound faith in the unity of nature.

Bohr's postulates (like any other postulates) cannot be justified logically or deduced from simpler ones. They remain arbitrary products of the human mind until experiment confirms their consequences. If confirmed they become theories, and the most apposite of these theories are called laws of nature.

We will confine ourselves to the above three consequences of Bohr's theory. In fact, there are many more and they all demonstrate the unquestionable power of these incomprehensible postulates. Bohr, of course, came to them along a somewhat different path than we have taken just now. When you climb an unfamiliar peak for the first time, you can hardly expect your path to turn out to be the shortest. Only from the summit, which commands a good view of all approaches, can you see the best way.

Post-Bohr Times

Bohr's postulates were startlingly unusual but they quickly won general recognition, and among their followers were many talented and keen researchers. If we were to assess the attitude of physicists toward them in those days we would find it to be a feeling of relief — relief from the continuous stress under which they had been before, trying to form a coherent jig-saw picture out of odd facts. Now all atomic phenomena appeared to be centred around an incomprehensible but simple model, which brilliantly explained some of them. Others awaited further refinement of the model.

In particular, Bohr's model provided a simple explanation of Kirchhoff and Bunsen's experiment. When light is passed through sodium vapour, consisting of atoms in the ground state, it excites some of them, each atom requiring the energy $E = h\nu$ to be excited. The frequency ν is exactly the frequency of the D line of sodium. And so the transmitted light no longer contains radiation with that frequency and on the screen we see a continuous spectrum with a dark D line in the yellow part of the spectrum. Conversely, when sodium atoms undergo a transition from an excited state to the ground state, they emit light with the same frequency ν as they absorbed earlier, i.e., the same D line, but now it is bright yellow.

Despite its successes, Bohr's theory was at first treated rather like a convenient model. Physicists were dubious about the existence of the energy ladder in the atom. Compelling evidence came in that same year (1913) from James Franck (German, 1882-1964) and Gustav Ludwig Hertz (German, 1887-1975), a nephew of the famous Heinrich Hertz. Like any good idea, Bohr's theory not only explained old facts, but also prompted ways of verifying itself.

Arnold Johannes Wilhelm Sommerfeld (1868-1951) of Germany, an eminent physicist and brilliant teacher, was one of the first in Europe not only to accept Bohr's theory immediately but also to refine it "...by following, as Kepler had once done in studying the planetary system, his inner feeling of harmony". He reasoned as follows: if an atom is similar to the solar system, then an electron there can travel either in a circle, as in Bohr's model, or in ellipses, with the nucleus being at one focus.

Ellipses with the same semi-major axis are described by the same *principal quantum number* n, because an electron in these orbits has the same energy. (Sommerfeld could prove this, but we will just take it on trust.) These orbits differ in their degree of ellipticity, which depends on the *orbital momentum* l. In keeping with Bohr's ideas, Sommerfeld assumed that for a given n the elliptical orbits must only have ellipticities such that their orbital quantum numbers l (which distinguish them) are integers $l = 0, 1, 2, \ldots, n-1$, i.e., there are n permissible ellipses.

But this is not all there is to the Bohr-Sommerfeld atom: if Einstein's theory of relativity is taken into account, electrons in elliptical orbits with the same values of n and various values of l will differ slightly in energy, and so the energy levels in the atom must be labelled with both quantum numbers n and l. For the same reason, the spectral lines produced by transitions of electrons between levels with different n must have a *fine structure,* i.e., they must split into several components. At Sommerfeld's request, Friedrich Paschen tested and confirmed this inference for the helium line of λ = 4683 angstroms, which corresponds to a transition from level $n = 4$ to level $n = 3$. Careful examination of a photograph of a helium spectrum revealed that the line is actually a set of thirteen closely spaced lines. This astonishing agreement at that time (1916) was compared to the calculations of Leverrier and Adams, who predicted the existence of the planet Neptune.

"The spatial quantization of Keplerian orbits is one of the most surprising consequences of quantum theory. In simplicity of derivation and results it is like magic," wrote Sommerfeld in 1916. (In that same year he introduced the term *quantum number* instead of Bohr's rule of whole numbers.)

But even the two quantum numbers n and l did not explain all the details of spectra. For instance if we place a radiating atom in a magnetic field, the spectral lines will split in an entirely different manner. The splitting was discovered by Pieter Zeeman (1865-1943) of Denmark back in 1896. His observations can be treated on the basis of the Bohr-Sommerfeld model as follows: an electron travelling in a closed orbit can be likened to a turn of the winding of an electric motor and likewise it orients itself in a magnetic field. But unlike the winding, the electron orbit can take up only certain definite orientations. These permissible orientations of the orbit are dictated by the *magnetic quantum number m,* which (again in the spirit of Bohr's ideas!) can only assume integral values $m = -l, -(l-1), ..., -1, 0, 1, ..., l-1, l$, or, as easily seen, $2l + 1$ values in all. This implies that in a magnetic field each level with given n and l will split further into $2l + 1$ sublevels, each of which is uniquely defined by the three quantum integers n, l, and m.

As Bohr's theory became more and more sophisticated, it gradually lost some of its initial elegance and graphic simplicity. It evolved into the *formal model of the atom,* which was only called

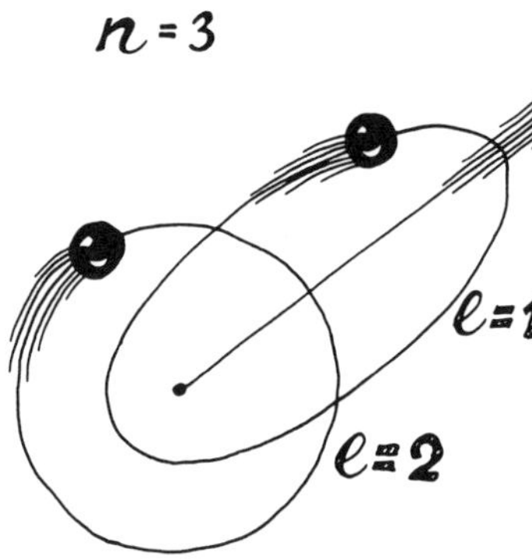

upon to catalogue the spectral terms, that is, the possible energy levels of the whole atom, without the need to specify what the electrons were doing. The word "quantization" thus lost its initial meaning of restricting the allowed states of motion of the electron and came to connote the formal assignment of the quantum numbers n, l and m to each energy level. The quantum numbers, however, do also define the allowed orbits of electrons in the atom and many scientists continued to think in this way. This mode of thinking helps to explain why external electric and magnetic fields, which, of course, influence the motion of the atomic electrons, immediately affect the spectra of atoms by altering and splitting the possible energy levels of the electron (and hence also the spectral lines).

Formal Model of the Atom

Like any other literary form, popular science is not without its constraints. As a rule they start to make themselves felt at the point when it is no longer possible to draw on concepts and images from everyday life to explain scientific facts. To overcome these constraints requires us to switch over to the formal language of science, however primitive that may be. To do otherwise makes the description of the science so long-winded and cumbersome that the mind ceases to comprehend it. Any attempts to avoid this step are bound to lead to a loss of information and a distortion of our understanding, leaving the very core of the science concealed. On the other hand, if you take the trouble to make an additional effort you will be rewarded with additional insights into the power of the logical constructs of the science and into the beauty of their consequences. As a rule,

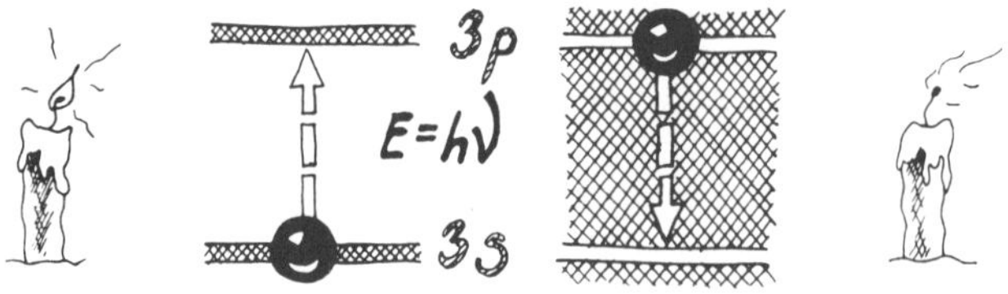

the technical difficulties that arise are no greater than those encountered by any student of chemistry. In almost no time at all he finds it much easier (and, most important, much clearer) to write the formula H_2O than to say each time "a molecule consisting of two atoms of hydrogen and one atom of oxygen".

The theory of spectra has adopted a language resembling chemical formulas: the principal quantum number n is denoted by the integers 1, 2, 3, ..., and the orbital moment l by the letters $l = s, p, d, f$, ... which correspond to the series of numbers $l = 0, 1, 2, 3, \ldots$. Therefore, a level with the quantum numbers $n = 3$, $l = 0$ will be $3s$, and with $n = 3$, $l = 1$, $3p$.

In an unexcited sodium atom, a radiating electron is in the state $3s$. The dark D line appears when the atom is excited to the state $3p$. In the reverse transition ($3p \rightarrow 3s$), the atom emits energy, and so the bright yellow D line appears.

What will happen when the radiating sodium atom is placed in a magnetic field? At first, according to Sommerfeld, the upper level $3p$ should split into three components, since $2l + 1 = 2 \times 1 + 1 = 3$, and the lower level should remain unchanged. Each of the lines D_1 and D_2 should thus have split into three components.

This conclusion was not borne out by experiment. In actual fact, the D_1 line splits into four components and the D_2 line into six. This phenomenon is a special case of the so-called *anomalous Zeeman effect.* To understand it we will have to return to the question we purposely avoided before — why even in the absence of a magnetic field does the D line of sodium consist of two closely spaced components D_1 and D_2?

This question intrigued Wolfgang Pauli (Austrian-Swiss, 1900-1958), one of Sommerfeld's students. His painful speculations led him in 1924 to conceive the idea of electron spin. He argued as follows: both lines, D_1 and D_2, correspond to the same transition from the level with $n = 3$ and $l = 1$ to that with $n = 3$ and $l = 0$. But still there are two lines! Consequently, there exists not one but two upper levels $3p$ with some kind of additional quantum number that distinguishes one from the other. The property that corresponds to this fourth quantum number S he called "the nonclassical two-valuedness of the electron" and assumed that it could only take on two values, +1/2 and – 1/2. Pauli thought that any graphic representation of this property was impossible.

But within a year George Eugene Uhlenbeck (1900-1988) and Samuel Abraham Goudsmit (1902-1978) thought up a graphic model to explain this property of the electron by assuming that it spins around its axis. The model was also a direct consequence of the analogy between the atom and the solar system; the Earth not only revolves in an elliptic orbit around the Sun, but also rotates about its own axis (the analogy had been noted by Compton already in 1921 but Pauli was firmly opposed to it).

Uhlenbeck and Goudsmit suggested that, in addition to orbital angular momentum l, the electron has also inherent to it the intrinsic angular momentum, or *spin,* S equal to 1/2. Combined with the orbital angular momentum l, this intrinsic momentum S can either increase or decrease it to yield the *total angular momentum j,* equal to either $j_1 = l - 1/2$ or $j_2 = l + 1/2$, depending on the mutual alignment of the vectors $\vec{l}$ and $\vec{S}$. If $l = 0$, then the total angular momentum and the spin coincide ($j = S = 1/2$).

Things have thus worked out splendidly. With the electron spin taken into account the level $3s$ in the sodium atom remains unchanged, since it corresponds to $l = 0$, but the $3p$ level with $l = 1$ splits into two, $3p_{1/2}$ and $3p_{3/2}$, *whose energies differ slightly. (There is the convention that the values* $j = l + S$ and $j = l - S$ are put as subscripts to the right of the term symbol, and so $3p_{1/2}$ corresponds to $j = 1/2$ and $3p_{3/2}$ to $j = 3/2$.) Accordingly, instead of one D line we observe two closely spaced spectral lines: D_1 corresponding to the transition $3p_{1/2} \rightarrow 3s_{1/2}$ and D_2 to the transition $3p_{3/2} \rightarrow 3s_{1/2}$.

Just as in the case of l, in a magnetic field each level with j splits further into $2j + 1$ components, which differ in their magnetic quantum number m. And so each of the levels $3p_{1/2}$ and $3s_{1/2}$ will split into two additional sublevels, and the $3p_{3/2}$ level into four. The result is the diagram of levels and transitions depicted in the figure.

The figure illustrates how the initial Bohr model was becoming more and more complex. It started with only one level, $n = 3$. When relativity was taken into account, the level split into two: $3p$ ($n = 3, l = 1$) and $3s$ ($n = 3, l = 0$). With the electron spin taken into consideration the $3p$ level splits further into two sublevels: $3p_{1/2}$ ($n = 3, l = 1, j = 1/2$) and $3p_{3/2}$ ($n = 3, l = 1, j = 3/2$). Finally, in a magnetic field we obtain a system of levels and appropriate transitions between them that accounts for the line pattern observed experimentally.

The hypothesis of the electron spin is one of the most profound hypotheses in physics. Its significance has not yet been comprehended completely. Pauli was, of course, right when he cautioned against the temptation to picture the electron as a kind of spinning top. At times the influence of spin on the workings and structure of the atom manifests itself in an inconceivable manner. One such feature of spin is described by Pauli's exclusion principle that states that *it is impossible for any two electrons in the atom to have the same set of quantum numbers n, l, m, and S.* Later in the book we will see that this principle alone provided a rational explanation for the periodic system of Dmitry Ivanovich Mendeleev and the idea behind the periodic law.

You may have noted already from the language of this section that the formal model is far poorer in descriptive imagery as compared with Bohr's model and that it is much more difficult to give a dramatic representation of it using everyday words and no formulas, however simple. Nonetheless, you may have sensed its power: it can explain and predict rather fine details of spectra. It helped to wrest a measure of order out of that formidable number of spectral lines that had been accumulated over the half-century. Now, any line in the spectrum of an atom could be

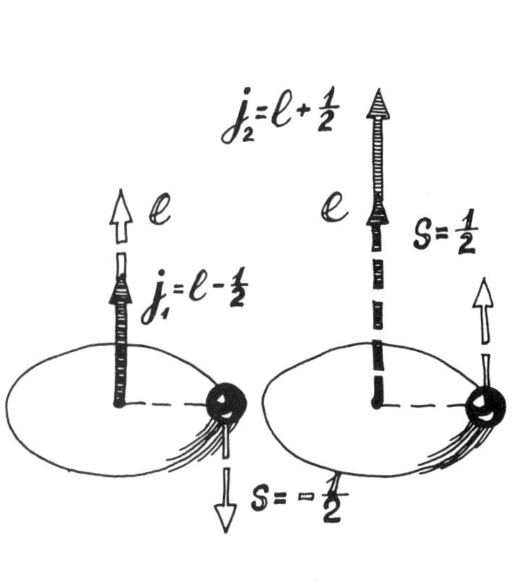

uniquely defined by the eight quantum numbers: four for the initial level of the radiating electron (n_i, l_i, m_i, S_i) and four for the final level (n_k, l_k, m_k, S_k). By 1925 this titanic labour had been completed. The hieroglyphics had been deciphered and the first, yet crude, picture of atomic architecture had been drawn.

Of course, even today, it is not an easy task to decipher the spectrum of an element. It requires expert knowledge. Likewise, you cannot learn to read even deciphered hieroglyphics at one sitting; besides, it is hardly necessary that everyone be able to do this. Physicists are no longer baffled by the long tables of spectral lines, just as biologists are no longer terrified by the millions of species of plants and animals after the taxonomic work of Carolus Linnaeus, Chevalier de Lamarck and Charles Robert Darwin.

With spectral lines things were as they had been with real Egyptian hieroglyphics: until the latter were deciphered they had been of genuine interest only to Egyptologists. But when the hieroglyphics and spectral lines *had* been deciphered, the former enabled the history of an ancient people to be read, and the latter revealed the atomic structure, and these insights were now of great interest to everyone.

The formal model of the atom, despite its triumphs, no longer satisfied the criterion of elegant simplicity bordering on obviousness (which so favourably distinguished Bohr's model). The formal model had become so complex that it aroused suspicion and a certain weariness, similar to that experienced by physicists before Bohr's

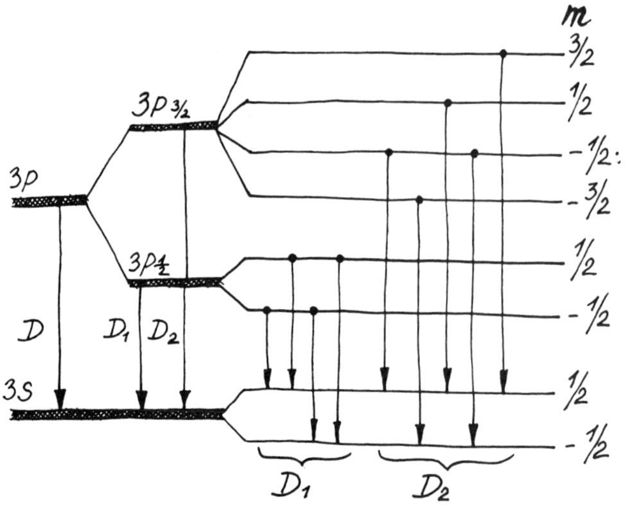

model. Moreover, all attempts to extend Bohr's model to include more complex atoms were failures. Bewildered physicists began to question everything, e.g., the validity of Coulomb's law, the applicability of electrodynamics and mechanics to atomic systems, and even the law of conservation of energy. Everyone understood that this crisis had arisen from formally fusing empirical data, principles of quantum theory and the remains of classical concepts that could not be rejected because nothing definite had been proposed to replace them. When dealing with quantum phenomena, physicists continued to employ classical concepts. But atomic objects did not have the properties that corresponded to these concepts, and so nature was asked what was essentially illegitimate questions. Or rather, questions in a language that nature did not understand. A search then began for a general principle from which the formal model and other aspects of atomic objects would follow. The general consensus was "let there be only one unified principle, comprehensible or incomprehensible."

As if in answer to that wish, the year 1925 saw the birth of quantum mechanics — the science of the motion of electrons in the atom. It was created by a new generation of physicists. By a freak of chance they were all nearly the same age; Werner Karl Heisenberg of Germany was born in 1901, Paul Adrien Maurice Dirac of England in 1902, and Wolfgang Pauli in 1900. Only slightly older were Louis de Broglie (b. 1892) of France and Erwin Schrödinger (b. 1887) of Austria. It was their good fortune to write down the images and concepts of atomic mechanics in the language of formulas.

Quantumalia

Niels Henrik David Bohr

Bohr's discovery, like any truly great insight, was difficult to make, yet easy to understand. The power of Bohr's ideas lies in their simplicity and accessibility. Their gist can be grasped by any intelligent layman. Bohr provided us with an image that was a clue to the world of bizarre *concepts* of quantum mechanics, an image that was destined to become a symbol of our age. And if we take into account that, for all its simplicity, the image correctly reflects the principal features of atoms, its outstanding status will become evident to us.

On 3 March 1972 the *Pioneer 10* Jupiter probe was launched. Among other things on board the spacecraft was a plaque with some pictures engraved on it. These pictures contained the most important information that the people of Earth wanted to communicate to other civilizations: the outlines of a man and a women, the place of the Earth within the Solar system, and a diagram of the hydrogen atom.

God often visits us, but most of the time we are not at home.

French proverb

Out of a hundred physicists taken at random, today hardly more than one or two have read Bohr's famous papers of 1913. But any one of them can explain in detail the ideas set forth in those papers. Moreover, Bohr's ideas are now no longer the property of science alone; they have become part and parcel of our culture, the acme for any theory.

"A man who is destined to endow the world with a great constructive idea has no need for the praise of posterity. His creative work has given to him a more significant blessing." These words of Einstein about Planck can just as appropriately be said of Bohr, and Einstein himself.

N. Bohr

In his twilight years Niels Bohr visited the Soviet Union and came to Georgia. One fine day he went on a picnic with a group of Soviet physicists to Alazany valley. Some of the local peasants were resting on the grass nearby, singing native songs and drinking wine. An inquisitive person, Niels Bohr came to the picturesque group and was welcomed with traditional cordiality. The Georgian physicists began to explain, "This is the famous scientist Niels Bohr...". But the *tamada* (Georgian toast-master) silenced them with a gesture and addressed his companions with the following toast: "My friends! The world's greatest scientist, Professor Niels Bohr, is our esteemed guest. He is the founder of modern atomic physics. His works are studied by schoolchildren all over the world. He has come to us from Denmark. Let us wish him and his companions a long life, much happiness and health. Let his native land live in peace and prosperity." The *tamada* 's speech was translated for Bohr in a low voice. When the *tamada* finished speaking, an old man got up, carefully took Bohr's hand into his own and kissed it. Then, another peasant stood up, filled a drinking horn, solemnly bowed to Bohr, and emptied it.

Niels Bohr had spent a lifetime amongst the paradoxes of quantum mechanics, but even he was struck by the unreal quality of this event. Tears of astonishment and gratitude brimmed his eyes.

Experimental Proof of Bohr's Postulates

Franck and Hertz's experiment closely resembled that of Kirchhoff and Bunsen; but instead of sodium they used mercury atoms and instead of light they used a beam of electrons whose energy they could vary. Franck and Hertz observed an interesting thing: as long as the energy of electrons was arbitrary, the number of electrons that had passed through the mercury atoms equalled the number of electrons in the initial beam. But when the electron energy reached 4.9 electron-volts, or 7.84×10^{-12} erg, the number of electrons that survived beyond the mercury dropped sharply — they were absorbed by the mercury atoms. At the same time, a bright violet line with a wavelength $\lambda = 2{,}536$ angstroms, or frequency $\nu = c/\lambda = 1.18 \times 10^{15}$ per second, appeared in the spectrum of the mercury vapour. The energy of a quantum of that frequency is $h\nu = 6.62 \times 10^{-27} \times 1.18 \times 10^{15} = 7.82 \times 10^{-12}$ erg, i.e., it is almost exactly equal to the energy expended by the electron. Evidently, this radiation is due to the reverse transition of the mercury atom from an excited state to the ground state.

Everything has its beauty but not everyone sees it.

Confucius

The picture observed was direct experimental evidence for both of Bohr's postulates: in the atom there are real stationary states; therefore, the atom is unable to absorb arbitrary portions of energy. Transitions of the electron between the levels are only possible in jumps; and the frequency of the emitted quanta is governed by the difference between the energy levels and is calculated by Einstein's formula $\nu = \Delta E/h$. This, of course, is all so easy now, but in 1913 even Franck and Hertz interpreted their experiment in an entirely different manner.

James Franck remembered fifty years later: "...since back then among physicists there predominated an open mistrust of attempts to construct a model of the atom at the level of knowledge of the time, only few took the trouble to read carefully a work devoted to this. Worthy of particular note is the fact that Gustav Hertz and myself were at first unable to perceive the enormous significance of Bohr's work..."

Lin-Yui, Chinese

Chapter Five

Teachings of the Ancients • First Attempts
Elements and Atoms • Table of Elements • The Periodic Law

Imagine that you have decided to study the life of a cell. You perform every possible experiment: you heat it, expose it to radiation, destroy it, and examine it in a microscope. Your knowledge of the cell however will not be complete unless you remember that a cell is a part of a living organism and only as such does it display its properties fully.

Something similar occurred in the science of the atom. Up till now we have deliberately tried to isolate the atom and have selected only those experiments that could throw some light on the properties of an individual atom. But long before evidence for the complex structure of the atom was forthcoming, Dmitry Ivanovich Mendeleev (1834-1907) had established that the atoms of the various elements composed a single organism — *the natural system of elements.*

The word "elements" came to be known by the letters *l*, *m*, and *n* (*el-em-en-t*). Words are made up of letters, and together they form an alphabet. Likewise, all substances in nature are built up of a small number of chemical elements, which themselves are members of one system — the alphabet of elements.

It must be said that chemists could never reconcile themselves to the notion that there is an independent collection of qualitatively different elements. They always tended, therefore, to think of elements as clusters of particles of the same kind. Such attempts began in ancient times and then developed along two different paths.

Democritus believed that all substances in nature were built up of atoms and that the properties of substances depended upon combinations of their component atoms. Aristotle, on the other hand, contended that all matter consisted of elements, which themselves were holders of specific qualities.

An echo of this ancient controversy has even reached our times: the word "atom" brings to mind something hard and massive, whereas the words "chemical element" conjure up thoughts of some pure quality, regardless of its holder. This perhaps explains why the teaching of chemical elements in earlier times developed quite independently of the idea of atoms. Later on, the two teachings became so interwoven that they could

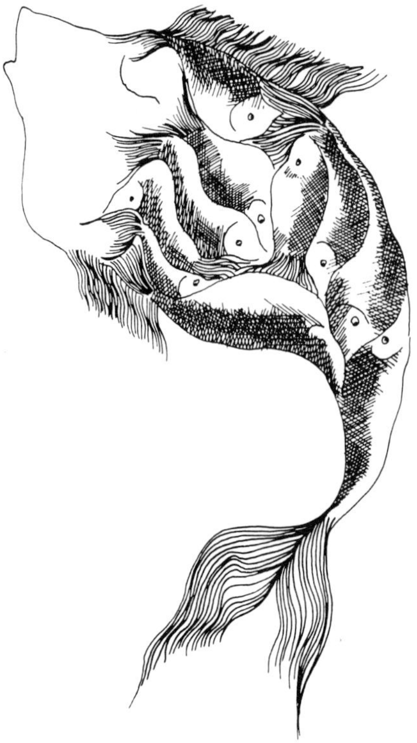

no longer be distinguished from each other. The psychological barrier between them has not yet been surmounted even to this day.

So far we have been dealing at length with the "physical origins" of the science of the atom. Now we will turn to its "chemical origins".

Teachings of the Ancients

Philosophers of the Ionian school of thought, one famous representative of which was Thales of Miletus (*c.* 640-546 BC), recognized only one basic element — water — "on which the Earth floats and which is the beginning of all things". Later Empedocles of Agrigentum (*c.* 490-430 BC) added three more elements — earth, fire and air. Aristotle (384-322 BC) supplied a fifth element, the spiritual one — *quinta essentia,* whose corruption "quintessence" has come down to us.

Something like that is found in ancient Hindu philosophy. But, in contrast to the Greeks, who understood elements to be *substances* that influence our sensory organs, in India the elements were conceived of as manifestations of the Universal Soul or Essence of the Universe. There were five such element-manifestations in Indian philosophy, one for each of the senses capable of perceiving them: ether for hearing, air for touch, fire for sight, water for taste, and earth for smell.

In the Middle Ages the idea of the elements was revived by alchemists, the most famous of whom were the Egyptian Zosimus (fl. AD 300), the Arab Geber (Abu-Mousa-Djaber ben Hayyan Ec Coufy of the eighth century) and Saint Albertus Magnus (Albert von Bollstadt of the thirteenth century). The alchemists (after Aristotle) thought of the elements as being *qualities* or "principles", not substances. So mercury was for them the "principle" of metallic lustre; sulphur, of combustibility; and salt, of solubility. They believed that these principles, if mixed together in the proper proportions, would give any substance in nature.

Examples of alchemical classifications of "elements" are given in the figures. So the fingers of the hand each represent a chemical substance, the fish in the middle being a symbol of mercury. In the other figure the hands point to a vase surrounded by symbols of seven metals.

The word "alchemy" generally conjures up attempts to change mercury into gold, to concoct the elixir of life and other miracles. This association is responsible for a condescending and scornful attitude to the entire history of alchemy. This history encompasses several centuries, however! And it is hard to believe that its carefully developed philosophy and practical recipes contain only plain nonsense. Few know that alchemists discovered alcohol, which alone has perhaps justified their existence. But their main merit is that by their blind experimentation they gradually amassed a heap of facts without which the science of chemistry would never have made its appearance.

First Attempts

In the seventeenth century alchemy and natural philosophy gave way to chemistry and physics. In 1642, the German natural scientist Joachim Jungius (1587-1657) published his *Disputes on the Principles of Matter,* which he concluded, quite in the spirit of the day, with the words: "Which principles should be recognized as primary for homogeneous bodies? — a question that can be answered only with honest, detailed and diligent observation, not by making guesses."

In 1661 appeared the celebrated book *The Sceptical Chemist* by Robert Boyle (1627-1691), in which he defined chemical elements as simple or primitive substances. This was the first and almost modern definition of an element: an element is primarily a substance, not a "principle", substratum, or idea. Well and good, but how to extract elements from natural substances and how to distinguish pure elements from their mixtures or compounds. For instance, Boyle himself supposed that water was almost the only pure element; on the other hand, he treated gold, copper, mercury and, sulphur as chemical compounds and mixtures.

Antoine Laurent Lavoisier (1743-1794) sided completely with Boyle on elements, but since he lived a century later, he went a step further; he learned to isolate elements from chemical compounds. Lavoisier seems to have been one of the first to use scales for scientific purposes and not just for preparing powders and mixtures. He started out with the assumption that *each element of a compound weighs less than the compound as a whole* — an assumption which seems trivial today, but which called for no little courage in the age of phlogiston. By consistently applying this principle, he compiled the first table containing about thirty elements.

The views of Lavoisier were so at odds with the generally accepted tenets of the day that zealous followers of phlogiston theory in Germany burnt his portrait in public. Lavoisier did not finish his investigations. Accused of high treason, he was guillotined in the Place de la Révolution on the afternoon of 8 May 1794, and his body was buried in a common grave. Next morning Lagrange said bitterly: "It took but a moment to sever that head, but it will perhaps take more than a century to produce another like it."

The next hundred years saw a good many new additions to Lavoisier's table. Especially noteworthy among the contributors was the "king of chemists", Baron Jöns Jakob Berzelius (1779-1848), who analysed over 2000 substances, determined to high accuracy the atomic weights of forty-six of the forty-nine elements known at that time, and discovered selenium, thorium, lithium, vanadium, and some other rare-earth elements. (Incidentally, he introduced in 1814 the modern symbols of the chemical elements, using the first letters of their Latin or Greek names.)

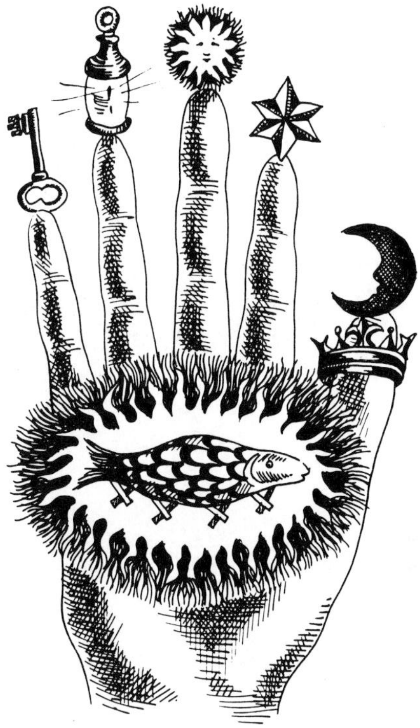

By the mid-nineteenth century about sixty elements were known — not as many as Democritus had expected but not so few that they could be regarded as being all independent. Chemists were becoming ever more confident that the set of elements formed some system, and they set out to search for such a system. As a matter of fact, this search had actually begun a long time before (when it was obviously premature) and had never ceased since. For example, as far back as 1786 Marne was convinced "that all that exists in nature is interrelated in a single continuous series" and that "...from the tiniest dust speck in a sunbeam to the holiest seraph, a whole stairway of creation can be erected..." In 1815 the English physician and chemist William Prout (1785-1850) elaborated Marne's idea concerning the relationship between the elements and proposed the simple hypothesis that all elements were formed as a result of successive condensation of the lightest element, hydrogen.

It is beyond the scope of this book to give a detailed discussion of all attempts at constructing a system of elements, undertaken at various times by Johann Wolfgang Döbereiner (1817), Max Jo-

seph von Pettenkofer (1850), John Hall Gladstone (1853), William Odling (1857), Beguyer de Chancourtois (1863), John Alexander Reina Newlands (1865), and especially Lothar Meyer (1830-1895). What is more important by far is to examine the ideas and driving forces behind these studies.

At the heart of any science lies the human capacity to wonder at things. And the existence of elements has always been and will always be a cause of wonderment. Isn't it strange that this whole world, bursting with colour, odours, sounds, and human passions, is built up of only several dozen elements. And these elements themselves are, as a rule, quite plain in appearance and in no way resemble the colourful world built of them.

In the minds of scientists, however, the feeling of wonderment is soon replaced by the pressing need to put their impressions into some kind of order. This purely human trait runs deep in each of us. A child is pleased when he succeeds in making a regular figure out of a pile of building blocks, a sculptor when he carves a statue from a chunk of marble, a musician when he arranges sounds into a melody.

Whenever we want to put things into some sort of order the question immediately presents itself: "On what basis?" If we have a boxful of numbered wooden blocks, we can easily put them into order by lining them up in increasing order of their numbers. Now, just imagine that instead of blocks you have test tubes with different chemical elements. The substances in the tubes have different colours and smells, they are liquid or solid, light or heavy. Which of these properties shall we use as the basis for our classification? For example, the tubes can be lined up on a shelf so that their colours form a rainbow. The show will be pretty but quite useless for the science of elements. Any classification only makes sense if it unveils fundamental properties or structural features (as does, for instance, the classification of the animal kingdom).

What in general is the use of any classification, besides satisfying our instinctive craving for simplicity? First and foremost, no science can exist without it. Nature has an infinite variety of aspects, and a scientist can only hope not to be lost in them if he learns to pick out the main features from heaps of detail.

Chemical elements have a great many properties. This is quite understandable; otherwise they could not make up such a complex world. The most important of their properties is the capacity to react chemically. It might appear that it is this property that should form the basis of their classification. But this is not so; there is no method for accurately measuring (or even rigourously determining) the reactivity of elements. This makes any such classification unreliable. Not to be arbitrary, a classification must be numerical, i.e., elements must be classified according to a property that lends itself to precise measurement.

But this is not the whole of the story; we can measure the density of elements with great accuracy, but we cannot use it as a basis for their classification, if only because they include gases, liquids and solids.

The many abortive attempts to find a system of elements helped scientists to realize, at least, that among the various obvious properties of elements that can be directly observed *not a single one* was suitable as a basis for their classification. Such a property — the *atomic weight of the element* — lies outside of the domain of chemistry; it belongs to physics. The moment this was first realized can be regarded as the starting point of the modern theory of chemical elements. This decisive step was taken by John Dalton.

Elements and Atoms

Among the men of science of his time John Dalton was a remarkable figure. Early in the nineteenth century society had come to believe in science and understand the secret of its power — science dealt with numbers, and numbers do not lie. The art of marking precise experiments was, therefore, more highly valued than any other skill in those times. Dalton was rather helpless in this art and so he came under severe attacks from the venerable scientists of his day. By his cast of mind Dalton was a typical theoretician, in the present-day sense of the word. And so one should not be unduly critical of some inaccuracies of his measurements for these measurements enabled him to express lucid and fruitful ideas that predetermined the development of chemistry for the next hundred years. In a nutshell, he devised a method of experimentally verifying the atomic hypothesis.

Dalton defined an element as a *substance consisting of atoms of a single species.* The atoms of various substances differ in mass and remain unchanged in all transformations of the substances; the atoms are only rearranged. Dalton wrote that destroying or creating an atom of hydrogen is no less difficult than adding a new planet to our solar system. He was not only a strong believer in the atomic hypothesis, but he also set out to search for its consequences, especially *observable* ones. He reasoned as follows.

Suppose that all elements consist of atoms. Then, say, 16 grammes of oxygen contain N atoms of oxygen. Suppose further that we burn hydrogen in this oxygen. We can easily show that we need 2 grammes of hydrogen to burn 16 grammes of oxygen to obtain 18 grammes of water. The facile answer is that each oxygen atom O combines with one hydrogen atom H to form one molecule of water HO. This is exactly what Dalton thought. Later on, Berzelius proved Dalton wrong: each atom of oxygen combines with *two* atoms of hydrogen so that the chemical formula for water acquires its familiar form H_2O. The most important idea here is that a *whole* number of hydrogen atoms combine with each oxygen atom. Therefore, if 16 grammes of hydrogen contain N atoms, then 2 grammes of hydrogen contain $2N$ atoms. This means that *one oxygen atom is sixteen times heavier than a hydrogen atom.*

It thus became possible to compare the masses of atoms of various elements. To this end, the concept of *atomic weight* was introduced, which is simply a number indicating how many times heavier an atom of a particular element is than the hydrogen atom. By definition the atomic weight of hydrogen is unity.

Dalton came to this conclusion in 1804-05, and in 1808 he published his famous book *A New System of Chemical Philosophy,* which marked the beginning of a new epoch in science. His findings were immediately confirmed by the English physician and chemist William Hyde Wollaston (who had discovered the dark lines in the solar spectrum). The figure shows Dalton's symbols for chemical elements. Some of these are now known to be compounds.

The last step was to learn to *determine* the atomic weights of elements. It was a good idea to begin with the simplest substances. The first to attract the attention of scientists were gases. Very soon, in 1809, the French chemist Joseph Louis Gay-Lussac (1778-1850), a former assistant to Berthollet, made a very important discovery. He found that the volumes of two gases that combine in a chemical reaction are always in a simple whole-number relation to each other. *Volumes,* not masses — an extremely important fact as we shall soon see. To obtain water, for instance, we must burn exactly two volumes of hydrogen in one volume of oxygen. This strongly suggests that equal volumes of gases contain equal numbers of atoms.

This was precisely the conclusion reached in 1811 by the Italian scientist Amedeo Avogadro (1776-1856), except that his formulation was more exact: equal volumes of gases contain equal numbers of *molecules.* As we now know, the molecules of most gases (hydrogen, oxygen, nitrogen, etc.) consist of two atoms (H_2, O_2, N_2). This explains the classical experiment on the combustion of hydrogen in oxygen. We know that from one volume of oxygen and two volumes of hydrogen we obtain two volumes of water vapour. Or in symbols

$$2H_2 + O_2 = 2H_2O\,.$$

What is the significance of Gay-Lussac's and Avogadro's discoveries and why have we spent so much time on these simple facts?

Let us take a closer look at the situation. Equal volumes of gases contain equal numbers of molecules. We know that at standard temperature and pressure 2 grammes of hydrogen occupy a volume

ELEMENTS
Wt
Hydrogen 1
Azote 5
Carbon 5,4
Oxygen 7
Phosphorus 9
Sulphur 13
Magnesia 20
Lime 24
Soda 28
Potash 42
Wt
Strontian 46
Barytes 68
Iron 50
Zinc 56
Copper 56
Lead 90
Silver 190
Gold 190
Platina 190
Mercury 167

of 22.4 litres. If we denote the number of molecules in this volume by N, then the same number N of oxygen molecules will occupy the same 22.4 litres. But their total weight will be 32 grammes instead of 2. It follows that each oxygen atom is sixteen times heavier than a hydrogen atom. This means that if we know the density of a gas, we can easily work out its atomic weight. The reality of the atomic hypothesis had never before been so evident. *Density* is a readily measurable quantity and we are accustomed to it since it affects our sense organs. It is odd that we can thereby measure in a simple experiment the *atomic weight*, a quantity that we cannot sense directly but that is, nevertheless, absolutely real.

The number N of molecules contained in 22.4 litres of any gas under standard conditions is now called Avogadro's number. This is one of the universal physical constants, like the velocity of light c and Planck's constant h. To determine N it is sufficient to know the absolute mass μ of one hydrogen atom. And since 22.4 litres contain two grammes of such atoms, then $N = 2/\mu$.

Avogadro's hypothesis was soon forgotten, and it was not until half a century later, in 1858, that it was revived by another Italian scientist, Stanislao Cannizzaro (1826-1910) — a most opportune development, since the chemists of that day could not come to any consensus. Each one only recognized his own table of atomic weights; the organic chemists did not trust the inorganic chemists, and the First International Chemical Congress, a meeting of the world's most famous chemists in Karlsruhe, Germany, in 1860, failed to reach any satisfactory agreement on these matters. (Although in a resolution dated 4 September 1860 the Congress did establish the difference between atoms and molecules.)

Only then, finally, were the atomic weights of the elements determined accurately enough for chemists to start thinking of some classification for the elements.

Table of Elements

A story has it that Mendeleev chanced upon his periodic system as follows. He had written out the names of all the elements known at that time on the back of his business cards and began arranging and rearranging them in various ways, in solitaire fashion. He had been doing this for a long time until he dozed off. It was during that short nap that the solution came to him. Perhaps this account is not fully authentic, but even those who regard it as the absolute truth should bear in mind that before that fortunate day, 1 March 1869, there had been many other days and nights, sleepless and frustrating.

What then was the stumbling block? Remember the example with the pile of numbered wooden blocks. They can be put into numerical order quite easily. But there are no labels with numbers on the chemicals; they are merely substances of various colours — some solid, some liquid, some gaseous. We only know that each one can be placed according to a *number* — its atomic weight. You can, of course, simply arrange all the elements in increasing order of their atomic weights, but this would be an occupation worthy of an apprentice, not a master-craftsman. To begin with, how do we know that all the elements have already been discovered? Without being sure that they have, what would be gained by arranging them in increasing order of their atomic weights?

The problem rather resembles a jig-saw puzzle in which some of the pieces have been lost and others have been distorted. It may still be possible to restore the whole picture, but this, of course, will be extremely difficult. This can be done, however, only if we try to conceive of the *picture as a whole* and not just hope that it will emerge by itself if we just rearrange the pieces at random.

Synthesizing power was one of Mendeleev's virtues. From the very beginning he conceived of the elements as parts of a single system rather than as a collection of random substances. In his quest for this system of elements, he did not restrict himself to their physical properties alone, i.e., to their atomic weights (though this was the basis of his classification); he also did not overlook their chemical behaviour.

In Mendeleev's days only sixty-three elements were known. In the table he drew up in 1869 only thirty-six of them complied with the principle of increasing atomic weights. For another twenty elements this principle was violated; as to the remaining seven elements, Mendeleev used his table to correct their atomic weights. He so trusted his system that he predicted the properties of five then unknown elements and left gaps for them in his table: scandium (Sc), germanium (Ge), gallium (Ga), technetium (Tc), and rhenium (Re).

These elements were indeed discovered soon afterwards: scandium (No. 23) in 1875, gallium (No. 31) in 1879, germanium (No. 32) in 1886, rhenium (No. 75) in 1925, but technetium (No. 43) was only synthesized in 1937.

We must admit that Mendeleev compiled his system in spite of the facts, rather than from the facts. He visualized the entire table and only took into account the facts that fitted. As in a children's picture puzzle, Mendeleev all of a sudden made out the clear-cut outlines of the pattern in the tangle of lines. Once discerned, it stays imprinted in your mind, no matter how hard you try to get rid of it — a well-known aspect of human psychology. And here Mendeleev displayed that trait of his intellect that distinguishes a genius from a talented person — his formidable intuition. This is a rare gift of nature that enables someone to see the truth through a hodgepodge of false facts.

Mendeleev's periodic table of elements put an end to the centuries-old controversy between the ideas of Aristotle and Democritus on the nature of elements. Along the horizontal rows of the table varies an unobservable property of Democritus's atom — its atomic number. Along the vertical columns the atoms are naturally grouped in families with similar chemical properties, such as valence, chemical reactivity, and so forth. These are observable properties; they affect our sensory organs and are akin to the ancient "innate properties" of Aristotle.

Lagrange once said, "Lucky Newton! The system of the universe can be established but once". Mendeleev has established the system of the chemical universe. Again, this can only be done once. Therefore, his name, like Newton's, will never be forgotten in the history of science.

The Periodic Law

When looking at Mendeleev's table a question arises and has always arisen: "What is this, just a convenient trick for memorizing the elements, or a fundamental law of nature?" The table tells much to a chemist, perhaps too much for us to cover in this book. We will only try to answer the key questions: assuming this is a law of nature, then:

What determines the arrangement of the elements in the table?

Why do their properties recur periodically?

What dictates the length of the periods?

Attempts were made to answer these questions for half a century by great minds ranging from Mendeleev to Pauli. Time and again the table was rewritten, cut up into pieces and glued together again on a plane or three-dimensional surfaces in all possible and even impossible manners. But, as so often in science, the clue to the phenomenon lay elsewhere — in the physical theory of the atom.

Even Mendeleev knew that the atomic weight determines the place of an element only approximately. Nevertheless, by some method known only to him, he did manage to achieve a proper arrangement of the elements in the table; then he assigned a number to each of them. But has this numbering any profound meaning? We could equally well number the pieces of our jig-saw puzzle, and then quickly and easily restore the picture each time we happened to scramble the pieces. This may, of course, be convenient, but has no particular significance because the numbers on the pieces are in no way related to what is painted on them.

Is there any deep intrinsic relationship between the chemical properties of an element and its *ordinal number* in the table? Or is this an external and arbitrary feature, like the numbers of the houses along a street? If this were actually so, it would be necessary to change the numbering each time a new element was discovered, as it is necessary to change the house numbers in a street each time a new building is erected. Put another way, is the number of an element just a convenient tag that helps to find the element in the table, or is it an intrinsic characteristic, one which in independent of any tables? The latter assumption appears to be closer to the truth: more than a hundred years since the table was compiled and not a single change has been made in the numbering of the elements.

The mystery of the stability of the table was only unravelled after the work of Rutherford. In 1913 the Dutchmen Antonius Van den Broek (1870-1926) wrote a short article in which he made the assumption that *the number of an element in the periodic table is equal to the nuclear charge of its atoms.*

In that same year, studying the X-ray spectra of various elements, Henry Gwyn Jeffreys Moseley (1887-1915), one of Rutherford's most brilliant assistants, proved this hypothesis. Moseley's work was a major development in physics even in those eventful years. He did not complete his work. On the bright day of 10 August 1915 on the Gallipoli peninsula during the Dardanelles campaign, Liaison Officer Moseley of a British field-engineer company was killed by a bullet that hit his head.

Scrutiny of his scientific legacy reveals that he believed that all the known elements were adequately placed in the table, and that vacant spaces awaited discoveries of new elements. There is always something inexplicably attractive about such categorical statements, especially those about the structure of the universe. The work of Moseley had put the finishing touches to the system of chemical elements. It only remained to understand its details.

Nature took pains to conceal its main secrets as far as possible from the eyes of natural scientists: the nuclear charge of the atom, securely shrouded by a coat of electrons, cannot be measured by any chemical or by most physical methods. To uncover this property of atoms scientists had to bombard them with such missiles as alpha particles. At the same time, it is precisely this well-hidden property that determines the structure of atoms and all the observable behaviour of the respective elements. To gain an intimate knowledge of what occurs within the atom, we must first get to its nucleus. (As in the Russian fairy tale about Kashchey the Immortal: high on a mountain there grows an oak, in the oak there is a chest, in the chest there is hare, in the hare there is a duck, in the duck there is an egg, in the egg there is a needle, and at the point of the needle is the death of Kashchey.)

In accordance with the laws of nuclear physics, the nuclear charge is numerically about one half of its atomic weight. Therefore, if we arrange the elements in increasing order of their atomic weights we will thereby arrange them more or less in increasing order of their nuclear charges. Mendeleev, of course, was unaware of the existence of nuclei, but he somehow sensed that atoms have some kind of additional property, more fundamental than atomic weight. And so he relied more on his intuition than on the atomic weights. It was as if he peered under the electron shell of the atoms, counted the number of positive charges in the nucleus, and then assigned this number to the element. (Later Van den Broek would call it "ordinal number" and Rutherford, "atomic number".) Obviously, the atomic number is an intrinsic characteristic of an element, and it is not assigned arbitrarily like the house numbers on a street. (To pursue further our analogy of the jig-saw puzzle, we can say that all the pieces turned out to be numbered beforehand. Only the numbers were concealed inside the pieces.)

Now, at last, we can give a precise definition of an element.

An element is a substance consisting of atoms with the same nuclear charge.

One more question remains to be answered: why does a *monotonic* variation of the nuclear charge of atoms lead to *periodic* variation of their chemical properties? And not only chemical, but also physical properties, such as density, hardness, and even the state of aggregation. The reason for the periodic variation of the properties should obviously be sought in the surrounding electron cloud, not in the nucleus. The first idea to come to mind is that the electrons are not randomly scattered around the nucleus, but are arranged in layers or shells. Starting a new shell coincides with the beginning of a new period and with an abrupt change in the chemical properties of elements. After Bohr's work the idea seemed quite natural — an idea that he was the first to put forward.

The foregoing, however, gives us no hint as to how to calculate the length of the periods. On the face of it the periods vary in length quite arbitrarily: two elements in period I, eight each in periods

1 Hydrogen (H) 1.008								
3 Lithium (Li) 6.939	4 Beryllium (Be) 9.012							
11 Sodium (Na) 22.99	12 Magnesium (Mg) 24.312							
19 Potassium (K) 39.102	20 Calcium (Ca) 40.08	21 Scandium (Sc) 44.956	22 Titanium (Ti) 47.90	23 Vanadium (V) 50.942	24 Chromium (Cr) 51.996	25 Manganese (Mn) 54.938	26 Iron (Fe) 55.847	27 Cobalt (Co) 58.933
37 Rubidium (Rb) 85.47	38 Strontium (Sr) 87.62	39 Yttrium (Y) 88.905	40 Zirconium (Zr) 91.22	41 Niobium (Nb) 92.906	42 Molybdenum (Mo) 95.94	43 Technetium (Tc) 98.91	44 Ruthenium (Ru) 101.07	45 Rhodium (Rh) 102.905
55 Cesium (Cs) 132.905	56 Barium (Ba) 137.34	57 Lanthanum (La) 138.91	58 Cerium (Ce) 140.12	59 Praseodymium (Pr) 140.907	60 Neodymium (Nd) 144.24	61 Promethium (Pm) 145.	62 Samarium (Sm) 150.35	63 Europium (Eu) 151.96
			72 Hafnium (Hf) 178.49	73 Tantalum (Ta) 180.948	74 Tungsten (W) 183.85	75 Rhenium (Re) 186.2	76 Osmium (Os) 190.2	77 Iridium (Ir) 192.1
87 Francium (Fr) 223.	88 Radium (Ra) 226.05	89 Actinium (Ac) 227.	90 Thorium (Th) 232.038	91 Protactinium (Pa) 231.	92 Uranium (U) 238.03	93 Neptunium (Np) 237.	94 Plutonium (Pu) 242.	95 Americium (Am) 243.
			104 Rutherfordium (Rf) 261.	105 Hahnium (Ha) 262.				

Modern Periodic Table

								2 Helium (He) 4.003
			5 Boron (B) 10.811	6 Carbon (C) 12.011	7 Nitrogen (N) 14.007	8 Oxygen (O) 15.999	9 Fluorine (F) 18.998	10 Neon (Ne) 20.183
			13 Aluminum (Al) 26.982	14 Silicon (Si) 28.086	15 Phosphorus (P) 30.974	16 Sulphur (S) 32.064	17 Chlorine (Cl) 35.453	18 Argon (Ar) 39.948
28 Nickel (Ni) 58.71	29 Copper (Cu) 63 54	30 Zinc (Zn) 65.37	31 Gallium (Ga) 69.72	32 Germanium (Ge) 72.59	33 Arsenic (As) 74.922	34 Selenium (Se) 78.96	35 Bromine (Br) 79.909	36 Krypton (Kr) 83.80
46 Palladium (Pd) 106.4	47 Silver (Ag) 107.87	48 Cadmium (Cd) 112.40	49 Indium (In) 114.82	50 Tin (Sn) 118.69	51 Antimony (Sb) 121.75	52 Tellurium (Te) 127.60	53 Iodine (I) 126.904	54 Xenon (Xe) 131.30
64 Gadolinium (Gd) 157.25	65 Terbium (Tb) 158.924	66 Dysprosium (Dy) 162.50	67 Holmium (Ho) 164.930	68 Erbium (Er) 167.26	69 Thulium (Tm) 168.934	70 Ytterbium (Yb) 173.04	71 Lutetium (Lu) 174.97	
78 Platinum (Pt) 195.09	79 Gold (Au) 196.967	80 Mercury (Hg) 200.59	81 Thallium (Tl) 204.37	82 Lead (Pb) 207.19	83 Bismuth (Bi) 208.98	84 Polonium (Po) 210.	85 Astatine (At) 210.	86 Radon (Rn) 222.
96 Curium (Cm) 244.	97 Berkelium (Bk) 245.	98 Californium (Cf) 246.	99 Einsteinium (Es) 253.	100 Fermium (Fm) 255.	101 Mendelevium (Md) 256.	102 Nobelium (No) 255.	103 Lawrencium (Lr) 257.	

II and III, eighteen each in periods IV and V, and thirty-two in period VI. But as long ago as 1906, Johannes Robert Rydberg noticed that the series of numbers 2, 8, 18, and 32 obey the simple formula $2n^2$ with $n = 1, 2, 3$, and 4. Explanation of this pattern had to await Pauli's discovery in 1924 of his exclusion principle.

Pauli reasoned as follows. The motion of the electron in an atom is, indeed, described by four quantum numbers, which we dealt with at length earlier in the book:

n — the principal quantum number, which takes on the values 1, 2, 3, ...;

l — the orbital quantum number, which for a given n can have the values 0, 1, 2, ..., $(n - 1)$;

m — the magnetic quantum number, which for given n and l can have the series of values $-l$, $-(l-1)$, ..., -1, 0, 1, ...,$(l-1)$, l;

S — the spin quantum number, which can have the values +1/2 and − 1/2.

Pauli's exclusion principle states that: *in a given atom no two electrons can have all four quantum numbers identical.*

Following Pauli, let us compute how many electrons the nth layer contains. The layer with $n = 1$ can only have $l = 0$ and $m = 0$, and S can be +1/2 and − 1/2, i.e., the first shell can only accommodate two electrons. Accordingly, the first period only includes two elements — hydrogen and helium.

In the next shell ($n = 2$), l may be either 0 or 1. For each l the magnetic quantum number assumes $2l + 1$ values, i.e., one for $l = 0$ and three for $l = 1$. And for each of these two states, two spins are possible: +1/2 and − 1/2, i.e., for $l = 0$ two electrons are possible, and for $l = 1$, six electrons. All in all, the shell with $n = 2$ can hold 2 + 6 = 8 electrons, exactly the length of the second period, from lithium to neon.

Likewise, we can easily calculate that the layer with $n = 3$ contains eighteen electrons, i.e., not more than $2n^2$ electrons in the nth layer.

Each period in Mendeleev's table begins with an alkali metal and ends with an inert gas. The chemical properties of these elements differ markedly. Now, we also understand the reasons for these differences. The inert gases, helium, neon, argon, etc., differ from all the other elements in that all of their shells are completely filled.

The atoms of alkali metals Li, Na, K, etc., which follow after the inert gases in the table, each contain one electron in the next higher shell. The bonds of these electrons with the nucleus are much weaker than in the atoms of the inert gases; therefore, the atoms of alkali metals readily lose these electrons and become singly charged positive ions: Li^+, Na^+, K^+, and so on.

Conversely, the atoms of fluorine (F), chlorine (Cl), bromine (Br), etc., lack one electron in the outer shell to make it resemble that of an inert gas. That is why the halogens, as these elements are called, so willingly accept an electron to form the negative ions F^-, Cl^-, Br^-, and so on. When an atom of sodium meets an atom of chlorine, the sodium yields its outer electrons to the chlorine, and so the ions Na^+ and Cl^- are formed. These ions attract each other, forming the molecule NaCl, that of ordinary table salt.

In the nineteenth century the numbers 2, 8, 18, and 32 evoked much perplexity and came to be known as "magic numbers". Various explanations were suggested. For instance, some argued that an octahedron is the strongest of the polyhedrons, and added that in Buddhist philosophy there are eight paths of good. But nobody really supposed that the answer could be so simple and rational.

Quantumalia

Atoms and People

Robert Boyle (1627-1691) was an outstanding personality. He was deeply influenced by the philosophy of Francis Bacon and his view that experiment is the principal yardstick of truth. This, perhaps, explains why Boyle established one of the first quantitative laws of physics, known now as the Boyle-Mariotte gas law. Curiously

D. Mendeleev

How wonderful that we have met with a paradox. Now we have some hope of making progress.

Niels Bohr

H. Moseley

enough, the style of his work was rather like that of scientists of our days. So he did not *write* scientific papers, but *dictated* them to his secretary; he did not *do* his experiments himself — instead he *entrusted* them to his assistant (he was fortunate, by the way, in having as his collaborator the subsequently famous Robert Hooke).

Boyle was the fourteenth child and the seventh son of a rich family. Since his early years he had suffered from kidney stones, which probably predetermined the tenor of his life. He never married, he was devoutly religious, and his friends who had known him over forty years said that he never mentioned God without making a reverential pause. For sixteen years, from 1661 to 1677, he was a director of the famous East India Company, and in this capacity his chief concern was the activities of the missionaries in the colonies. About one third of his scholarly works were on theology. He financed translations of the Bible into Turkish, Arabic, and Malayan, and even into the languages of the American Indians. At the same time Boyle was one of the founders of the Royal Society.

John Dalton (1766-1844) was born into the family of a poor weaver in the County of Cumberland in the north of England. He was educated at the village school. When the boy was twelve, the schoolmaster resigned and Dalton himself reopened the school, first in his home and later at a Friends' Meeting House in the village, where he taught for the next two years. This fact in itself is extraordinary, to say the least, but no comments or recollections to this effect by his contemporaries have come down to us.

After a year of farming, at the age of fifteen, he left home and joined his elder brother Jonathan. The brothers opened a school and taught there for twelve years until 1793, the year young John was invited to teach physics and mathematics at the New College in Manchester. He spent six more years there. In Manchester he joined the Manchester Literary and Philosophical Society, where from time to time he presented his scientific papers. His first report was on colour blindness, from which he himself suffered and which is now known as Daltonism. He lived the rest of his life in Manchester and died there on 27 July 1844, after having been paralysed for the previous seven years.

Dalton came from a family of Quakers, members of the Society of Friends, an extremely strict protestant sect. This perhaps accounts for some features of his character. He lived a calculated life, his daily routine never varied. His neighbour could tell the time to the minute by noting when Dalton took outdoor measurements on his thermometer and barometer. His day's work finished at nine o'clock in the evening and, after dinner, he sat quietly with his family, smoking his pipe, making only a brief remark from time to time. On Sundays, in his Quaker dress — knee breeches, grey stockings, and buckled shoes — he would twice go to public prayer assemblies, although he never discussed his religious views.

Dalton was no reader and boasted that he could carry his library on his back, and that even of these he had not read half. One of his biographers wrote that as in all self-made men, the desire to learn of somebody else's accomplishments was less developed in him than his firm belief in the accuracy of what he himself had discovered.

I have hardly ever known a mathematician who was capable of reasoning.

Plato

His contemporaries did not remember him as a likeable character. One of them recalled in later years that "his aspect and manner were repulsive... his voice harsh and drawling, his gait stiff and awkward". Yet his fellow members of the Literary and Philosophical Society elected him their president in 1817. Towards the end of his life he enjoyed world-wide recognition; in 1822 he was elected Fellow of the Royal Society, and in 1830 he was one of the eight foreign associates of the French Academy; he replaced Sir Humphry Davy, who had died the previous year.

Subsequent generations are totally indifferent to the personal shortcomings of a scientist. They only remember his ideas. This, it seems, is one of the reasons why man makes progress.

2. Ideas

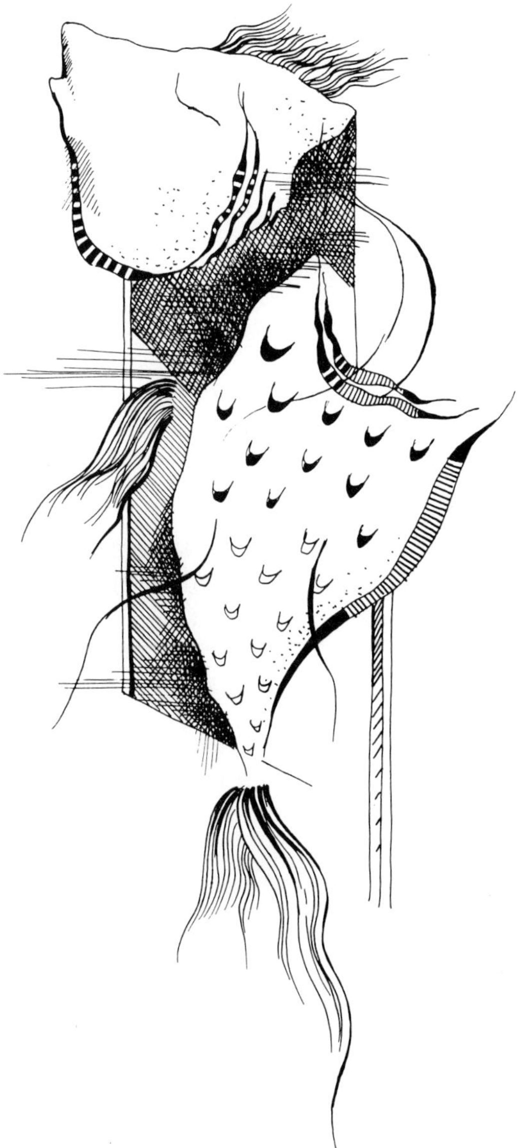

To see a world in a grain of sand,
And a heaven in a wild flower:
Hold infinity in the palm of your hand,
And eternity in an hour.

William Blake

Science is built up with facts, as a house is with stones. But a collection of facts is no more a science than a heap of stones is a home.

Henri Poincaré

Neptune's horse, Roman

Chapter Six

Contemporaries Comment on Bohr's Theory
Phenomenon, Image, Concept, Formula
Heisenberg's Matrix Mechanics

The meaning of words is determined by tradition and custom, but their true import only follows from context. Always and everywhere: in science and in art, in technology and in politics. When confronted with new phenomena man names them with old names, but now he gives them another meaning, a meaning that can only be understood if one knows the origins of new notions and their relation to past ones. This trend to distinguish the esoteric meanings of words from the customary ones produces scientific jargon, which as a rule is at odds with literary language. Scientific amateurs run to the other extreme; they accept all its statements literally, absolutely unaware of the complex system of conventions inherent in each scientific statement. This often leads to misunderstandings, which amuse physicists and distress the amateurs.

At the turn of the century physicists uncovered a new world, the world of the atom. They were stunned by the wealth of novel phenomena, and they rushed to invent names for them, although they did not quite understand what should be read into their new names. When Bohr pronounced the words "stationary state" and "quantum jump" for the first time, hardly anybody, including himself, could explain what these words really meant.

We began our story of quantum mechanics by saying that this science is the science of the structure and properties of quantum objects and phenomena. It became clear from the outset what the words "quantum object" mean, one of these objects being the atom. We still cannot quite give an unambiguous definition of the concept "atom", although we know much more about it now than at the beginning.

Experiments gradually changed speculative images into more complex, less visualizable, and yet more authentic views of the atom. Men of science succeeded in proving that the atom is a real entity, although it turned out to be a far cry from the atom of Democritus. They learned that it consists of the nucleus and electrons. They found that it can emit radiation. They established that its emission is caused by the motion of the

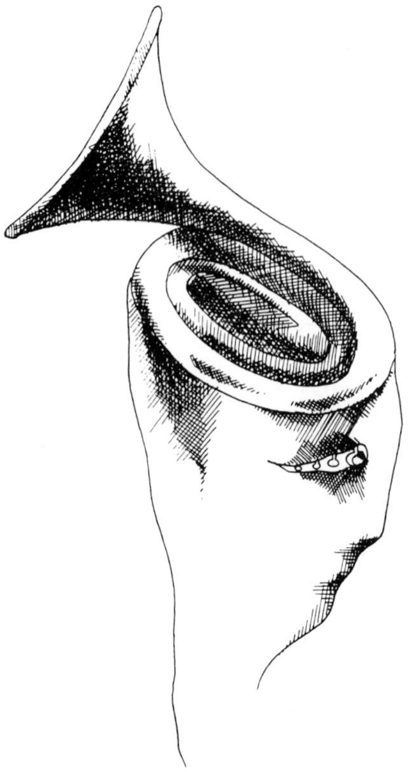

atomic electrons. And now they had to uncover the laws governing this motion. And so quantum mechanics came on the scene.

The foundation stone of this revolutionary science was laid by Niels Bohr. His postulates were an upheaval in the comfortable world of classical physics, and a blessed remedy that brought order to the unspeakable chaos of experimental evidence that was mounting in the new physics.

But science takes nothing for granted, even Bohr's postulates. Physics faced a dilemma — either to reject them or to eliminate their contradictions.

Contemporaries Comment on Bohr's Theory

In 1949 Albert Einstein wrote about the earliest days of quantum mechanics: "All my attempts to adapt the theoretical foundations of physics to the new facts were completely unsuccessful. It was exactly as if the ground was slipping away from under our feet and we had no firm soil that we could build on. It always seemed a miracle to me that this vacillating and quite contradictory basis turned out sufficient to enable Bohr — a man of brilliant intuition and keen perception — to find the predominant laws of spectral lines and the electron shells of atoms, including their significance for chemistry. This still seems to be a miracle to me. It is a manifestation of the highest form of musicality in the realm of thought."

"Perhaps this is madness, but it has method," commented the young Heisenberg on Bohr's postulates in 1920. Forty years later he changed his tune: "The language of Bohr's images is the language of poetry, which is only partially related to the reality it depicts and which must never be taken literally... The postulates of Bohr are like brushes and paints, which in themselves make up no picture yet, but which help to create one."

It is always easier and safer to evaluate discoveries with the advantage of hindsight. The task of contemporaries is always more difficult. They know as yet too little to form a correct judgement of the advantages and disadvantages of a theory. Whatever the successes of Bohr's hypothesis, it made his contemporaries unhappy. What they wrote and said at that time is both astounding and instructive to us now.

"If this is correct, it signifies the end of physics as a science" (Einstein, 1913).

"I am confident that this teaching is fatal for the healthy development of science" (Sir Arthur Schuster, 1913).

"The atom exists eternally. This we know beyond question. But do we now understand this? No, we do not. And we conceal our lack of understanding by the likewise incomprehensible quantum conditions. The emission process is an event of regeneration of a demolished atom. Its mechanism is not understood by us. And again we conceal our lack of understanding by another incomprehensible quantum condition, Bohr's second hypothesis... The whole method of Bohr is based on quantization, which is a blind, scarcely logical way of reasoning. It is based on what could be called formal intuition" (Dmitry Sergeevich Rozhdestvensky, 1919).

"The theory of quanta is similar to other victories in science; for some months you smile at it, and then for years you weep" (Hendrik Anthony Kramers, 1920).

"In their present form, the quantization laws are, to a certain extent, theological in nature, entirely unacceptable to a naturalist. Thus, many scientists are rightfully indignant over these *Bauern-Regeln* (peasant laws)" (Paul Sophus Epstein, 1922).

"We are immeasurably far from such a description of the atomic mechanism that would allow us to retrace, for instance, all the motions of the electron in the atom or to understand the role of the stationary states...

The theory of quanta can be likened to a medicine that cures the disease but kills the patient" (H.A. Kramers and Gilles Holst, 1923).

"This all is very beautiful and extremely important, but, unfortunately, not very clear. We understand neither Planck's hypothesis of oscillators nor the forbiddenness of nonstationary orbits. Nor do we understand how, eventually, light is formed according to the theory of Bohr. It is beyond doubt that the mechanics of quanta, the mechanics of the discrete, is yet to be created" (Lorentz, 1923).

"Physics is a blind alley again. In any case, it has become too difficult for me, and I would prefer to be a comedian in the cinema, or something like that, and hear no more about physics" (Pauli, 21 May 1925).

Otto Stern reminisced many years later that in those days Max Laue and he had vowed that they would give up physics if "in that nonsense of Bohr's there is at least something". Lorentz lamented that he had not died five years previously, when there was still a measure of clarity in physics. Even Bohr himself in those days suffered from a "feeling of melancholy and hopelessness".

This unanimous discontent is incomprehensible to anyone unacquainted with the structure and methodology of modern physics. To understand the stir produced by Bohr's ideas requires a knowledge, at least in broad outline, of the intrinsic logic of the natural sciences.

Phenomenon, Image, Concept, Formula

What shocks a layman who first leafs through a textbook on quantum mechanics is the sheer number of formulas and equations. If he takes the trouble to delve deeper into the subject, he finds quite soon that formulas are a necessary part of the science of the atom, but not the most difficult. It is more difficult by far to understand *what* lies behind the formulas or, to use the parlance of physicists, to "understand the physical meaning of the formulas". We should not exaggerate the difficulties, but, since they are there, we should not forget them. Some of them are associated with the fact that many words we have used throughout our lives acquire an unusual connotation in quantum mechanics.

We start to know nature first with our senses. A child touches a wooden toy horse with his fingers, listens to his mother's voice, and sucks her breast. In short, from the earliest days of his life, he finds himself in a world of *phenomena,* which produce *images* in his mind. At this point, these phenomena and images have no names. Bit by bit, step by step, he begins to recognize *words* that correspond to the phenomena and images. Quite soon he finds that the same words engender different images for different people. Then he finds, perhaps to his surprise, that there are words (or phrases) that are linked but indirectly to images, though they are begotten by them. These are *concepts*.

Concepts are a distillation of collective experience; they are purposely deprived of details inherent in specific images, and can therefore serve as a means of communication between different persons. But, again, concepts are not totally unambiguous, if only because they give rise to different images in different minds. Even in everyday life this is often a cause of misunderstandings. In science this is even more dangerous; scientific results claim to have an objective meaning and to be independent of whims or wanton opinions. Therefore, each concept in science is supported by a formula, or a set of symbols and numbers, and strictly defined rules to handle them. It is this that enables scientists of different countries and generations to communicate with one another. The sequence

phenomenon → image → concept → formula

summarizes what I have just said: the phenomenon on the outside produces an image, which is ultimately generalized into a concept, which is made precise by a formula.

Man observed various phenomena: waves at sea, the concentric circles produced by a stone thrown into a pond, the propagation of light and the vibration of a string. They all produced some images in his mind. Gradually, he noticed that all these different phenomena have something in common, namely they are all associated with some *periodic processes* whose hallmarks are the phenomena of interference and diffraction. Thus a new concept — the *wave* — was formed. To make it entirely unambiguous, man associated it with four parameters — the amplitude A, the velocity of propagation v, the wavelength λ, and the frequency ν.

Likewise, when a physicist uses the concept of a *particle*, he does not visualize a specific image such as a billiard ball or a speck of dust. It is enough for him to know that a particle is a specific object whose internal structure is of no interest to him, but which has mass m, velocity v, momentum $p = mv$, and a trajectory of motion that the physicist can follow.

The *trajectory*, or path, is yet another concept required in order to define the concept "motion of the particle". At first, this process of definition may seem infinite — in order to define one concept we have to use another concept, which again must be defined, and so forth. But this is not so. In physics there are several primary concepts, ones that can be defined without reference to others, namely by giving exact recipes for measuring the quantities that correspond to these concepts. Such concepts are the time t, coordinate x, charge e, and so on.

The trajectory of the motion of a particle $x(t)$ is said to be specified if at each instant of time t we can indicate its coordinate x. To do this, we must either measure the coordinates x_i at times t_i or calculate them. Measurement is the concern of experimental physics; calculation, of theoretical physics. The latter, by the way, can only be used to solve a problem if the physical laws of the motion of the particle are known.

A physical law is a constant relation between the appropriate quantities that characterize the phenomenon; usually it is written in mathematical symbols in the form of equations. Each group of phenomena is ruled by a set of laws of its own: one set for mechanics (Newton's equations) and another set for electrodynamics (Maxwell's equations.) But concepts, laws, formulas, etc., taken together, constitute an exact science.

Any established science should be consistent. This means, in particular, that each concept within the framework of this science can only be employed in a strictly defined sense. This may be hard to accomplish, but there are no two ways about it since scientists, like all other people, communicate with one another by means of words and not formulas. They only need formulas to record the results so that other physicists can understand them.

For centuries mechanics was quoted as an example of a full-blooded coherent science; it had even come to be known as classical mechanics. Mechanics is the science of the mechanical motion of bodies. Its laws are obeyed by almost all the visible movements in nature, be it the fluttering of a butterfly or the motion of planets. The classical perfection of mechanics had long mesmerized men of science. They endeavoured to employ it to explain all forms of motion in nature, not just mechanical.

"Everyone unanimously agrees that the task of physics is to explain all phenomena of nature using the simple laws of mechanics," wrote Heinrich Hertz as late as 1894, on the very threshold of the revolution in physics.

Motion is one of the most complex concepts of physics. The images it conjures up in our minds are as diverse as the rustling of leaves and the charge of an infuriated rhinoceros. Even the most exotic pictures of motion, however, have something in common — the displacement of some objects in relation to others with respect to the passage of time. If we bring in the concept of trajectory the concept of motion becomes more concrete, perhaps because it again becomes dramatic. Since our earliest years the only idea of motion ingrained in us is that of mechanical motion, and so we tend to see all other kinds of motion in terms of the concept of trajectory. This naturally becomes impossible, for instance, when we attempt to conceive of electrical motion. We could, of course, when thinking of a high-voltage transmission line or a telephone trunk line imagine that the wire is the "trajectory" of the electric signals. Such a mental picture would, however, have no practical purpose; the electromagnetic waves are not a liquid flowing through the wires.

It is even harder to define the concept of motion in quantum mechanics. What is more, the very day that this fact was established can be regarded as the birthday of modern quantum mechanics.

Heisenberg's Matrix Mechanics

When the euphoria over the first successes of Bohr's theory was gone, physicists all of a sudden soberly realized the simple truth: Bohr's scheme was inherently contradictory. There was no denying it. This was the reason for Einstein's pessimism, as well as for Pauli's despair. Time and again, researches found that the electron's motion within the atom defies the laws of electrodynamics. It does not fall into the nucleus, nor does it radiate unless the atom is excited. This was all so shocking that people could not grasp it: the electron, which had "come from electrodynamics", suddenly no longer obeyed its laws. Any attempts to find a way out of this vicious circle always led to the same conclusion — Bohr's atom cannot exist.

But nature does not care about our mental constructs. Contrary to logic, atoms are stable and, as far as we know, they have existed forever. If the laws of electrodynamics cannot explain the stability of the atom, so much the worse for them. Hence the motion of electrons in the atom obeys some other laws. It was agreed later on that Bohr's postulates had been fortunate guesses of fundamental laws, unknown at the time. These fundamental laws were later named the laws of quantum mechanics.

Quantum mechanics is the science of the motion of electrons in the atom. Initially it was called precisely that — atomic mechanics. And one of the founding fathers of this science was Werner Karl Heisenberg (1901-1976).

At Bohr's invitation Heisenberg came to Copenhagen in the spring of 1925 from Göttingen, where he had worked as an assistant to Max Born after his graduation from the University of Munich. (He had studied physics there under Arnold Sommerfeld.) In Denmark Heisenberg found himself in the midst of heated scientific arguments and among people who had dedicated their lives to physics. Half a year was spent in work and endless discussions about the same things — why doesn't the electron, an object of electrodynamics, obey its laws within the atom; or why do Bohr's illogical postulates have such surprising power; finally, what does the very concept of "motion" mean in this case?

Heisenberg's painful reflections begot a stunning conjecture, which after a while became his firm belief: the motion of the atomic electron cannot be pictured as the motion of a small ball along a trajectory. Not because an electron is not a ball, but because it is something more sophisticated. One simply cannot follow the motion of this "something" as closely as one can follow the motion of a billiard ball. Therefore, when we attempt to follow the trajectory of an electron in an atom, we ask nature illegitimate questions, such as those asked in ancient times: "What is it on which the Earth rests?", "Where does the end of the world lie?", and later "Which side of the world is up and which is down?"

Heisenberg contended that the equations by which we want to describe motion in the atom must contain only those quantities that can be measured experimentally. Experiments demonstrated that the atom is stable and that it consists of a nucleus and electrons, and that it can emit radiation if disturbed from its equilibrium. The radiation can have fixed wavelengths, and, if Bohr was to be believed, they are emitted when an electron jumps from one stationary orbit to another. Bohr's scheme, however, gave no indication as to what happened to the electron during the jump, so to speak, in its "flight" between the two stationary states. But all, including Heisenberg, were seeking an answer in this very question. One fine day, however, it struck Heisenberg that the electron simply cannot at any moment in time be "between" stationary states — it simply does not possess such a property!

Well, what property *does* it possess? There was some property that Heisenberg did not have a name for yet, although he was sure that it ought to depend only on the electron's *position* before and after the jump.

Up until that time everyone had been using the equations of electrodynamics to find the hypothetical trajectory $x(t)$ of an electron in the atom, a trajectory that would have a continuous dependence on time and that could be expressed as the series of numbers $x_1, x_2, x_3,...$ — the positions of the electron at $t_1, t_2, t_3,...,$ respectively. Heisenberg held that there was no such trajectory in the atom and that, instead of a continuous curve $x(t)$, there was a set of discrete numbers x_{nk}, where n and k are the numbers of the initial and final states of the electron.

This important and astounding statement can be illustrated by a simple analogy. Imagine a fly walking across a chessboard. You could, if you wished, trace its path in detail by recording its lo-

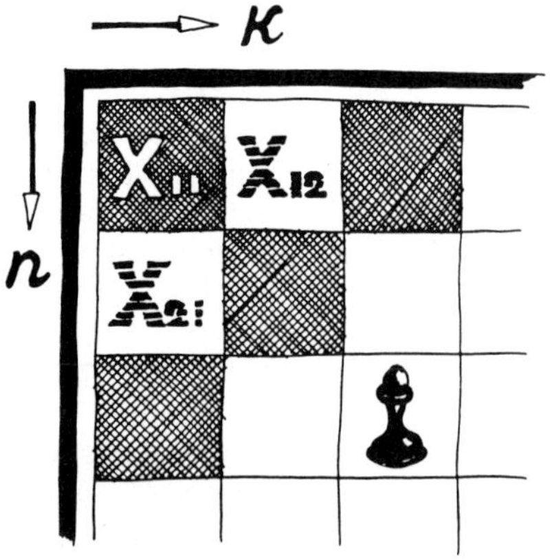

cation x_i at each instant of time t_i. From these measurements you could then readily plot the curve $x(t)$, i.e., the trajectory of the fly's motion. Or you could merely write down the squares through which the fly passed. You would, thus, also obtain some information about the fly's displacements, although as is easily seen, it would not be as complete as before from the viewpoint of classical mechanics.

Now imagine that you were playing chess on the same board and you had decided to make the traditional move e2 to e4. The resultant situation on the board would be absolutely the same, whatever the path you had taken to move the pawn. This is obvious: the rules of chess have no relation whatsoever to the rules of mechanics, and so the concept of trajectory is irrelevant here.

Heisenberg surmised that the "rules of the atomic game" also do not require a knowledge of any trajectory. And so he conceived of the states of the atom as being an infinite chessboard, each square of which has a number x_{nk} written in it. To be sure, the values of these numbers depend upon the location of the square on the "atomic board", i.e., on the number n of the row (initial state) and the number k of the column (final state) — at their intersection lies the number x_{nk}.

It is no surprise that given a record of a chess game one can reproduce it at any time even many years later. Such a record generally includes no information about how long the game lasted, what emotions the players experienced, and exactly how they moved the pawns and other chessmen. But this all is of no significance if we are only interested in the course of the game itself.

In much the same way, if we know the numbers x_{nk}, a record of sorts of our "atomic game", we know enough about the atom to predict its observable behaviour — its spectrum, the number and velocity of electrons ejected from the atom by ultraviolet radiation, and much more. The numbers x_{nk} cannot be called the coordinates of the electron in an atom. They replace them or, as they began to say later, *represent* them. But even Heisenberg did not understand at first what these words meant.

Instead of the table x_{nk} you may feel that one might as well draw any figure, say a cube, and assume that it represents the motion of atomic electrons. Max Born soon realized that the table of x_{nk} is not just any table but a *matrix*.

What does the term mean? Mathematics deals with quantities and symbols, and each symbol obeys definite rules. For example, ordinary numbers can be added or multiplied, and the result of these operations does not depend upon the order in which they are carried out. So

$$5+3=3+5 \quad \text{and} \quad 5\times 3 = 3\times 5 .$$

But mathematics also deals with more complicated things: negative and complex numbers, matrices, and so on. Matrices are tables of quantities of the type x_{nk} for which there are rules of addition and multiplication that have nothing to do with the rules governing ordinary numbers. For example, just like ordinary numbers, matrices can be added and subtracted in an arbitrary order. But the product of two matrices is dependent on the order in which they are multiplied together, i.e., $x_{nk}\cdot p_{nk} \neq p_{nk}\cdot x_{nk}$. For example the product

$$\begin{pmatrix} 1 & 0 \\ 0 & -1 \end{pmatrix}\begin{pmatrix} 0 & 1 \\ 1 & 0 \end{pmatrix} = \begin{pmatrix} 0 & 1 \\ -1 & 0 \end{pmatrix}$$

clearly differs from the "reverse" product

$$\begin{pmatrix} 0 & 1 \\ 1 & 0 \end{pmatrix}\begin{pmatrix} 1 & 0 \\ 0 & -1 \end{pmatrix} = \begin{pmatrix} 0 & -1 \\ 1 & 0 \end{pmatrix}.$$

This rule may seem strange, and even suspicious, but it is not arbitrary, by any means. This rule is precisely what distinguishes matrices from other quantities. Mathematicians, of course, used matrices long before Heisenberg. To their surprise, however, they found these strange mathematical objects with their strange properties corresponding to something real in nature. It is to Heisenberg and Born's credit that they surmounted this psychological barrier and found a correspondence between the properties of matrices and the details of the motion of electrons in the atom. This correspondence gave rise to a new, *atomic quantum matrix mechanics.*

Atomic — because it describes the motion of electrons in the atom.

Quantum — because the concept of the quantum of action h is central to this description.

Matrix — because the required mathematical tool kit consists of matrices.

In the new mechanics each characteristic of the electron — the coordinate x, momentum p, and energy E — is associated with the corresponding matrix: x_{nk}, p_{nk}, and E_{nk}. And it is for these matrices (not for ordinary numbers) that the equations of motion, known from classical mechanics, are written. The only requirement is that all the operations with x_{nk}, p_{nk}, and E_{nk} be in strict accordance with the rules of matrix mathematics.

Max Born also established that the quantum-mechanical matrices of x_{nk} and p_{nk} must conform to the *commutation relation*

$$x_{nk} \cdot p_{nk} - p_{nk} \cdot x_{nk} = i\hbar$$

where $i = \sqrt{-1}$, and $\hbar = h/2\pi$.

In the new mechanics this commutation relation played the same role as Bohr's quantization conditions in the old mechanics. Just as Bohr's conditions identify stationary orbits, so Heisenberg's commutation relation singles out from the set of matrices only the quantum-mechanical ones. It is no coincidence that h is present in both Bohr's quantization conditions and the commutation relations. It is an indispensable member of all the equations of quantum mechanics. In fact, the presence of h infallibly tells us that these equations must be quantum-mechanical.

The new equations derived by Heisenberg resembled neither those of mechanics nor those of electrodynamics, and so they could in no way violate them. In terms of the new equations, the state of an atom is considered to be completely specified if all the numbers x_{nk} and p_{nk} are known, i.e., if the matrices corresponding to the coordinate and momentum of the electron are known.

Notice that in our discourse we have never mentioned the concept of the "motion of the electron in the atom". We simply do not need it any more. According to Heisenberg, motion does not mean the travel of the ball-type electron along some orbit around the nucleus. Motion is the change in the state of the system "atom" in time, and it is described by the matrices x_{nk} and p_{nk}. The issue of the motion of the electron in the atom became clear, and the question of the stability of atoms automatically lost its significance. In the new picture the electron is at rest in the unexcited atom and thus it should not radiate.

We could go on discussing the consequences of Heisenberg's mechanics without using any formulas. This, however, would be just as unnatural as attempting to retell a musical piece in words. To gain deeper insights into the workings of quantum mechanics requires a knowledge of mathematics, especially matrix mathematics — in short, it requires some command of a physicist's trade. There is nothing mystical or inscrutable about matrices; it is much easier to master them than, say, Latin. But, like music, it takes some effort. Otherwise, the unpleasant after-taste of incomplete knowledge will even spoil the pleasure that one derives from the interplay of the images and concepts of any deep science.

Physicists welcomed the coming of Heisenberg's matrix mechanics with a tremendous sigh of relief: "Heisenberg's mechanics returned my joy of life and my hope. Though it does not solve the riddle, I believe we can move forward again," wrote Wolfgang Pauli on 9 October 1925. Soon afterwards he had an opportunity to prove this. He applied the new mechanics to the hydrogen atom to obtain the same formulas that Niels Bohr had arrived at on the basis of his postulates. Of course, new troubles lay waiting just around the corner, but those were just teething troubles — not a blind alley.

Quantumalia

The Foundation of Physics

W. Heisenberg

The basic concepts of physics, such as length, time, mass, charge, and so on, cannot be defined with words unambiguously for two reasons. First, these are primary concepts that cannot be reduced to simpler ones; second, physics is a quantitative science, which means that from the very beginning concepts have to be correlated with numbers. There is only one way to make these concepts well-defined: to give a precise recipe for measuring the corresponding quantities.

We have already defined the concept of "length". One metre is equal to 1,650,763.73 wavelengths of the orange-red line of ^{86}Kr (a krypton isotope with mass number 86) in a vacuum. This spectral line, which has been accepted as the standard for length, corresponds to the transition of an electron in a krypton atom from the $2p_{10}$ to the $5d_5$ level. One metre thus defined is about one 40,000,000th of the Paris meridian, which was initially, in 1800, accepted as the length standard.

The unit of mass is 1 kilogramme, which is defined as the mass of a specially shaped cylinder made in 1789 of platinum-iridium alloy (its height and base diameter are 39 millimetres). This mass is about that of a litre of distilled water at a temperature of 4°C.

To define the unit of time requires some stable cyclic process, such as the revolution of the Earth about the Sun. So 1 second is one 31,556,925,9747th of the tropical year, which is the interval between two successive identical positions of the Earth in its rotation with respect to the stars. The length of the tropical year is, however, slowly changing (by 0.5 second per century) due to the precession of the Earth's axis and other disturbances. The tropical year 1900 was specified as the standard — or, more precisely, the year that began at midnight on 31 December 1899.

With the passage of time it became clear that, just as with the unit of length, it is better to base the unit of time on spectroscopic measurements. In 1967 the Thirteenth General Conference on Weights and Measures redefined the second as the duration of 9,192,631,770 cycles of radiation associated with the transition of an electron between two hyperfine levels of the ground state of a caesium isotope with mass number 133 (caesium frequency standard).

Conversion to the atomic standards of length and time was inevitable. And not only because spectroscopy is the most precise branch of physics. The main reason is the exceptional stability of atomic standards; they are affected neither by temperature, nor by pressure, nor by any cosmic catastrophe — unlike the original standards. (The standard metre, for instance, is kept under a bell-glass at constant temperature in a steel cabinet placed in a deep vault whose three keys are in the custody of three different officials. Additional precautions are also taken.) With the second, the situation is even worse. If some celestial body were unexpectedly to pass through our solar system, the period of revolution of the Earth around the Sun would irreversibly change and so would the duration of the second. The atomic standards know no such dangers; they are stable and invariable as the atom itself, whose properties they rely on.

The three quantities — the metre, the kilogramme, and the second — belong to the Systèm International d'Unites or International System of Units (known as SI) and they are sufficient to describe all kinds of mechanical motion. Electromagnetic theory involves two more fundamental quantities: the charge e and the velocity of light c. To describe atomic phenomena a knowledge of Planck's constant h is also required.

A. Einstein

Founded in 1875 to measure precisely the fundamental physical constants, the International Bureau of Weights and Measures holds a General Conference on Weights and Measures every six years. These conferences carefully analyse the conditions under which standard quantities, such as temperature, pressure, and altitude above sea level are to be measured, as well as all the details of the instruments needed for these measurements.

Note one important feature of such measurements — it is only rarely that one quantity can be determined independently of others. In other cases it is necessary to use the laws of physics. For example, if the velocity v of a particle is constant, it can be determined by measuring the distance Δx covered by the particle during the time Δt:

$$v = \frac{\Delta x}{\Delta t}.$$

Conversely, the unit of length can be defined by the formula

$$l = c \times \Delta t,$$

where $\Delta t = 1/c$ seconds and $c = 299{,}792{,}458$ metres per second is the velocity of light in a vacuum. This was accepted in 1983.

Scientists have to admit that they do not know how to bring about the major advances of science. There are no rules or methods that allow one to specify what needs to be done now to make possible a great step forward in a few years time.

Thomas Gold

This simple example illustrates the fact that fundamental constants are all, in a sense, interrelated. There is a special branch of physics, called metrology, whose task is to make consistent determinations of all such constants, taking due account of all the available data relating to their measurements.

But the most difficult issue is that of the applicability of concepts found by such methods. It is quite evident that the units — the metre, the kilogramme, and the second — have been selected in such a manner that a person can easily comprehend them. So, 1 metre is the height of a five-year-old child, 1 kilogramme is the mass of a big round loaf of bread, and 1 second is the time elapsed between heartbeats. But do these concepts retain their previous meanings in the realm of very large or very small distances, masses, or time intervals?

As yet there is no general answer to this question. For instance, we saw that the concept of size is no longer applicable to the electron. In atomic theory the concept of "motion" was replaced with a new one, but the previous concepts of "length", "mass", and "time" are still valid. This means that at least the distance 10^{-10} metre, the mass

10^{-27} kilogramme (the atomic size and mass), and the time interval 10^{-17} second (the revolution period of an atomic electron) can still be understood in the ordinary sense.

A similar problem arises in astronomy when we try to comprehend the mass of other galaxies and how far away they are. This is perhaps an even more difficult task than conceptualizing the theory of elementary particles. In fact, no one can assert light-heartedly that he fully understands the words "1 billion light years". In theory it is quite simple — this is the distance covered by light in 10^9 years, i.e., 10^9 (years) $\times$ 3.15×10^7 (seconds in a year) $\times$ 3×10^8 metres per second = 10^{25} metres. But how can we grasp, or at least intuitively sense, what is behind this symbol?

Pascal was right: "Man dangles between two infinities".

Anubis, Egyptian

Chapter Seven

Louis de Broglie • Matter Waves
Optical-Mechanical Analogy
Schrödinger's Wave Mechanics

Janos Bolyai (1802-1860), a Hungarian officer, developed non-Euclidian geometry at the age of twenty-three. He was quite pleased with his discovery until he found out that somewhere on the border between Europe and Asia one Nikolai Ivanovich Lobachevsky (1793-1856) had published the same ideas several years earlier. Then Bolyai's life turned into a nightmare. He began to believe that everybody was spying on him; he became hard and suspicious and blamed everyone, even his father who had devoted his whole life to the same problem. The old Bolyai was not as gifted as his son, but he was wiser and more humane. On his deathbed he said, "Don't despair, my son; when spring comes, all the violets burst into bloom at the same time."

For the science of the atom such a spring came in 1925: within only two years the violets of a new science, quantum mechanics, bloomed. Its basics have not changed ever since: in this way a volcanic island sometimes appears in the midst of the ocean and remains there unchanged for centuries. The island in the ocean and the spring come as surprises only to those who have not experienced subterranean quakes, and who passed by the swollen buds without pausing to take notice. In Part One we tried to sense the vague tremors, to detect the almost imperceptible flow of the sap that ushered in the spring of quantum mechanics.

Heisenberg's paper, the one that contained the ideas of *matrix mechanics,* appeared in the autumn of 1925. This was the first consistent theory to explain the stability of the atom. But (quite in keeping with the laws of spring) only half a year later Erwin Schrödinger constructed still another mechanics, *wave mechanics,* which at first glance bore no resemblance to matrix mechanics, but accounted for the structure of the atom equally well. It turned out that both matrix and wave mechanics are just two different forms of representation of the same theory, quantum mechanics.

Louis de Broglie

Heisenberg was born in the year in which Planck

published his celebrated work. When he was finishing his gymnasium, his native Germany was at war with all of the world — with Russia, the native land of Mendeleev; with England, the native land of Rutherford; and with France, where in 1892 Louis Victor Pierre Raymond, Prince de Broglie (1892-1987), a descendant of kings and a future Nobel Prize winner, was born. Like so many at that time, de Broglie fought in the war, and only after the war did he begin working in the laboratory of his elder brother Maurice, who was studying the X ray spectra of elements. This explains why workers in his laboratory not only knew Bohr's work well but were also aware of all the latest developments in atomic physics.

Louis de Broglie was preoccupied with the very same questions: "Why are atoms stable? And why doesn't the electron radiate when it is in a stationary orbit?" Bohr's first postulate defined these orbits as those that satisfy certain "quantum conditions":

$$mvr = n\hbar ,$$

where r is the radius of the orbit, v the velocity, and m the mass of the electron. De Broglie wanted to find a reasonable justification for this condition, i.e., to explain it in terms of other, more common concepts. In other words, he tried to understand its physical meaning.

When confronted with incomprehensible things we usually resort to analogies. And so did de Broglie in his search for a way out of the atomic dead-end. He figured out that these difficulties were akin to those that had arisen with the contradictory properties of light. The situation with light became decidedly muddled in 1923, when Arthur Compton carried out his famous experiment showing that the scattering of X rays by electrons in no way resembles the scattering of sea waves — it rather resembles the collision of two billiard balls, one being the electron of mass m and the other a quantum of light of energy $E = h\nu$. After Compton had preformed his experiment, and together with Peter Joseph William Debye (1884-1966) had explained it, there was no doubting any more that in nature there really exist light quanta, i.e., photons with energy $E = h\nu$, momentum $p = h\nu / c$, and wavelength $\lambda = c / \nu$.

Neither de Broglie nor his contemporaries could explain the meaning of the phrase "light quanta correspond to a light wave". On the other hand, they had no grounds to question the experiments that had demonstrated that under certain conditions light behaves as a wave with length λ and frequency ν, and under other conditions it behaves as a flux of particles, photons, with energy $E = h\nu$ and momentum $p = h / \lambda$ (earlier these particles were called *corpuscles*). Three years or so later all would understand that this phenomenon was just a special case of a universal *wave-corpuscle*, or *wave-particle,* duality in nature, but in those days de Broglie had to grope his way through *terra incognita.*

Matter Waves

De Broglie sincerely believed in the unity of nature, as all great scientists had believed before him. He could not admit, therefore, that light was something extraordinary, unlike anything else in nature. And so de Broglie supposed that not only light but *all bodies in nature must possess both wave and particle properties at the same time.* Therefore, in addition to light waves and particles of matter, there must also exist in nature quanta of light and waves of matter.

To make such a simple, forceful statement takes much courage and self-confidence. To understand it is even more difficult; this requires an open mind accustomed to abstract thinking. And to picture this is hardly possible — nature, accessible to the perception of our five senses, does not produce images helpful to us in this effort. Indeed, when you hear the word "particle" any number of things may come to your mind — a grain of sand, a billiard ball, or flying stone — but you would hardly think of sea waves or the vibration of a string. To a normal person these are such contradictory images that it would seem unnatural to lump them together.

Any account of the birth of a new physical theory is inherently inaccurate, even its founder's. Such accounts almost always rely on concepts that did not exist at the time the theory was being produced. For present-day physicists the notion of the "matter wave" conjures up complex images that have no parallel in the world around us. This image takes shape gradually, after years of dealing with the mathematics of quantum mechanics and its problems. The image is hard to describe in words. Clearly, de Broglie's mind in those days just could not have conceived of such a sophisticated image. To explain his reasoning we will make use of a fitting analogy, namely the image of a wave produced by the vibration of a string.

It is well known that when we strike a taut string, the sound it produces will depend on the tension and length of the string. The mechanism behind this is also well known: the vibration of the string is transmitted to the air, and we do not perceive the vibration of the string itself, but of the molecules of air. There is, however, a simple relationship between them. For instance, if we hear the note "la" in the first octave the string is vibrating at a frequency of $\nu = 440$ hertz, i.e., it makes 440 vibrations per second. Since the velocity of sound through air is $v = 344$ metres per second, the wavelength of sound will be $\lambda = v/\nu = 78$ centimetres.

When the string vibrates we hear the fundamental tone, the one due to the vibration of the string as a whole. There are, however, additional vibrations, called overtones. This complicates the picture, and the string now has "nodes", i.e., points that remain stationary in the process of vibration. However complex the vibration, one condition is always met: the length of the string accommodates a whole number of half-wavelengths $\lambda/2$. For the fundamental tone the length of the string is exactly half a wavelength $\lambda/2$. For the first overtone it is equal to two half-wavelengths, between which a stationary "node" is located, and so forth.

The rest is relatively simple. Let us make rings of our strings and imagine they are orbits of the electrons in the atom. And now we mentally replace the motion of the electron in these orbits with waves that "correspond to the electron", an analogy de Broglie thought made sense. Further assume that the motion of the electron will be stable if and only if the orbit can accommodate a whole number n of "electron waves" λ. We thus obtain the simple condition

$$2\pi r = n\lambda.$$

De Broglie compared this condition with Bohr's first postulate

$$2\pi mvr = nh$$

and obtained the "electron wavelength"

$$\lambda = \frac{h}{mv}.$$

Voila! Simple, isn't it! But the simplicity of this is like that of Planck's formula $E = h\nu$, Bohr's postulates, or Newton's law of universal gravitation, i.e., it is the simplicity of genius. Discoveries like these seem to be simple because they require the simplest of concepts. But they change the very foundations of our thinking. In the history of the evolution of man's intellect only a few such breakthroughs are known. You can never fully comprehend *how* they have been accomplished. They are always miracles that even their initiators are incapable of explaining. They can only echo Newton's words "I've known it for years".

De Broglie was thirty when he found his formula. But he had begun his search for it twelve years before, when his brother Maurice had returned from Brussels, there he had been the secretary of the First International Solway Congress on Physics — the congress of 1911 where Planck had reported on the evolution of "the hypothesis of quanta". The significance of those discoveries and his brother's first-hand accounts of meetings with the greatest physicists of that time had made such a deep impression on the younger de Broglie that he was not able to forget them even during the war. Years of deep reflection finally paid off in 1923 in the form of the hypothesis of matter waves. The new hypothesis enabled de Broglie to give a new definition to the concept of a "stationary orbit": an orbit whose length accommodates a whole number of "electron waves" $\lambda = h/mv$.

If this is actually so, then the problem of the stability of the atom has been solved, since in its stationary state the electron would resemble a string vibrating in a vacuum without friction. Such vibrations are not damped and thus the electron can stay in a stationary state forever if there is no external influence.

To come up with a new hypothesis is always the hardest part — the process has no logic. But once the hypothesis has been stated the laws of logic enable us to extract from it all its consequences. The main one is obvious: if "matter waves" exist they can be observed and measured. Within four years they would actually be discovered and their reality would be proved with the rigour common in physics.

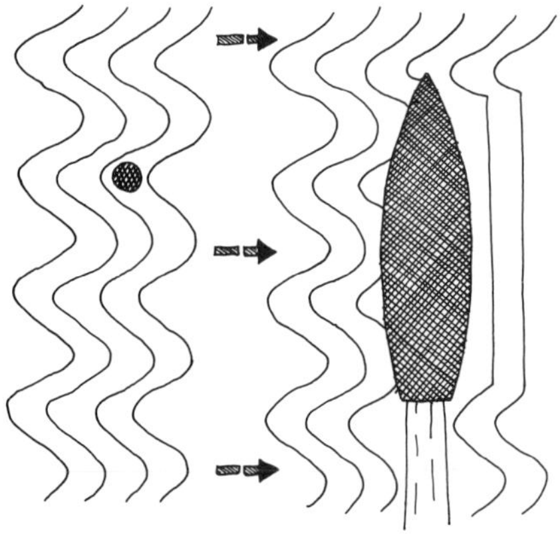

De Broglie wrote down his formulas in 1923, two years before the works of Heisenberg and Schrödinger. The simplicity of the formulas and the elegance of the idea were reminiscent of Bohr's postulates. And just like Bohr's postulates, de Broglie's ideas did not yet lead to a full-blown atomic theory, because they had not been written in the language of equations. When Heisenberg created matrix mechanics, he thereby translated Bohr's ideas into the language of precise formulas and rigourous equations. De Broglie's ideas became the basis for the wave mechanics to be produced by Schrödinger.

Optical-Mechanical Analogy

When we speak about atoms or quanta we time and again return to the properties of light. No wonder — light embodies nearly all the properties of present-day physics. Let us now take a closer look at its properties. We will return to the days of Sir Isaac Newton and recall his arguments with Christiaan Huygens on the nature of light.

It is common knowledge that in empty space light propagates in a straight line. Physics textbooks generally show this by drawing straight lines connecting a source of light with the eye of the observer, i.e., the imaginary trajectory of the light beam.

This trajectory has the same meaning and evokes the same images as the trajectory of a moving particle. That is why in Newton's time light was thought to be a flux of very small particles. Of course, the path of these "particles of light" (like that of ordinary particles) would be bent, for instance, when passing from air into water, but the concept of the trajectory is still relevant. In everyday life the notion of the light beam is useful and does not lead to confusion. It helps us to avoid ill-fated encounters with cars in the streets, to determine the positions of stars in the sky, and to design cameras.

Advances in experimental physics expanded the lay person's horizons and revealed new properties of light. It was found that light loses its customary properties entirely when passing around a "very small obstacle". Physics is a quantitative science, and such indefinite statements are of no use to the physicist. "Small" — well, but in comparison with what?

Huygens conceived of the propagation of light as the vibrations of a "luminiferous ether". This is reminiscent of the circles that are formed when we throw a stone into a pond or of endless crests of sea waves. Physicists no longer doubted the legitimacy of these images after the work of Maxwell and Hertz, who proved that light is just a form of electromagnetic oscillation.

Let us now recall what we mentioned in the first chapter of the book: the main characteristic of any wave process is its frequency or wavelength. This statement can now be couched in more definite terms: "a beam of light loses its customary properties if the size of the obstacle is commensurate with the wavelength of light". Under such conditions, light fails to propagate in a straight line, resulting in the phenomenon of diffraction. Moreover, some of the waves in the beam begin to interact, i.e., to reinforce or to cancel one another or, to use the parlance of modern physics, to interfere. Both phenomena, *diffraction and interference,* produce a pattern on a screen, which is quite difficult to understand in Newtonian terms. On the other hand, the wave theory of light explained this pattern perfectly, and so it won out.

As time went on the wave properties of light became so commonplace that they turned into a sort of standard for all wave processes. Henceforth, any process that displayed the tell-tale signs of interference and diffraction was recognized as having a wave nature. This explains why de Broglie's hypothesis on matter waves was immediately accepted by physicists as soon as they saw the first photographic evidence of the diffraction of electrons.

For the present generation of physicists, however, this is no longer an article of faith, but rather a basic fact useful even in some applications in technology.

For the theory of wave optics to become absolutely consistent one road-block had to be removed — we still perceive a beam of light as a beam, not as a wave. How can we explain this fact from the viewpoint of wave optics? The answer was given by Augustin Jean Fresnel (1788-1827); it can now be found in any textbook on physics. It turned out that because of interference all the waves from a source of light cancel one another with the exception of those that lie along a narrow ray originating from the source and terminating at the observer's eye. The "width" of this ray is one half the wavelength of the light, i.e., about 2×10^{-5} centimetres. If we neglect the thickness of this "light channel", we obtain the trajectory of the light beam, which we all know so well in our everyday life.

We even know how to construct it. First draw a line through all the wave crests, or, as one says in physics, mark the *wave front.* Then draw a line from the light source perpendicular to the wave front. The result will be the trajectory of the light beam. The front near an obstacle distorts, and as a result the trajectory of the beam bends. The ray will thus bend around the edge of the obstacle, causing diffraction.

In 1834 William Rowan Hamilton (1805-1865), the world-famous professor of astronomy at Dublin, was working on a problem that was incomprehensible to his contemporaries. He was trying to prove that the formal analogy between the trajectory of a particle and the trajectory of a light ray had a rigourous mathematical meaning. We have already seen that in physics, corresponding to the concept of a law of motion, are formulas — the equations of motion. The formulas for waves and particles differ considerably. In one case we solve the equations to find the trajectory of the particle; in the other, to find the shape and velocity of the wave front. But we also know that in geometrical optics we can draw the trajectory of a light ray if we know the motion of its wave front.

Hamilton proved that in mechanics we can do the opposite. We can replace the trajectory of a particle with the motion of the front of a certain wave. Or rather, we can write the equations of motion in mechanics in a form such that they will fully coincide with the equations of geometrical optics which describe the propagation of a light ray ignoring its wave properties. Hamilton proved

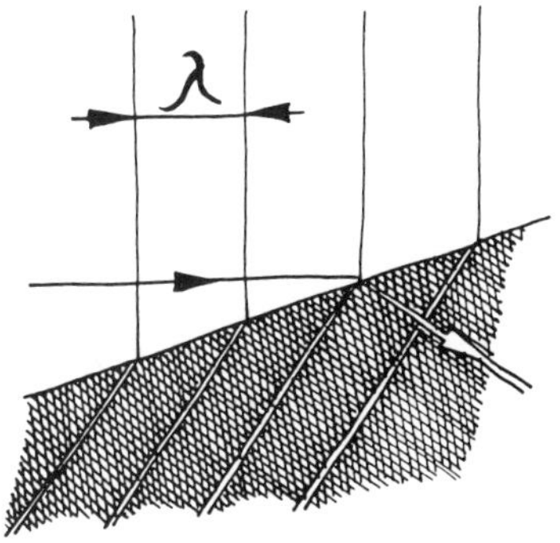

thereby the validity of the *optical-mechanical analogy,* i.e., that the motion of a particle along a trajectory can be represented as a light beam if its wave aspects are ignored.

Schrödinger's Wave Mechanics

In 1911 Erwin Schrödinger (1887-1961) graduated from the University of Vienna, which was still steeped in the traditions of Christian Doppler, Armand Fizeau, Ludwig Boltzmann, and in general the spirit of the classical age in physics. These traditions were soundness in the study of natural phenomena and a leisurely interest in them. By 1925 he was a professor at Zurich University. Although he was no longer young, he still had the youthful urge to understand the most vital problem of his day, namely the architecture of the atom, and the way electrons travel in it.

In late 1925 Schrödinger read in one of Einstein's articles words of praise for de Broglie and his hypothesis. This was enough for him to take seriously the matter-wave hypothesis and to start working on it. Looking back now, half a century later, we can easily understand his reasoning. First of all, he recalled Hamilton's optical-mechanical analogy. He knew that it had been proven only in the limit of geometrical optics, i.e., when we can ignore the wave aspect of light. Schrödinger went one step further to assume that the optical-mechanical analogy remains valid even for wave optics. This means that *any motion of a particle is always similar to wave propagation.*

As with any fundamental discovery, Schrödinger's hypothesis had no logical premises. But,

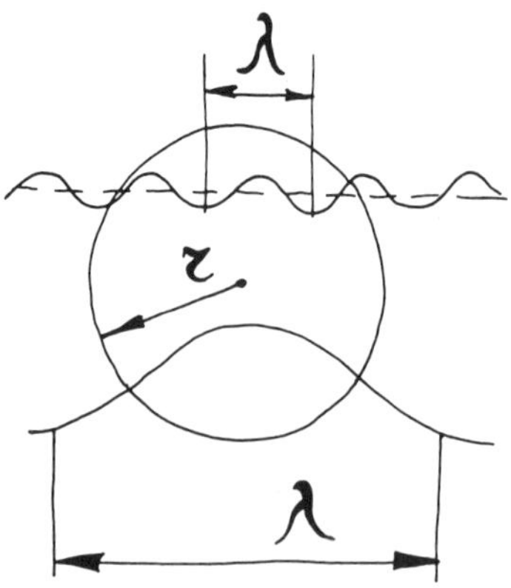

like any true discovery, it did have logical *consequences.*

First of all, if Schrödinger's ideas were correct, the motion of particles should display wave properties when dealing with space scales commensurate with the wavelength of these particles. To a great extent this concerns the motion of an electron in an atom. A comparison of the formulas of de Broglie $\lambda = h/mv$ and Bohr $mvr = h/2\pi$ shows that the atomic radius $r = \lambda/2\pi$ is about one-sixth of the electron's wavelength λ. If this length is taken to be the size of the atomic electron, it becomes immediately clear that an atomic electron cannot be viewed as a particle, because we would have to suppose then that the atom is constructed from particles that are larger than the atom itself. An immediate and somewhat unexpected consequence of this is Heisenberg's postulate, which we have already discussed in the previous chapter: there is no such thing as the trajectory of an electron in the atom.

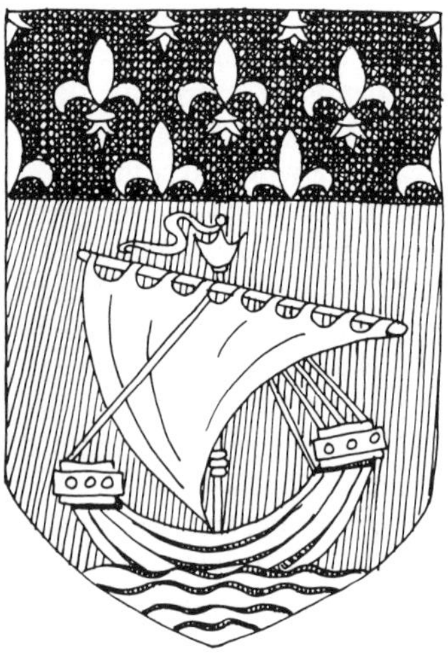

How can an object possibly move about within something of smaller size, and along some kind of trajectory at that. This clears up the problem of the stability of the atom, since electrodynamics only forbids the electron to travel in the atom along a trajectory; it is not responsible for the phenomena connected with other kinds of motion. This all implies that electrons exist in the atom not as particles but as waves, whose nature was not quite clear at first, even to Schrödinger himself. What *was* clear to him was that whatever the nature of these electron waves, their motion must obey a *wave equation.* Schrödinger derived such an equation. It looks like this:

$$\frac{d^2\Psi}{dx^2} + \frac{2m}{\hbar^2}\Big(E - V(x)\Big)\Psi = 0 .$$

The equation says absolutely nothing to those who see it for the first time. It induces curiosity or even a nebulous feeling of instinctive objection (without serious grounds for the latter).

But isn't the design on the accompanying figure as baffling as Schrödinger's equation? Why then do we accept it without a feeling of resistance? It somehow comforts us to know that this design is simply the coat of arms of the City of Paris. Only the most finicky of readers will try to find out why it looks the way it does and not otherwise. As in Schrödinger's equation, each line and symbol in these armorial bearings are not without meaning. At the top are fleurs-de-lis, or lilies, which first appeared as a heraldic emblem of the French royal family and of France toward the end of the fifth century, after the victory of Clovis I over the Huns on the River Lys. (Tradition has it that on their way home the soldiers decorated their helmets and shields with white lilies, or "li-li", which in Gallic means "white-white".) The ship in the lower field has outlines resembling those of the Ile de la Cité, the island in the River Seine, which was the home of the ancient Gallic tribe called the Parisii, from which the City of Paris got its name. The shape of the escutcheon resembles a sail — a symbol of the main occupation of the ancient inhabitants of Paris. It is not difficult, you see, to decipher the coat of arms, although it is really significant only to the inhabitants of Paris.

Now let us approach Schrödinger's equation in the same manner. At first we simply accept it as a symbol of quantum mechanics, as a coat of arms of the quantum land through which we are travelling, and make an attempt to understand why it looks that way. Some features in this "coat of arms" are already old acquaintances: m is the mass of the electron, $\hbar$ is the Planck's constant h divided by 2π, E is the total energy of the electron in the atom, $V(x)$ is the potential energy, and x is the distance from the nucleus to the electron. It is perhaps somewhat more difficult to understand the symbol of the second derivative d^2/dx^2, but this cannot be helped for the time being. For the moment we will just have to remember that this is the symbol for the second derivative of the function Ψ. This circumstance makes Schrödinger's equation not an algebraic equation, but a differential one.

The most difficult thing here is to grasp the meaning of the Ψ function (read as psi function). Suffice it to say that even Schrödinger misinterpreted the meaning of this function at first. We will also understand this a bit later in the book. For the moment we will just assume that the psi function somehow represents the motion of the electron in the atom. Not in the same manner as Heisenberg's matrices x_{nk} and p_{nk}, but nevertheless it does, and quite adequately at that. So adequately that it makes the solution of all problems in quantum mechanics simpler and quicker than in terms of Heisenberg's matrices.

Physicists were quick to appreciate the advantages of wave mechanics — its universality, elegance, and simplicity. Ever since they have almost abandoned matrix mechanics.

Quantumalia

The Life of Boscovich...

Man is the measure of all things.
Protagoras

Roger Joseph Boscovich (1711-1787) is now known only by a narrow circle of experts, but early in the nineteenth century he was a celebrity, and his theory of the atom influenced the thinking of such men of science as Faraday and Maxwell.

Boscovich was born and spent his childhood at Ragusa (now Dubrovnik) on the Dalmatian coast. He was the eighth of nine children and the youngest of six sons in a family of well-to-do merchants. Those were the days when any vocation was considered significant and enjoyed recognition only if it was sanctified by the Church or associated with it. Boscovich had studied in the local Jesuit College from the age of eight, and when he turned fourteen his parents sent him to Rome, his mother's native city. After two years as a novitiate he was admitted to the *Collegium Romanum.* There he excelled in mathematics, physics, and astronomy. In 1736 he published his first paper on the Sun's equator and period of rotation. At the age of twenty-nine he became an instructor and at thirty-three a priest and a member of the Society of Jesus. For fourteen years he taught physics and mathematics, studied the aberration of light and the shape of the Earth, and mapped the Vatican. In 1760 the St.Petersburg Academy elected him a foreign member.

Boscovich was also a poet: in 1779 he dedicated a poem to Louis XVI, in which he predicted that Louis XVI's reign would be devoid of solar eclipses. The brilliant qualities of his many-sided nature won him admission to the most eminent church, academic, and diplomatic coteries of Europe.

In 1757 he went to Vienna as a member of the embassy, and there during a period of eleven months he wrote his *Theoria Philosophiae Naturalis,* a treatise he had been pondering over for twelve years. Following his return from Vienna, he set out on a four-year tour of Paris, London, and Constantinople. Afterwards he lectured, worked in the Milan observatory, and made enemies everywhere owing to his independent views. In 1772 he found himself in Venice, penniless. His friends obtained for him a post in Paris, where he lived for ten years before returning to Italy in 1783, to publish his works and to die. Late in 1786 he showed symptoms of mental disorder, which developed into pathological melancholia. After an abortive attempt to take his life he went insane, and on 13 February 1787 he ended all the complications of his turbulent life.

L. de Broglie

...and His Atom

Of the few who in the eighteenth century believed in atoms Boscovich was the only one who did not think of atoms as tiny solid balls. His views were rather closer to ours than all the atomic theories of the nineteenth century.

His arguments against the hypothesis of incompressible ball-shaped atoms were as follows: it does not account for the crystalline structure and elasticity of bodies, the melting of solids, the evaporation of liquids, and especially chemical reactions between substances constructed out of these round, solid, impenetrable spheres.

Boscovich pictured the atom as being the centre of forces that vary with the distance from that centre. Near the centre the forces are repulsive, which corresponds to the repulsion of atoms when they are brought into close proximity or collide. As we move away from the centre, the repulsive forces first decrease, then vanish and finally become attractive. According to Boscovich, solids change into liquids, and vice versa, exactly at the moment when the forces change their sign. Still farther away from the centre, the forces again become repulsive, causing liquids to evaporate. Lastly, far away from the atom the forces are always attractive in accordance with Newton's law of universal gravitation.

It is only when the cold season comes that we know the pine and cypress to be evergreens.

Chinese proverb

Each of Boscovich's atoms thus "extends to the very boundary of the solar system"; and since centres of forces can neither be destroyed nor created, the atoms of Boscovich were eternal, just as those of Democritus. It was this part of Boscovich's theory that was especially congenial to Faraday. We can readily see its analogy to Faraday's idea of the lines of force of the electromagnetic field. And later on, at the turn of the twentieth century, Lord Kelvin once again turned to the atom of Boscovich in his attempt to explain the nature of radioactivity.

Boscovich's atom is a good deal closer to the modern atom than Democritus's. Like the modern atom, it has no definite geometrical dimensions. On the other hand, it helps one's understanding of the variety of crystalline shapes and chemical transformations in which these atoms take part. The atom of Boscovich is clearly a speculative scheme, which is based neither on experiment nor mathematics, only on common sense and keen observations of nature. Boscovich wrote: "There are indeed certain factors regarding the law of forces of which we are all ignorant — they are concerned with intervening arcs and other things of that sort. This all, however, far surpasses human understanding, and He alone, Who founded the universe, had the whole before His eyes."

Quantum mechanics enables us to derive the law governing the variations of the force between two atoms without any arbitrariness or any references to divine prevision. We can use this law of forces to predict the spectrum of the hydrogen molecule and calculate the energy needed to separate one hydrogen atom from another. We can even foresee the results of mixing hydrogen with, say, chlorine, and what will change if we expose this mixture to ultraviolet radiation.

We are able to determine the shape of crystals and even to predict the colour of chemical compounds. When I say "we", I mean of course those conversant with the fairly complex mathematical tools of quantum physics. Anyone, however, who is the least bit acquainted with its images can readily understand many of the features of the structure and properties of matter.

Paul Ehrenfest (1880-1933)

Science needs prophets, but it also needs apostles. In addition to solitary geniuses who alter its course, science also needs torch-bearers who are able to keep its fire alight and kindle it in the hearts of neophytes. These devotees create around themselves an atmosphere of intellectual energy and inspiration in which talents quickly blossom and gifted minds find fertile soil. Such torch-bearers were Arnold Sommerfeld in Germany, Paul Langelvin in France, and Leonid Mandelshtam in Russia.

P. Ehrenfest

Another of that race was Paul Ehrenfest. He was born and raised in Vienna and studied first at the University of Vienna under Boltzmann and then in Göttingen under Felix Klein. When he had finished his education, he lived for five years in Russia. Then, in 1912, he accepted the invitation of Lorentz to succeed him as professor of theoretical physics at the University of Leyden. There, every Tuesday, for twenty years, Ehrenfest held a seminar, which drew all the eminent and celebrated scientists who in a quarter of a century reconstructed the very foundations of physics. It was at this seminar that the hypothesis of electron spin was born, and its midwife and godfather at the same time was Ehrenfest. He initiated and "added fuel" to the famous polemics between Bohr and Einstein. He lived in the centre of the "physical events" of his day and himself was a major contributor.

Ehrenfest was a man of rare warmth. Among his frequent visitors were Bohr, Planck, Heisenberg, Pauli, and Schrödinger. In a letter to Ehrenfest, Einstein wrote: "We have been created by nature for each other. I have an even greater need of your friendship than you have of mine." But something broke his spirit, and on 25 September 1933 he committed suicide.

Ehrenfest left physical ideas that have outlasted the living memories of his students and friends. He bridged the gulf in the minds of his contemporaries that separated quantum phenomena from classical ones. He proved the theorem that can be summarized as follows.

We have stated repeatedly in our narrative that the equations of quantum mechanics differ markedly from those of classical mechanics. Therefore, the motion of quantum objects can neither be described nor represented by classical concepts and images, just as it is impossible to trace on the terrestrial globe all the movements of a passenger crossing the Atlantic on an ocean liner. But, no matter how the waves rock the ship and what the passenger may be doing during his voyage, he is still basically travelling along a set course.

Where the telescope ends, the microscope begins. Which of the two has the grander view?

Victor Hugo

In the quantum world things are similar. Admittedly, we cannot visualize quantum motion. Admittedly, we cannot clearly understand what the position and momentum of an electron is. But we do know for sure that the *mean values* of quantum quantities obey the equations of classical mechanics. This is the essence of the *principle of correspondence,* which was formulated in 1918 by Bohr and proven in 1927 by Ehrenfest.

A while later everybody realized that Ehrenfest had proved something more than that in the limit quantum mechanics becomes classical mechanics. In fact, at all times it was tacitly (and in many cases openly) understood that the dynamical laws of classical mechanics are primary, true laws, and that the statistical laws of quantum mechanics are second-rate laws, ones we have to employ because atomic objects are especially complicated. The work of Ehrenfest showed that such thinking is actually prejudice, because the equation of motion of classical mechanics are a limiting case of the more general equations of quantum mechanics.

Sphynx, Egyptian

Chapter Eight

Schrödinger's Equation
The Meaning of the Ψ Function
The Image of the Atom
Quantum Truth

"A naturalist, leaving the realm of direct sensory perceptions behind him with the aim of discovering more relationships, can perhaps be likened to a mountain-climber who wishes to reach the summit of the highest mountain to view the countryside stretched out before him in all its diversity. The mountain-climber must also leave the fertile populated valleys. As he ascends, he gains an ever wider view of the neighbourhood but, at the same time, the signs of life about him become less frequent. He eventually reaches a dazzling bright world of ice and snow where there is no more life, and breathing is almost impossible. Only by crossing this region can he reach the peak. But when he finally reaches the summit, a moment comes when he gains complete command of the view of the whole locality spreading before him. Then perhaps, he would not feel so far removed from life...."

These words of Heisenberg give one a good feeling for the qualitative leap that occurred in the minds of people when they crossed over from the observation of phenomena that directly acted on their sensory organs to the study of atomic objects. This breakthrough, which occurred at the turn of the century, is so important that we shall illustrate it again with a specific example.

Let us return to the example of a taut string that produces a sound. You hear the sound, you see the vibrating string, and you can touch it with your hand. From this an *image* is formed in your mind of the physical phenomenon occurring before you. The *concept* of a "wave process" comes later, when observing other similar processes. To make this concept unambiguous it is fixed using a *formula,* which is an equation that enables the whole process of string vibration to be predicted. We can check this prediction by *experiment,* e.g., by recording, for instance, the vibration of the string on motion picture film.

We have once more traced the classical chain of cognition:

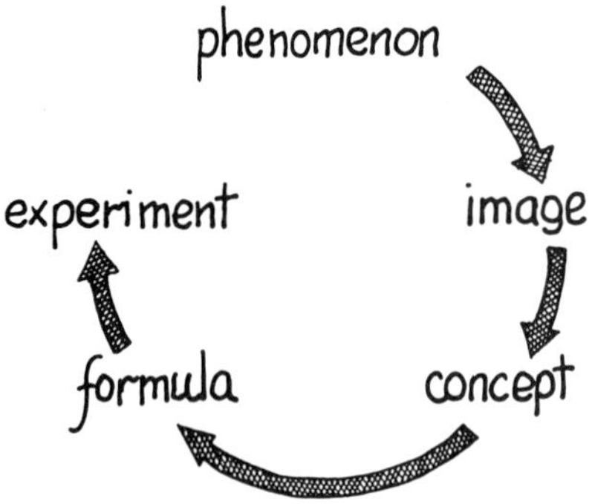

It lies at the basis of knowing in physics. The last link in the chain — experiment — checks how correctly we imagine a phenomenon as a whole with only a partial knowledge of it.

But this simple scheme breaks down completely when we try to answer the question "What is the atom?" simply because the entity "atom" does not affect our organs of perception, and so they cannot give us any "image of the atom", however approximate. Therefore, at first the "atom" was a purely speculative thing unrelated to our sensory organs, and for twenty centuries it remained just a curious hypothesis, one that had no advantage whatsoever over other hypotheses on the structure of matter.

The true history of the atom began with the advent of science, which taught people to trust the readings of instruments and not merely to rely on their perceptions. People used instruments to observe how substances behaved when dissolved, when an electric current was passed through the solution, when substances were heated, illuminated and subjected to many other types of conditions. Men of science *studied* these phenomena, not merely observed them. They *measured* the temperature of bodies, the wavelength of the light that they emitted, and many other features that we are already familiar with. They recorded the results of their measurements in the form of *numbers.* The numbers replaced direct sensations scientists previously relied upon. *Numbers* were the only things that they came to trust when they began to study things that they could not physically perceive. Supplied with these numbers they began to seek relationships between them and to write them down in the form of *formulas.*

But people communicate with one another by means of words, not formulas. When they want to speak about new relations in nature, they have to invent *concepts* that correspond to these formulas. Even though these concepts are often quite extraordinary, people become accustomed to them and learn to apply them correctly and even create images for themselves that they associate with the new concepts. This reverses the conventional chain of cognition, which now becomes

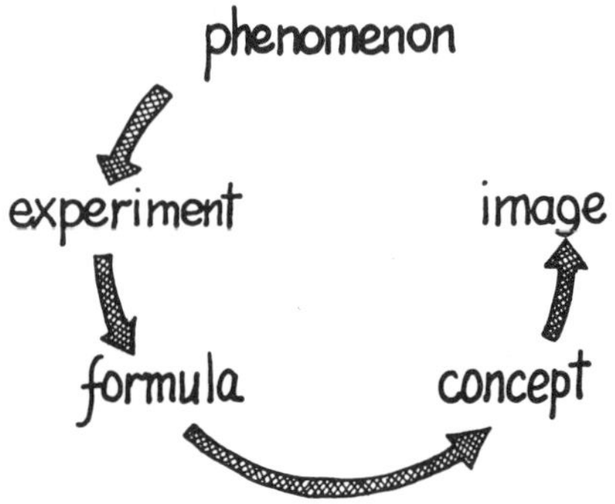

Now the image is the last, not the first, step in the sequence.

This chain can readily be traced back through the history of the atom. Fraunhofer, Kirchhoff, and Bunsen discovered that each atom emits a strictly definite set of spectral lines (the phenomenon) and that each spectral line corresponds to a number — the wavelength λ (the experiment). Balmer, Rydberg, and Ritz found simple relationships (the formula) between these numbers, and Bohr showed that their formulas follow from a single principle, which was named quantization (the concept). Eventually, from these experiments, formulas, and concepts emerged an image — Bohr's atom.

Experiments continued; fresh numerical and factual evidence mounted, and some of the new facts could no longer be accommodated by old formulas, concepts, and images. And then came quantum mechanics — a unified principle from which followed all the previous empirical formulas and apt conjectures.

Thus far we have learned quite a lot about the experiments of atomic physics and the concepts that have to be employed to cater for them. What we want now is a bit more: at this new, higher level of knowledge to create an *image of the atom.* To do this we must concern ourselves, at least cursorily, with the formulas of quantum mechanics. This is a necessary condition — after all, the beauty of logical constructs in science is much more important than the effects of stunning associations produced by its simple consequences.

Schrödinger's Equation

All the foregoing will have convinced us that an electron is not a point, it does not occupy a definite position in the atom, nor can it move in some orbit. We have thus been initiated to the rather vague idea that the atomic electron is "smeared out". Schrödinger succeeded in catching this vague idea quite precisely in the unambiguous language of formulas. Schrödinger's equation, just like any other fundamental law of nature, cannot be deduced rigorously from simpler ones. It can only be guessed. (Later in his life Schrödinger confessed that he himself still did not fully understand how he had done this.) But it is one thing to guess an equation, and quite another to learn to apply it, to find out what all the symbols in the equations mean and what phenomena in the atom they reflect.

We have already written out Schrödinger's equation earlier:

$$\frac{d^2\Psi}{dx^2} + \frac{2m}{\hbar^2}\Big(E - V(x)\Big)\Psi = 0$$

and explained the symbols that appear in it: $\hbar$ is Planck's constant h divided by 2π, m is the electron mass, E is the total electron energy in the atom, and $V(x)$ is the potential energy of the interaction between the electron and nucleus separated by the distance x. But the meaning of the wave function Ψ still remains obscure. To perceive it, we return to the analogy of the vibrating string.

Its equation, well known in classical physics, is

$$\frac{d^2u}{dx^2} + \left(\frac{2\pi}{\lambda}\right)^2 u = 0\,.$$

Several of the solutions of the string equation, i.e., the functions $u_k(x)$, are shown at the left-hand side of the figure. These are ordinary, familiar sine curves and their meaning is obvious: they depict the *shape of the string* at some instant of time, a snapshot of its vibration. The mode of string vibration depends on the number of nodes

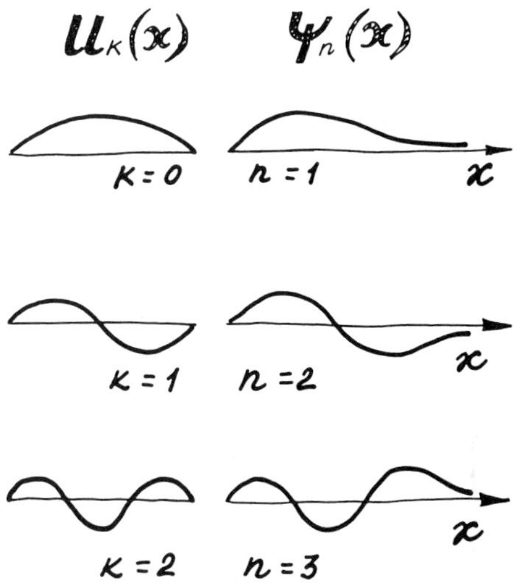

k, i.e., the number of points that remain stationary in the process of vibration. Corresponding to these nodes are an infinite set of solutions $u_k(x)$, which differ in k. It is important that there exist no other intermediate vibrational modes, besides those denoted by the subscript k.

Schrödinger's equation differs only slightly in shape from the string equation. To see it more clearly, we introduce the notation

$$\lambda(x) = \frac{h}{\sqrt{2m\,(E - V(x))}}\,.$$

And thus Schrödinger' equation becomes similar to the string equation

$$\frac{d^2\Psi}{dx^2} + \left(\frac{2\pi}{\lambda(x)}\right)^2 \Psi = 0\,.$$

If $V(x) = 0$, i.e., the electron moves about freely far away from the nucleus, then the energy E is just its kinetic energy, i.e., $E = mv^2/2$, and hence its wavelength

$$\lambda(x) = \lambda = \frac{h}{mv}$$

is constant and is equal to de Broglie's wavelength. Now Schrödinger's equation coincides in form exactly with the string equation. In the atom the electron interacts with the proton by the Coulomb law; therefore $V(x) = -e^2/x$, where e is the electron or proton charge. In this case the "electron wavelength"

$$\lambda(x) = \frac{h}{\sqrt{2m\,(E - V(x))}}.$$

has no definite value and varies from point to point with x. On the other hand, in the theory of vibrating strings such a case is also known: if instead of a uniform string we have a non-uniform one, i.e., one with all sorts of weights and inhomogeneities, then its vibrations will be described exactly by such an equation. Its solutions will only remotely resemble regular sine waves, although they will retain the main property of the previous solutions: they will still have stationary nodes by which these solutions can be numbered.

We have thus seen that formally Schrödinger's equation is identical with the loaded string equation, but their solutions, of course, have different meanings. Mathematically, the equations of quantum mechanics are even simpler than the equations of hydrodynamics or electromagnetism. Their complexity lies elsewhere: in the concepts we associate with the quantities in the equations.

Now look at the figure, where alongside the sine curves of the string $u_k(x)$ are given the solutions $\Psi_n(x)$ of Schrödinger's equation for the hydrogen atom. The two sets of curves look very much alike. And, even if in the atom there occur no real vibrations similar to the motions of the string, this does not make the analogy less useful.

The analogy enables the solutions $\Psi_n(x)$ to be numbered with integers n in the same manner as the solutions $u_k(x)$ are numbered with the integers k, and there are no other solutions apart from these *eigenfunctions* of Schrödinger's equation. Furthermore, the integer n is exactly that mysterious *quantum number* by means of which Bohr numbered the electron orbits in his atom. It is none other than the number of nodes of the wave function plus one: $n = k + 1$.

The first of Bohr's postulates deliberately ordained that electrons move only in those orbits in the atom that meet the quantum condition $mvr = n\hbar$. The principle worked well, but it was unnatural for physics, and so it evoked a complicated mixture of delight and discontent in the physical community of the day. Schrödinger's condition is clearer: no matter how intricate the motion undergone by an electron within the atom, it must still lie within the atom. Consequently, the Ψ function, which "represents" this motion, must be concentrated near the nucleus. And so from this single and natural *boundary condition* it follows unequivocally that Schrödinger's equation does not always have solutions; there are solutions only for certain values (*eigenvalues*) of the energy E_n to which the eigenfunctions $\Psi_n(x)$ correspond. The eigenvalues E_n of the energy of the electron in the hydrogen atom can be found by solving Schrödinger's equation with the potential $V(x) = -e^2/x$.

These discrete energy values

$$E_n = -\frac{me^4}{2\hbar^2}\cdot\frac{1}{n^2}$$

for *stationary states* are numbered by the integers n. It is easily seen that these energies are precisely the energies of the electron in stationary orbits in the Bohr atom, thus making Bohr's postulates redundant, while all the other positive results of his model remain with us.

At once these consequences of Schrödinger's theory won over many physicists by their simplicity and naturalness; physicists put their trust in Schrödinger's equation and then concentrated on the last point: the significance of the function $\Psi_n(x)$ itself. Well, if the function $u_k(x)$ depicts the shape of the vibrating string, *what* then does the Ψ function depict?

The Meaning of the Ψ Function

This is one of the most puzzling questions in quantum mechanics; even Schrödinger answered it incorrectly at first. But his answer was so convenient and so close to truth that we will make use of it for the time being.

In the atom an electron does not exist as a particle. It spreads out to form a kind of cloud. The shape and density of that cloud are determined by the function $\rho(x)$. This function of x equals the square of the function Ψ:

$$\rho(x) = |\Psi(x)|^2.$$

To illustrate this idea let us look at a watermelon and try to plot its density $\rho(x)$ as a function of the distance x from the centre of the watermelon. The function $\rho(x)$ is clearly constant throughout, decreasing slightly near the outside (the rind is lighter than the flesh), and then dropping abruptly at the boundary. A glance at the figure is sufficient even for a person who has never seen a watermelon to tell how it is arranged inside. True, this will give him no clue as to its taste, colour, or fragrance, or a thousand other minor features that distinguish one watermelon from another.

When we make attempts at penetrating into the atom we all find ourselves in the position of someone who has never seen a watermelon but wishes to picture one by using the function $\rho(x)$. For an atom we can calculate this function from Schrödinger's equation and then, using it, plot the distribution of the electron cloud in the atom. The pictures thus obtained are counterparts of the pictorial representation of the atom that we are all unconsciously striving for.

Shown on page 104 are such pictures; they are two-dimensional representations of the hydrogen atom reconstructed from the functions $\rho(x)$ obtained from Schrödinger's equation. This is the modern image of the atom towards which we have so long been travelling and to which we must now become accustomed. Later in the book the image will only change slightly; perhaps even not the image itself but our attitude to it.

We have now covered the roughest stretch of ground, and so we can sum up at leisure what we have achieved. To this end, we return — at a higher level now — to the question "What is the atom?"

The Image of the Atom

Recall Thomson's model of the atom: a large positive sphere with tiny negative electrons floating around in it. In reality, everything turned out to be exactly the other way round: in the centre of the atom lies a tiny positive nucleus surrounded by a negative cloud of electrons. The shape of this cloud is not arbitrary — it is determined by the laws of quantum mechanics. This shape is not, of course, a sphere with sharply defined boundaries, but generally the unexcited atom of hydrogen is very much like a sphere (this guess of Democritus's appeared to be correct).

Now, excited atoms differ in shape from a sphere, and the more so the more excited the atom is. When we excite an atom we expend energy on the restructuring of its electron cloud. Corresponding to each shape of the cloud is its own, definite energy. Therefore, for an atom to be transferred from one shape to another requires a fixed amount of energy, the quantum $h\nu$, in keeping with Bohr's second postulate.

So far we have only been concerned with the hydrogen atom. In fact, it is the only atom that physicists of today know in great detail and for which they can conceive a plausible image. In more complex atoms the electron cloud on the whole is also more or less as shown in our figures. The configuration of the clouds was computed with only moderate accuracy in the works of the Soviet physicist Vladimir Fock (1898-1974) and the English physicist Douglas Hartree (1897-1958). The task presents enormous computational difficulties and can only be properly handled by modern computers.

When speaking about the shapes of bodies we normally assume that the bodies have dimensions. This is not always so: a billiard ball has both shape and size, but if we turn, say, to a cloud, it is difficult to speak of its dimensions, although its shape raises no doubt. The most stunning consequence of the new model of the atom is that *the atom has no definite geometrical dimensions,* dimensions in the sense we generally attach to this concept when viewing, say, a billiard ball. Since an atom has some outline, we can, of course, single out a part where the density of the electron cloud is greatest and call this part its size. This definition is legitimate and we will use it (we often speak about the size of the atom); but at the same time we should remember that, strictly speaking, we cannot define the dimensions of the cloud — this is always a matter of convention.

Now, this consequence of quantum mechanics alone enables us to account for many of the observable properties of bodies. For instance, the diversity of geometries of crystals should not astonish us: after all, from identical bricks are built an infinite variety of houses, but we do not consider it strange that a brick is simply a brick and not a miniature house. Bodies surrounding us have colours, odours, and dimensions, but the atoms of which these bodies are made up have none of these qualities. Likewise, they have no definite shape. Only the laws of quantum mech-

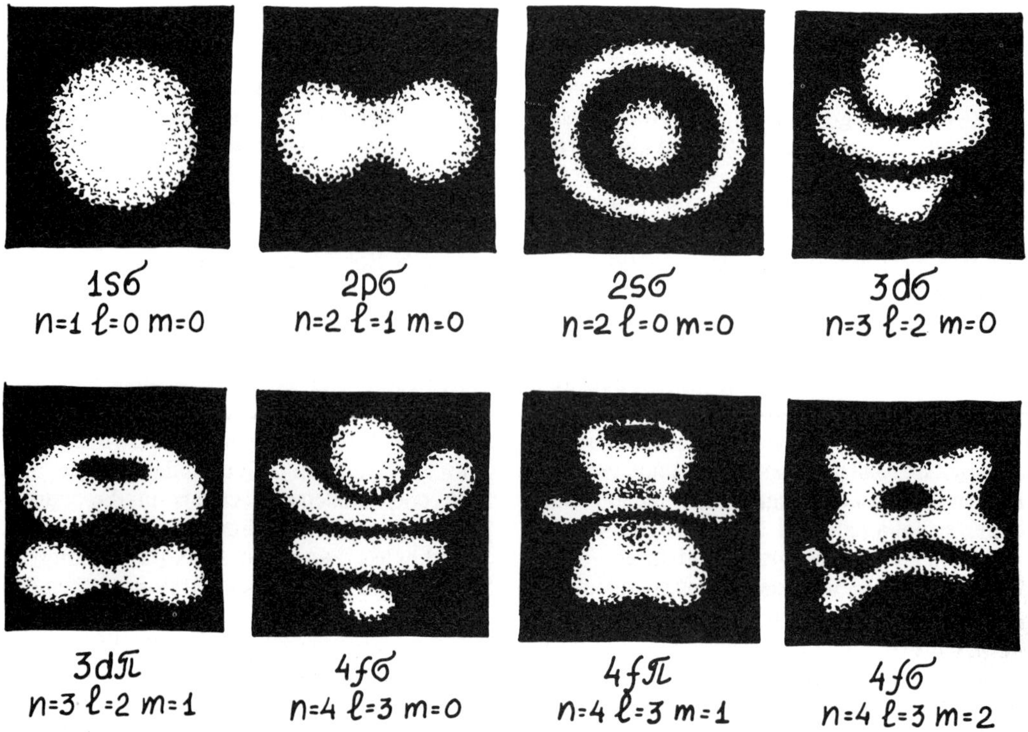

anics that govern that shape are invariable.

But why is the atom, which does not even have dimensions, so stable? This too should not astonish us: after all, the Earth does not rest on three whales but, nevertheless, poised in the void it has kept its orbit invariable for millions of years. The secret of this stability is in the motion and in the invariability of the dynamical laws that govern this motion. These same reasons account for the stability of the atoms, although the laws governing the motion of electrons are a far cry from those of celestial mechanics.

It is worth noting, in all fairness, that quantum stability is much more reliable than the dynamic stability of classical mechanics: a disturbed atom restores its structure but the Earth would never again be in the same orbit if were to be disturbed by the sudden intrusion of some foreign body from space.

The atoms of various elements vary widely in nuclear mass and charge. But how are we to distinguish two atoms of the same element? For watermelons the question is ridiculous; nobody has ever seen two exactly identical watermelons. To distinguish one brick from another is already much harder; if they are broken the task is somewhat simpler.

The same is true of atoms. If their masses and nuclear charges are equal, they can only be distinguished by the shape of their electron clouds, which depends on the degree of excitation of the atoms. But all unexcited atoms of the same element are indistinguishable from one another, just like bricks coming from the same mould. For atoms the role of the mould is played by the dynamical laws of quantum mechanics, which are invariable and the same for all of them.

The "portraits" shown here of the atom represent our level of knowledge about it today. This is the latest image of the atom that has come to replace the models of Democritus, Thomson, and Bohr. But even these "portraits", of course, should not be taken too literally: they are by no means "photographs of atoms", like the photographs of a vibrating string. Instruments, however sophisticated, tell us nothing about the distribution of electron density within the atom because the very act of measuring will destroy the atom (even a watermelon must be cut open to make sure that it is ripe). All the same, we have good reason for believing in the picture thus obtained: it enables us to give a consistent account of the experiments that have led us to this image of the atom.

We should no longer be surprised that the alpha particles in Rutherford's experiments flew unimpeded through billions of atoms as through a void. So the Earth, when it passes through the tails of comets, is never deviated from its orbit. The origins of spectral lines should also be clear to us now: in the atom the shape of its electron cloud simply changes in a jump, and as a result a quantum of energy is emitted. Now this also explains the splitting of spectral lines in an electric field (Stark effect) and in a magnetic field (Zeeman effect): the electron cloud is charged and its many shapes change somewhat in the fields, by splitting into closed "subshapes"; so does the energy of the quanta needed to transform the cloud from one shape to another, and also the wavelength of the spectral line that corresponds to that quantum. Using the equations of quantum mechanics we can support this simple qualitative reasoning by exact calculations and show that it agrees with the experimental data.

We might go on to use this new model to analyse the mounting evidence of atomic physics. But right now it is more important for us to look into another matter: why are we so sure that the image of the atom we have found is the true one?

Quantum Truth

Now, what kind of truth are we to discuss? And what is taken to be truth in quantum mechanics? If we were talking about a watermelon, everything would be quite simple. We would say, for instance, that if we only knew the density distribution inside the watermelon we would not know the entire truth about the object. Only when we could see, touch, and finally eat the watermelon could we say that we know sufficiently about it. For the majority of people this knowledge would now be quite adequate, but not for scientists. They would begin examining watermelon flesh under the microscope and this would tell them that the watermelon consists of cells. They would then say that the cells are made up of molecules and the molecules of atoms. We have thus completed the circle. To know everything about the watermelon we must again answer the question "What is the atom?"

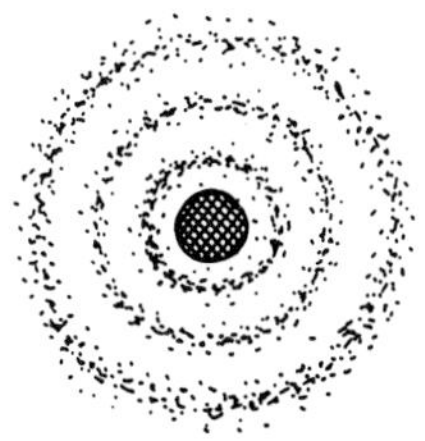

In actual fact, things are not as bad as they might seem. The concept of the "watermelon" had taken shape many centuries before any science emerged and it does not depend on any past or future scientific achievements, since it is only based on our sensations. This concept can only change if people all of a sudden develop another, sixth, sense. Assuming this to be unreal we can lightheartedly assert that we know the entire truth about a watermelon if we put it to the test of our five senses. (Remember how you buy a watermelon. First, you select a likely one from the heap, then you take it in your hands; sometimes you hold it up to your ear to hear the slight cracking sound it makes when you squeeze it, and finally you cut out a small wedge to taste.)

Is this approach good for the concept of the "atom"? After all, we piece together our image of the atom from endless experiments, and in principle each of them adds something new to our knowledge. We cannot just stop and say "No more experiments. We've already constructed for ourselves an image of the atom and any further experimenting can only spoil it." On the contrary, we happily accept each new piece of experimental evidence, especially when it does not fit into the earlier picture. It was such evidence that helped us to reject the idea of atoms being solid spheres and to devise a more refined model. Why are we now so sure that the current image is ultimate and true?

We must admit that this is something physicists are not at all sure of. They can, however, say honestly and with dignity: "Over the last hundred years not a single experiment has yielded results inconsistent with the picture we have constructed. Therefore, it is more appropriate to speak of its fruitfulness rather than its truth, i.e., to speak of how well it helps us to explain and predict the details of atomic effects." And here a startling fact comes to light: it is no longer critical for us to know "just what an atom actually looks like". It is sufficient for us to grasp the equations of quantum mechanics and the rules for using them. We can then predict everything: how a body changes in colour when heated, what spectral lines it emits in the process, and how their wavelength changes if we place the body in an electric or magnetic field. We can predict the shape of crystals, their specific heat, and electrical conductivity. Finally, we can build an atomic power station and an atomic ice-breaker and they will operate without fail. And all this without the slightest reference to the true shape of the atom.

Such arguments are behind the reasoning of a school of thought, whose major proponent was Heisenberg, that maintains that quantum mechanics can do without any graphic images whatsoever. One may well have objection to such an extreme view, but one cannot turn it down arbitrarily. These "extremists" answer the eternal question "What is an atom?" laconically: "The atom is a system of differential equations". Like it or not, there is much truth in this jest. In comparison with a whole watermelon, an "atom of watermelon" is exceedingly meagre in terms of its properties. But these few properties are contradictory and we cannot fuse them into reason without overstraining our logic and common sense except through the equations of quantum mechanics.

Quantum mechanics is a mathematical scheme that enables us to calculate physically measurable characteristics of atomic phenomena.

If this alone were the aim of physics, we could regard the task of constructing the mechanics of the atom as completed. Physics, however, is called upon to give us more than that, namely a rational picture of the world. It is impossible to accomplish such a grand programme with formulas and numbers alone. We must also find images and formulate suitable concepts. This is of especial interest to non-physicists, who neither know nor understand the formulas of quantum mechanics. The language of images and concepts is the only language that can usher them into the deepest recesses of matter. Since the times of Democritus we have advanced a great deal along this road and have constructed for ourselves a more or less satisfactory picture of the atom. But for this picture to be perfect several finishing touches are still needed.

Quantumalia

Compton's Experiment

Imagine that you are standing before a mirror dressed in a green sweater and suddenly see your reflection in the mirror "dressed" in a red sweater. You will, obviously, rub your eyes at first, and if that does not help, you will go to see an optician. Because what you see is "just impossible". Green light has a wavelength $\lambda = 5500$ angstroms. When this light meets an obstacle in its path (the mirror), it is reflected but can by no means change its wavelength and become, for instance, red ($\lambda = 7500$ angstroms). But this is exactly what Compton observed. When he allowed a beam of X rays of wavelength λ to strike a target, he found that the scattered X rays have a longer wavelength λ' than incident X rays!

To explain this marvel we recall Einstein's hypothesis about light quanta, which he proposed for the photoelectric effect. According to that hypothesis, instead of X rays with a wavelength λ and frequency $\nu = c/\lambda$ we must picture a flux of particles, or quanta, with the energy $E = h\nu$ and impulse $p = h\nu/c$. When the quanta collide with atomic electrons in the target, they knock out some of the electrons (expending the energy P), accelerate them to the velocity v (expending the additional energy $mv^2/2$), and are themselves scattered with lower energy $E' = h\nu'$ and lower frequency, i.e., larger wavelength λ. By energy conservation,

$$h\nu = h\nu' + P + \frac{mv^2}{2}.$$

E. Schrödinger

If the atom absorbs the light quantum completely ($E' = 0$), we will observe the ordinary photoelectric effect, and Compton's equation becomes Einstein's:

$$h\nu = P + \frac{mv^2}{2}.$$

We can carry out both experiments in a Wilson cloud chamber, where we can trace the path of each ejected electron and thereby visualize the collision of a light quantum with an electron.

In that case, what could prevent us from seeing ourselves in a red sweater in the mirror? It is the same quantum laws which forbid an electron to absorb arbitrary portions of energy. An electron in a stationary orbit in the atom can absorb only a quantum that can transfer it from one stationary state to another (remember Franck and Hertz's experiment), or knock it out from the atom (experiments by Lenard, Stoletov, and Millikan). The energy of "green quanta" (2.5 electronvolts) is much too small to knock out an electron from the atom ($P \sim$ 10 electronvolts). And so the quanta elastically bounce off the atoms of the mirror (with no loss of energy) and become no "redder" in the process. X rays are 5000-10,000 times more energetic, and the phenomena they undergo are different. So they are not reflected at all by the mirror but instead pass through it, stripping electrons from their atoms on their way.

As a matter of fact, even the simple process of reflecting green light from a mirror is more complicated than the picture presented above. But there is another difficulty, and a vital one: in our orderly picture, where instead of light waves we have only quanta of light, there is no room for the experiments of Friedrich, Knipping, and Laue, who discovered the diffraction of X rays and thereby demonstrated their wave aspect. How are we to reconcile these incompatible conceptions of ray-waves and ray-quanta? By using the ideas of quantum mechanics.

"The world is strange", said Jeremy. "As compared with what?" asked the Spider.

George Macdonald

Electron Diffraction

As with many other discoveries in physics, electron diffraction was first observed "by chance", although, as Louis Pasteur liked to repeat, "chance only favours the prepared mind". In 1922 the Bell Telephone Laboratories engaged Clinton Joseph Davisson (1881-1958) and his colleague to study the reflection of electron beams

It is not entirely true that the scientist goes after the truth, it goes after him.

Soren Kierkegaard

from the surface of metals. Suddenly the experimenters noted some anomalies. After the work of de Broglie, in 1925, one of Max Born's students, Walter Maurice Elsasser (b. 1904), suggested that these anomalies were due to electron waves. Davisson came to Europe in 1926 and showed his plots to Max Born and James Franck at Göttingen and to Douglas Rayner Hartree at Cambridge. They unanimously recognized them as being de Broglie waves, although their arguments had no success with Davisson.

Soon after his return to the United States, there occurred a failure of his installation: the vacuum tube burst and the crystal inside it, which was hot at the time, changed its structure when exposed to atmospheric oxygen. This regrettable happening turned out to be an unexpected success: after it the spectra of reflected electrons resembled the spectra of scattered X rays, whose wave nature was no longer questioned at that time. Davisson continued his experiment and in 1927 he, together with Lester Halbert Germer (1896-1971), confirmed beyond doubt that the matter waves associated with electrons are a reality.

Sir George Paget Thomson (1892-1975) approached the problem from another angle. From the very start he regarded de Broglie's hypothesis with great sympathy and, soon after Davisson's sojourn in England, he set out to devise experimental methods to prove it. In England, after the works of Crookes and J.J. Thomson, experiments with cathode rays had become an invariable and customary element in education. It was perhaps for this reason that G.P. Thomson first turned to them in his search for new experimental tools. Almost at once he found a suitable experimental set-up in Aberdeen, which had been used by the student Alexander Reid. (He was killed soon after in an automobile accident at the age of twenty-two.) Within two months they obtained excellent photographs of electron diffraction patterns, which seemed to be replicas of X ray diffraction patterns. This was to be expected since in their experiments electrons were accelerated by a potential V = 150 volts (roughly the usual voltage of the city mains). The wavelength of such electrons is about 1 angstrom = 10^{-8} centimetre. This is commensurate with the wavelength of X rays and atomic dimensions.

On 30 April 1897 J.J. Thomson made a report at the Royal Society about his cathode-ray studies. This day, if you wish, may be taken to be the birthday of the electron, the first elementary particle in physics. Ironically, thirty years later, in May 1927, his son G.P. Thomson proved that the electron is a wave.

And both were right, and both were awarded Nobel Prizes for their discoveries.

Siren, Greek

Chapter Nine

Wave-Particle Duality • Uncertainty Relation Complementarity Principle

Early in the 1920s the physicists Max Born and James Franck and the mathematician David Hilbert organized in Göttingen a "seminar on matter", where even long before the work of Heisenberg and Schrödinger they started using the term "quantum mechanics". The seminar drew renowned scientists of the day and young physicists who were to enjoy world fame later in their lives. Almost every session Hilbert opened with the question "So, gentlemen, like you, I wish somebody would tell me exactly — what is an atom?"

Today we know more about atoms than all the participants at the seminar, but still we are not adequately equipped to answer Hilbert's question. We know now many facts of quantum physics but we are short of concepts to interpret these facts correctly.

Thanks to Niels Bohr, even now, after so many years have passed, the word "atom" calls forth in our minds a tiny planetary system with a nucleus and electrons. It takes some effort to remember that the atom also has inherent wave aspects. Now, as before, the ideas of "electron-wave" and "electron-particle" exist side by side in our mind, and subconsciously we try to get rid of one of them. "Particle or wave?" — the question haunted the physicists of the 1920s: desire for definiteness is deeply ingrained in man.

By the spring of 1926 the situation in atomic physics was peculiar: two quantum mechanics existed separately and independently, their premises differing widely. Heisenberg, following Bohr, was sure that the electron was a particle, and derived his matrix equations accordingly. Schrödinger, on the other hand, could only derive his differential equation when he, together with de Broglie, believed in the wave-like behaviour of the electron.

Heisenberg demanded that the equations only contain quantities that can be measured directly, such as the frequencies and intensities of spectral lines. Accordingly, he excluded from the theory the concept of the "trajectory of the electron in the atom" as a quantity that is unobservable in principle. Schrödinger too had no use for

the concept of the trajectory, but he wrote down his equation for the Ψ function, which also cannot be measured.

Experiment — the ultimate judge in all controversies — was at first decidedly on the side of matrix mechanics. Faraday's experiments showed the indivisibility of the elementary electric charge, and the experiments of Crookes and Thomson provided more supporting evidence for this. Only a particle could possess this property. Millikan's experiments and the photographs of traces of electrons in a Wilson cloud chamber dispelled the last doubts that anyone might still have had. The concept of an electron as a particle, however, conflicted sharply with the astounding stability of the atom, since the planetary atom is unstable. Bohr devised his postulates precisely in order to explain the stability of the atom assuming that the electron is a particle.

De Broglie and Schrödinger went a different way and showed that the stability could be explained in the most natural way by assuming that the electron was a wave, not a particle. This hypothesis was soon confirmed by the direct experiments of Davisson, Germer, and G.P. Thomson, who discovered that the electron is capable of interference and diffraction.

It is customary to believe experiments. But how can you possibly believe two experiments if they exclude each other? The history of physics had known of similar situations; nonetheless, this case was so unusual that nobody suspected at first that the two systems of mechanics comprised a single entity, and so everybody was out to prove the validity of one and the falsity of the other. Bitter disputes were going on between the advocates of each of the two theories; some stressing the priority of matrix mechanics and others the mathematical simplicity of wave mechanics. These debates were stopped at the beginning of 1927 by Schrödinger, who proved that both systems of mechanics are *mathematically equivalent.* To physicists this meant that the systems were also equivalent physically, and that they were dealing with one and the same mechanics — *the mechanics of the atom* — although written in different forms. This also suggested the validity of the original premises of the two systems, namely the particle aspect of matrix mechanics and the wave aspect of wave mechanics.

Wave-Particle Duality

The more scientists learned about the atom, the less categorical became the questions they put to nature. In the times of Planck and Einstein physicists wanted to know "What is a ray of light, a wave or a flux of particle-quanta?" After the work of de Broglie they kept on trying to clear up the question: "What is the electron, a wave or a particle?" Gradually, and with great difficulties did the simple idea crop up: "But why *or*? Why must these aspects — wave-like and particle-like — exclude each other?" Upon second thoughts it turned out that there are no good reasons for the dilemma. The only reason why the dilemma remained unresolved for so long was the perennial pattern of our thinking: we always try to comprehend new facts in terms of old concepts and images.

There is still another difficulty, a psychological one. Our everyday experience tells us that the smaller an article, the simpler it generally is. For example, out of thirty-three matreshkas (wooden dolls painted like Russian peasant women with successively smaller ones fitted inside) the innermost, thirty-third matreshka is the simplest; a billiard ball is considerably simpler than the terrestrial globe, and the whole always consists of simpler parts. When, sitting by the seashore, Democritus divided his apple, he may have imagined the atom in any conceivable way, but hardly as being more complex in structure than the whole apple. And it is not, of course. But it so happens that some properties are evident in small objects and unnoticeable in larger ones. Likewise, when we divide up matter (which, by tradition, we imagine to be made up of particles) its new, wave properties do not *appear* — they just *manifest themselves* — we have simply not noticed them before.

We encounter phenomena of this kind much more often than we realize. A billiard ball and the terrestrial globe are both spheres and, in this, they are similar. But many people had suffered before the Earth was accepted as a sphere by everybody. A billiard ball, on the other hand, has always been round, even to the perpetrators of the Spanish inquisition. The key thing here is the scale of the phenomenon in relation to that of its observer. The Earth, just as each of the electrons it includes, has a wave-like aspect. If, however, we were to describe its motion using Schrödinger's equation, then given that its mass is 5×10^{27} grammes and its orbital velocity is 3×10^{6} centimetres per second, we would have to ascribe to this "particle" a de Broglie wavelength of 4×10^{-61} centimetre, a number small beyond comprehension. This fact alone, however, does not warrant any statements that the Earth possesses no wave-like properties. After all, we cannot use dividers and a ruler to measure its curvature, but still the Earth is round.

These examples are to help us to grasp the outcome of the reflection upon the "wave *or* particle" problem: the question of "either wave or particle" was formulated incorrectly: an atomic object is *both* a wave and a particle at the same time.

Moreover, *all* bodies in nature possess wave and particle properties at the same time, and these properties are but different manifestations of a unified *wave-particle duality*. This idea occurred to Bohr, Kramers, and John Clarke Slater as far back as 1924. They co-authored a work in which they definitely claimed that the wave nature of the propagation of light, on the one hand, and the absorption and emission of light in quanta, on the other, are experimental facts; they should be used as the basis for any atomic theory, and no explanations should be sought for them.

This indisputable, though unusual, unity of "wave-particle" properties is reflected in Planck's formula ($E = h\nu$) and de Broglie's formula ($\lambda = h/mv$). The energy E and mass m are characteristics of a particle; the frequency ν and wavelength λ are features of a wave process. The only reason why we do not notice this duality in our everyday life is the smallness of Planck's constant. Even if this is a chance circumstance, we have to take it into account.

If we lived in a world where Planck's constant were commensurate with customary scales, our lives would differ markedly from the ones we live now. Objects in it would have no definite boundaries, they could not be moved about arbitrarily, and it would be impossible to plan any appointments beforehand. Fortunately such a world is only a hypothesis, since we are quite unable to alter the value of Planck's constant at will. It is always unchanged and very small. But atoms are also very small; therefore they belong to this unusual world. We will now try to understand its uncommon logic in much the same way as Gulliver had to become accustomed to the ways and thinking of Lilliputians.

Uncertainty Relation

Suppose that we have become so inspired with the idea of the indivisibility of the "wave-particle" aspects that we now wish to express this idea in the precise language of formulas. These formulas should establish the relationship between numbers that correspond to the concepts of "wave" and "particles". In classical mechanics these concepts are miles apart and refer to disparate phenomena. In quantum mechanics wave-particle duality compels us to deal with both concepts simultaneously and to apply them to the same object. This is not easy; we have to pay for it, and pay dearly.

This became quite clear in 1927 when Heisenberg figured out that the concepts of "particle" and "wave", as applied to quantum objects, can only be defined strictly when defined separately.

In physics "to define a concept" means "to indicate a method of measuring the quantities to

which this concept corresponds". Heisenberg contended that it is impossible to measure both the coordinate x and the momentum p of an atomic object simultaneously and accurately. If we recall de Broglie's formula $\lambda = h/mv = h/p$, we find that it implies that we cannot determine the position x of an atomic object and its wavelength λ simultaneously and with absolute precision. It follows that the concepts of "wave" and "particle" have a restricted meaning when used *simultaneously* in atomic physics. Heisenberg found a numerical measure for this constraint. He showed that if we know the position x and momentum p of an atomic object (for instance an electron in an atom) with uncertainties Δx and Δp, we cannot improve our accuracies indefinitely; we can only do this as long as we satisfy the following inequality, called the *uncertainty relation*:

$$\Delta h \cdot \Delta p \geq \frac{1}{2}\hbar .$$

This constraint is very small, but it does exist and this is of fundamental importance.

The uncertainty relation is a strict law of nature, which is in no way related to some imperfections of our devices. It states that it is *in principle impossible* to determine simultaneously both the position and momentum more accurately than is allowed by the inequality $\Delta x \cdot \Delta p \geq \hbar/2$. This is impossible just as it is impossible to exceed the velocity of light or to reach the absolute zero of temperature. It is as impossible as being able to lift oneself by one's hair or to return to yesterday. And any references to the omnipotence of science are out of place here. The power of science lies in its capacity to discover, understand, and make use of the laws of nature, not in violating them.

This may seem a little strange. We are used to believing that science is omnipotent and that its vocabulary does not include the words "it is impossible". It is remarkable, however, that the greatest triumph in any scientific field is attained precisely at moments when such exclusion principles are established. When it was said, "It is impossible to build a perpetual motion engine", thermodynamics made its appearance. As soon as it was found that it is "impossible to exceed the velocity of light", the theory of relativity was born. And only when it was understood that various properties of quantum objects cannot be measured simultaneously with arbitrary accuracy did quantum mechanics acquired its final form.

The uncertainty relation is directly opposed to intuition and everyday experience. Heisenberg explained the matter by discarding still another idealization of classical physics — the concept of observation. It turned out that in quantum mechanics it had to be revised, along with the concept of motion.

The bulk of his knowledge of the world man acquires by means of his sight. This path of perception has dominated man's whole system of knowing. Almost everyone, on hearing the word "observation", thinks of a person attentively looking at something. When you stare at the person you are talking to, you are absolutely sure that even if your glare is one that chills the blood, you can do no physical harm. It is, in essence, this idea that lies behind our notion of observation in classical mechanics. Classical mechanics sprang from astronomy. But in astronomy no one doubted that when we observe a star we do not affect it in any way, and so nobody doubted this for other kinds of observation as well.

The concepts of "phenomenon", "measurement", and "observation" are closely related to one another but they are by no means identical. The ancients observed phenomena; this was their method of knowing nature. From their close observations they then derived consequences by pure speculation. From then on the conviction has evidently taken root that phenomena exist independently of their observation.

We have stressed repeatedly that modern physics differs from classical physics mainly in that it relies on experiment instead of speculation. Today's physics does not deny that phenomena in nature exist independently of observation (and of our mind, as well). But it asserts that these phenomena become objects of observation only after we can indicate a precise method of measuring their properties. *In physics the concepts of "observation" and "measurement" are inseparable.*

Any measurement is essentially interaction between the instrument and the object to be studied. And any interaction disturbs the initial states of both instrument and object. Therefore, a measurement gives us information about the phenomenon that is distorted by the interference of the instrument. Classical physics assumed that all such distortions could be taken into account, and that the "true" state of the object could be established independently of measurements. Heisenberg showed this assumption to be a delusion: *in atomic physics "phenomenon" and "observation" are inseparable from each other.* An "observation" is also a phenomenon of sorts, and not the simplest one at that.

Like many other things in quantum mechanics this statement sounds unusual and arouses our unconscious protest. We will nevertheless try to comprehend it, or at least to get some feeling for it.

We know from our everyday experience that the smaller the object we are dealing with, the easier is it to disturb its state. We know of nothing smaller in nature than atomic objects — the atom, the electron, the nucleus. We cannot determine their properties by simply using our will-power. In the long run, we are compelled to measure the properties of atomic objects by means of the objects themselves. Under such conditions the instrument is indistinguishable from the object being measured.

But why then is it impossible that in the process of measuring an atomic object one should exert only an insignificant influence upon the other?

The fact is that both of them — instrument and object — belong to the same quantum world and therefore their interaction obeys quantum laws. The main feature of quantum phenomena is their discrete nature. In the quantum world nothing is "very nearly" — interactions there occur only by a quantum, a whole one or none at all. We cannot influence a quantum system "slightly"; up to a definite moment it will not feel this action. But once the degree of influence has increased sufficiently for the system to perceive it, the system jumps to a new (also quantum) state or simply disintegrates.

Observation in quantum mechanics is rather like tasting than seeing. "The proof of the pudding is in the eating", was a favourite motto of the founders of quantum mechanics. Just as we cannot partake again of a pudding after we have eaten it, so we cannot endlessly refine our data on a quantum system: it will be destroyed, as a rule, by our first measurement. Heisenberg was the first to understand this stern fact; he also succeeded in expressing it in the language of mathematics. The uncertainty relation is one of the most important formulas of quantum mechanics; it is, as it were, a focus of the most significant features of quantum physics. After the uncertainty relation was discovered, it became necessary to revise not only the foundations of science but the theory of cognition as well. It took Bohr to do this. He was fortunate in combining the powerful intellect of a true scientist with the philosophical insight of a true thinker. At one time he had created a system of images for quantum mechanics and now, fourteen years later, he was painstakingly moulding the system of its concepts. After Bohr's work it became clear that both the uncertainty relation and wave-particle duality are only special cases of the more general *principle of complementarity.*

Complementarity Principle

The principle that Bohr named *complementarity* is one of the most fundamental philosophical and scientific ideas of our time, which can only be compared to the principle of relativity and the concept of a physical field. Its generality does not enable us to reduce it to any single statement — it must be mastered gradually, with reference to specific examples. It might be well to begin (as Bohr did) with the analysis of the process of

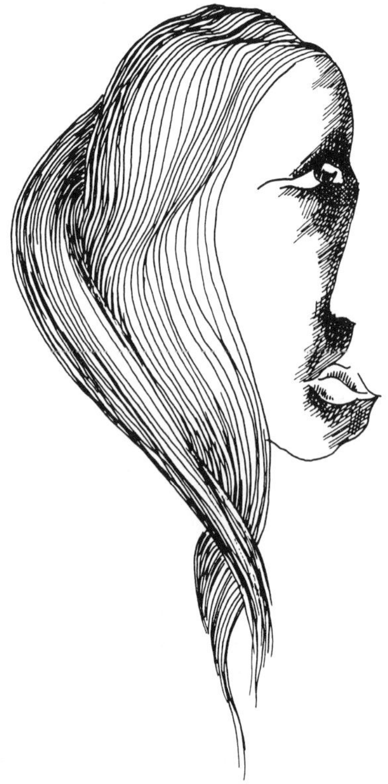

measurement of the momentum p and coordinate x of an atomic object.

Bohr noticed a very simple and clear fact: the position and momentum of an atomic particle can be measured neither simultaneously nor using the same instrument. In fact, to measure the momentum p of an atomic particle without changing it appreciably requires an exceedingly light mobile "instrument". But its very mobility makes its position extremely uncertain.

When we speak into a microphone, the sound waves of our voice are converted into vibrations of a thin diaphragm. The lighter and more mobile the diaphragm, the more faithfully it reproduces the vibrations of the air; and, on the other hand, the more difficult it is to determine its exact position at each instant of time. To measure the coordinate x we must therefore take another, exceedingly massive instrument, one that will not stir when hit by a particle. But no matter how its momentum changes in that case, we will not even notice this. This is the simplest experimental illustration concerned with Heisenberg's uncertainty relation: the same experiment cannot yield both characteristics, x and p, of an atomic object. For this purpose two measurements will be required, and two essentially different instruments, whose properties are complementary.

Complementarity is the word and the thinking that became commonplace in physics due to Bohr. Before Bohr all agreed that the incompatibility of the two types of instrument is a sure sign that the properties to be measured by them are inconsistent. Bohr denied this inflexible reasoning and explained: yes, these properties are incompatible all right, but both are equally needed to provide a full description of a quantum object, and so they complement rather than contradict each other.

You encounter situations like this everywhere. You may remember how in your childhood you used a lens to set some paper and dry grass on fire. This is one characteristic of sunlight — it carries energy in the form of photons. But when you pass sunlight not through a lens but through a prism, then you will see a spectrum. This is another, complementary characteristic of sunlight — it consists of waves of different lengths. The lens and the prism are different devices, which make it possible for us to observe various physical phenomena that characterise different complementary aspects of one quantum object. These properties cannot be observed at the same time, in the same experiment, using the same instrument, but they are each just as necessary for us to perceive the essence of sunlight in its completeness.

The above reasoning concerning the complementarity of two incompatible measurements can be illustrated by a simple analogy. Assume that you wish to find out what is inside a "black box" that has been designed in a special manner, namely with a camera obscura attached to it. In contrast to the ordinary camera, this camera has two holes and, consequently, two photographic plates on the walls of the camera opposite the holes. As long as the holes are closed, you know nothing at all about the object within the box; for you it simply does not exist. By opening the holes one at a time you will obtain two plane projections of the object on the photographic plates. Each of them, taken separately, is insufficient but both are equally necessary to reconstruct a three-dimensional picture of the object.

The two different projections of the object correspond to two different, complementary types of measurements in quantum mechanics. Clearly, we cannot make both measurements simultaneously because if we open both holes at the same time, in addition to the image formed by the object exposed to the required hole, light from the other, complementary hole will fall on each plate, and so both images will be spoiled. It is also clear that if the object is very small, then even the first measurement, or observation, will disturb its initial state, for example we may displace it or turn it. And this means that we will obtain on the other plate a distorted projection, not the true one. Under these conditions a three-dimensional picture can only be reconstructed with an error, which is better, nevertheless, than a plane picture, even if accurate. Quantum mechanics contends that to obtain a "three-dimensional" portrait of an atomic object two of its "plain projections", i.e., two complementary measurements, e.g., of position and momentum, are quite sufficient.

Now if we translate this reasoning into the language of abstract concepts we will have something like the following. An atomic object is a "thing in itself" until we have indicated a method for observing it. Various properties of an object require different methods, which are sometimes incompatible with one another. In actual fact, the concepts "object" and "observation" are just convenient abstractions needed to describe the more general concept of "experimental situation". Physical sciences study not objects by themselves but concrete realizations of experimental situations, which we will call "phenomena". Experimentally, any phenomenon is an ordered set of numbers that are the results of measurement of the response of the object to the action of a measurement of an given type. By selecting different, complementary measurements we change the experimental situation; by realizing a situation we affect different characteristics of the object. Finally, by observing the consequences of this action we obtain various sets of numbers, that is we study different phenomena. And although it is impossible to study complementary phenomena simultaneously, in one experiment, they nevertheless characterize a single quantum object and both are required for a complete description of its behaviour.

It has always been important to know what questions we ask nature. When we put questions to quantum nature we must be especially cautious, because the kind of question will predetermine the way in which nature will make the division into two parts, system and observer. The complementarity principle states that there are at least two qualitatively different ways of making this division. In the cognition chain in modern physics:

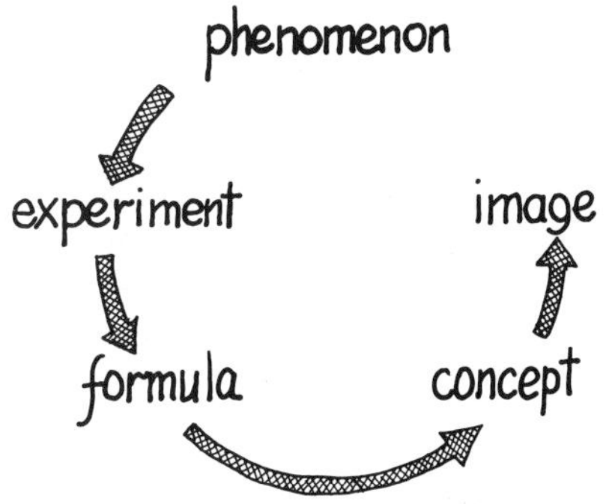

the complementarity principle primarily affects the system of concepts of quantum mechanics and the logic of its inferences. The fundamental notions of formal logic include "the rule of the excluded third", which states that of two contrary statements one is true, the other is false and there can be no third. Classical physics has given no grounds to doubt this rule, since there the concepts of "wave" and "particle" were really opposed and essentially incompatible. It turned out, however, that in quantum physics both are equally suitable for describing the behaviour of the same object; moreover, to obtain a full description, it is necessary to use them simultaneously. People brought up in the traditions of classical mechanics took these requirements as something contrary to their common sense, and there was even some talk that in atomic physics the laws of logic are violated. Bohr explained that the situation had nothing whatsoever to do with the laws of logic, it had rather sprung from the careless way in which, without any reservations, classical concepts had been applied to quantum phenomena. But such reservations are indispensable, and Heisenberg's uncertainty relation $\Delta x \cdot \Delta p \geq \hbar/2$ is just a formulation of them in the language of formulas.

The reason why our minds have difficulties in grasping complementary concepts is a profound one, and quite understandable. We cannot perceive an atomic object directly, using our five senses. Instead, we employ accurate and sophisticated devices, quite recent inventions. We need words and concepts to explain our findings, but they had been in use long before quantum mechanics appeared on the scene and are in no way adapted to their new usage in quantum mechanics. But we cannot do without them — there are no two ways about it: we master language and all other fundamentals much earlier than we find ourselves confronted with physics.

Bohr's complementarity principle is a successful attempt to reconcile the limitations of the established system of concepts with the advances in our knowledge of the world around us. This broadened the horizons of our thinking by explaining that quantum mechanics changes not only concepts but also the very formulation of questions concerning the essence of physical phenomena. (At one time Pauli even proposed to name quantum mechanics the "theory of complementarity" — in analogy to the theory of relativity.) But the significance of the complementarity principle oversteps by far the domain of quantum mechanics. Its true value for the whole system of human knowledge became evident much later, when attempts were made to apply it to other branches of science. The legitimacy of this generalization may well be debated, but there is no denying its fruitfulness in many cases, even ones far from physics.

Bohr liked to give an example from biology; it concerned the life of a cell, whose role is quite similar to that of the atom in physics. Just as an atom is the smallest part of a substance that still retains its properties, so a cell is the smallest part of any organism that still represents life in all its complexity and uniqueness. For science, to study the life of a cell means to reveal all the elementary processes that occur in it and to understand how their interaction leads to that peculiar state of matter — life.

But this combination of analysis and synthesis is impossible. In fact, to grasp the workings of the cell we observe it through a microscope, first an ordinary one and then an electron one; we heat the cell; we pass an electric current through it; we irradiate it, and break it down into its components. But the more intently we study the life of a cell, the more we will interfere with its functions and natural processes occurring in it. In the end, we will destroy the cell, and we will learn nothing about it as an integral living organism.

Nevertheless, to answer the question "What is life?" requires a combination of analysis and synthesis. These processes are incompatible but not contradictory. They are complementary, and the fact that we need both is just one reason why no answer exists as yet to the question of what life is.

As in a living organism, in an atom the main thing is integrity, the integrity of its "wave-particle" aspects. The finite divisibility of *matter* is the idea behind the finite divisibility of atomic *phenomena...* and the finite divisibility of the *concepts* to describe these phenomena.

One often hears that a question properly put is half the answer, and this is more than a witticism. A properly put question is one concerning the properties of a phenomenon that it actually possesses. Therefore, the question already contains all the concepts that are needed for the answer. An ideally put question has just "yes" or "no" for an answer. Bohr showed that in reference to an atomic object the question "Wave or particle?" was put incorrectly. An atom does not possess such properties *separately,* and so the question cannot be answered simply by "yes" or "no". Likewise, there is no answer to the question "Which is the larger, a metre or a kilogramme?" or to other similar questions.

A quantum object is neither a particle nor a wave nor both simultaneously. A quantum object is *something entirely different,* not simply the sum total of the properties of waves and particles — just as a melody is not just the sum of the sounds that constitute it and a centaur is not just the sum of a horse and a man — these are fundamentally different things, new entities. Quantum "entities" are imperceptible to our senses but they are real, all the same. We have no images and perception organs with which to picture adequately the details of this reality. However, the power of our intellect, aided by experiment, permits us to perceive them. The two complementary properties of quantum reality cannot be divided without destroying in the process the completeness and unity of that phenomenon of nature which we call the atom; just as we cannot cut a centaur into two parts and have the horse and the man as separate entities.

When Heisenberg discarded that idealization of classical physics, the notion of the "state of a physical system independent of observation", he thereby anticipated one of the consequences of the complementarity principle, since "state" and "observation" are complementary concepts. They are incomplete when taken separately, and so they can only be defined jointly, through each other. Strictly speaking, they do not exist independently: we always observe not something in general, but invariably some state. Conversely, any "state" is a thing in itself until we devise a method for its "observation".

"Wave" and "particle", "state" and "observation" are certain idealizations that have no relation to the quantum world, although they are needed to comprehend it. To obtain a full description of the workings of quantum effects, simple classical pictures have to merge in a harmonious manner, that is to become complementary. But within customary logic they can coexist peacefully only if the scope of their applicability is mutually restricted.

Bohr gave much thought to these and related problems, and he was led to conclude that it is a general rule rather than the exception that *any truly profound phenomenon of nature cannot be defined uniquely using the words of common language, and that the definition requires at least two mutually exclusive complementary concepts.* It follows that, provided we retain common language and logic, thinking in complementarity modes sets limits to the formulation of concepts that correspond to truly profound phenomena. Such definitions are unambiguous although incomplete or complete although ambiguous, since they include complementary notions that are incompatible in customary logic. Examples of such concepts are "life", "quantum object", "physical system", and even "cognition of nature".

It has long been known that science is only one of the two methods of knowing the world around us. Another — complementary — method is art. The existence side by side of art and science is in itself a good illustration of the complementarity principle. You can devote yourself completely to science or live exclusively in your art. Both habits of life are equally valid but, taken separately, are incomplete. The backbone of science is logic and experiment. The basis of art is intuition and insight. But the art of the ballet, say, requires mathematical accuracy and, as Pushkin wrote "... inspiration in geometry is just as necessary as in poetry". These two domains of human endeavour complement rather than contradict each other. True science is akin to art; real art always includes elements of science. In their utmost manifestations they are indistinguishable and inseparable, like the "wave-particle" aspects of the atom. They reflect dissimilar complementary aspects of human experience and only when taken together do they give us a complete idea of the world. Unfortunately, we do not know the "uncertainty relation" for the conjugate pair of concepts "science and art", and so we cannot assess the degree of damage we suffer from a one-sided perception of life.

However, this analogy, like any other for that matter, is neither complete nor rigourous. It only gives us some idea of the unity and inconsistencies of the entire system of human knowledge.

Quantumalia

Duality and Uncertainty

In the early days of quantum mechanics even good physicists would joke bitterly that they had to treat the electron as a particle on Mondays, Wednesdays, and Fridays, and as a wave on the other days of the week. With his inimitable humour Bohr wrote in 1924: "Even if Einstein sends me a telegram informing me that the reality of light quanta has been proved once and for all, even then it will reach me only owing to the existence of radiowaves."

"It is all rather paradoxical and confusing," wrote Davisson in 1928 in his famous articles with the typical name "Are Electrons Waves?", "We must believe not only that there is a certain sense in

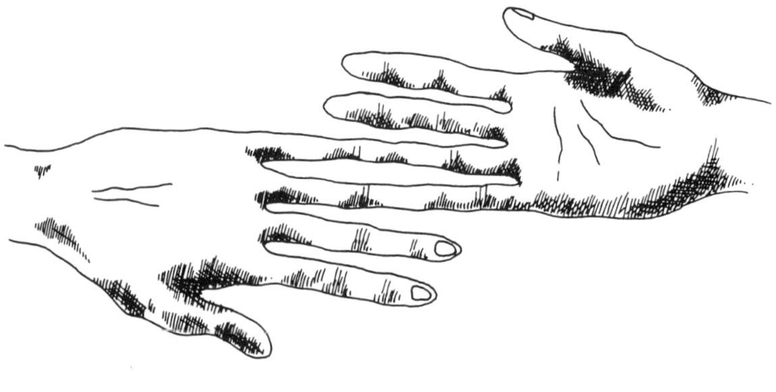

which rabbits are cats, but that there is also a certain sense in which cats are rabbits."

This thinking led to many paradoxes, which can be avoided if we make a point from the very outset not to separate the "wave-particle" aspects of the electron. Only then will Heisenberg's uncertainty relation no longer be something strange and incomprehensible and turn into a simple consequence of wave-particle duality.

Shape without form, shade without colour. Paralyzed force, gesture without motion.

Ebenezer Elliot

It has long been known in wave optics that no particle can be seen in any microscope whatsoever if its size is less than half the wavelength of light to which it is exposed. No one thought it strange: waves of light exist by themselves, and a particle by itself. But when it was found that a particle could also be ascribed a wavelength, this fact of wave optics became an uncertainty relation: a particle cannot localize itself with an accuracy greater than one half of its own wavelength.

Poets and the Complementarity Principle

The principle of complementarity as such, without reference to physics, is an ancient invention. It is quite a well-known category of dialectic logic and has been repeatedly proposed by various philosophers throughout the ages. Aristotle said, for instance, that "harmony is a blending and combination of opposites" and Hegel's triads might come in handy with our quantum-mechanical concepts.

When it comes to atoms language can be used only as in poetry.

Niels Bohr

It would be instructive to recall here how the complementarity principle was rediscovered by poets for their own purposes. In 1901 Valery Bryusov wrote a paper entitled "The Truth" in which we find the following: "Thinking requires plurality, irrespective of whether it is a fragmentation of my *ego* or appears as something external. Thought and, more generally, life comes into being from the juxtaposition of at least two principles. The unified principle is nonexistence; the unity of truth is no-thought. There would be no space if there were no right and left; there would be no morals if there were no good and evil...

What is valuable in truth is only what can be questioned. 'There is the Sun' — no doubt about it... This is truth but it has no value in its own right. No one has need of it. No one would go to the stake for such a truth. Or rather, this is not a truth but merely a *definition.* 'There is the Sun' is only a special phrase instead of 'Such an object I call the Sun'.

Truth only acquires a value when it becomes part of a plausible world outlook. But at the same time it becomes disputable; it can at least become the subject of an argument... Moreover, a valuable truth necessarily has the right to the exactly opposite truth that corresponds to it. In other words, an opinion directly opposite to a truth is, in turn, also true...."

You can never have the use of the inside of a cup without the outside. The inside and the outside go together. They're one.

Alan Watts

It is perhaps appropriate here to recall the words of Pascal who wrote: "All the principles of the Pyrrhonists, Stoics, atheists, etc., are the truth, but their conclusions are false because the opposite principles are also the truth."

It is remarkable that many of these statements are nearly word for word the same as those of Bohr's. It is not generally known that Bohr arrived at his principle of complementarity "from philosophy" rather than "from physics". The idea of complementarity struck roots in his mind in his adolescent years under the influence of the Danish philosopher Kierkegaard. Over the years it matured and refined itself, until in the long run it found a worthy application in quantum physics.

Cernunnos, Celtic

Chapter Ten

Heads Or Tails and Target Shooting
Electron Diffraction
Probability Waves • Electron Waves
The Atom and Probability
Probability and Atomic Spectra
Causality and Chance, Probability and Certainty

Just imagine that on the Trans-Siberian Express, somewhere between Novosibirsk and Krasnoyarsk, you become acquainted with an agreeable person. Further, a year later you run into him or her in the centre of London. No matter how glad you are to meet him or her again, first of all you are astonished because you know from experience how unlikely such a happening is.

When we employ the words "probably", "most probable", "in all probability", and "improbable", we never realize how strictly defined are the concepts corresponding to them. There are no such liberties in science, because there the concept of "probability" only makes sense when we can compute it.

This is not always possible. For example, we cannot predict the probability of the next chance meeting with an acquaintance in the harbour of Singapore: the laws governing the behaviour of people are all too complex. And so, all the textbooks on probability invariably explain the laws of chance with reference to the example of the tossing of coins.

Heads or Tails and Target Shooting

Any *event,* whose probability can be worked out, is one of the outcomes of a series of *tests.*

We agree that if a test has several outcomes, the total probability that at least one event will occur is *unity.* This is just a general convention or definition, the condition follows from nowhere. We too will abide by the tradition. Therefore, the statement "the event will occur with a probability of unity" means that it is certain.

It follows that the probability of any one outcome is always less than unity. In the example with a coin, each test — the toss of the coin — may only have two outcomes, the coin will fall either heads or tails up. (We have excluded the extremely rare event that the coin will end up standing on edge.) If the coin is "fair" it is logical to assume that both outcomes of a toss are *equally probable.* One is thus led to conclude that the probability of the outcome "heads", say, is 1/2. With equal ease we can calculate the probability of having a "three" when we cast a die. It is clearly equal to 1/6.

Examples can easily be multiplied, but they will be similar in many respects.

First, each subsequent test (the toss of a coin) is independent of the previous one.

Second, the outcome of each test is a *random event,* that is we don't know (or cannot take into account) all the causes of one outcome or another.

The latter is of especial importance. In fact, a coin is not an atom; its motion is governed by the well-known laws of classical mechanics. Using them we could actually foresee all the details of the motion of the coin and predict how it would fall, "heads" or "tails". We can even draw its trajectory. No doubt, this is all very difficult. We would have to take into consideration the air resistance, the exact shape of the coin, the elasticity of the floor on which it falls and many other important details. And, which is of prime importance, we must accurately specify the *initial position* and *velocity of the coin.*

It is not always possible, however, to take into account all the conceivable factors involved. For instance, in the case of the coin we never know its initial position and velocity with sufficient accuracy. Any, even the slightest, fluctuation can reverse the outcome of the toss. Then we can no longer be sure that in a given toss the coin will fall heads up. We can only say that the probability for the coin to fall heads up in *any* toss is 1/2.

The simple examples we have just discussed do not explain why probability is so important for quantum mechanics. Before turning to the issue, however, let us get acquainted, at least in outline, with the principal laws of the theory of probability. The laws of chance (notice the strange union of these two words!) are just as rigourous as other mathematical laws. They do, however, have certain unusual features and a clearly delineated domain of application. For instance, we can readily verify that in a large number of tosses heads will occur in about half the cases and the law is obeyed the more faithfully the more tests we carry out. Nonetheless, this accurate knowledge will be of no help in predicting the outcome of each individual tossing. This is what distinguishes the laws of chance: the concept of probability is valid for an *individual* event and we can *work out* a number that corresponds to this concept. But it can only be measured when *identical* tests are *repeated* a great number of times.

It is very important that tests should be truly identical, i.e., indistinguishable from one another. Only then can the measured value, the probability, be used to assess each individual random event, which is one of the possible outcomes of the test.

The unusual features of the laws of chance have a natural explanation. In fact, coin tossing is quite a complicated process. We do not wish or do not know how to study it in all of its complexity, and are only interested in the end result of a test. Such disregard for the details takes its toll — we can only predict with certainty the average result of numerous identical tests, and for each random event we can only indicate its likely outcome.

It is widely believed that a probabilistic description of motion is less complete than a strictly causal, classical one, with its concept of a trajectory. Classically, this is really so. If we forego some of the rigorous requirements of classical mechanics (for example, a knowledge of the initial coordinates and momenta), the classical description immediately becomes useless. The probabilistic approach takes its place and, under the new conditions, it will be complete because it will yield all the information about the system that can possibly be found experimentally.

In coin tossing we deliberately do not want to know the initial position and velocity of the coin and so we rely wholly on chance. In a shooting gallery, on the other hand, we always try to hit the bull's-eye. But try as we may, we never know what section of the target each bullet will hit. After our shooting exercise is over we find that our hits are grouped into a rather regular oval, which is commonly called the "ellipse of dispersion". Its shape is dependent on many things.

For all the bullets leaving the rifle to hit the same point of the target they would all have to have the same initial coordinate x_0 and velocity v_0 (or momentum p_0). And this is only possible if you aim perfectly and if each cartridge has exactly the same powder charge. Neither of these is generally achievable. Therefore, the bullet-holes in the target are always distributed according to the laws of chance, and so we can only speak about the probability of hitting the bull's-eye or the nine-ring of the target, but we can never be sure of anything beforehand.

As with the coin, this probability can be measured. Suppose that we have made 100 shots and have hit the bull's-eye (ten points) 40 times; the nine-ring, 30 times; the eight-ring, 15 times, and so on, down to zero. Then the probability W of hitting a 10, a 9, an 8, etc., will be

$$W_{10} = 40/100 = 0.4,\ W_9 = 0.3,$$

$$W_8 = 0.15, \text{etc.}$$

We can even construct the ellipse of dispersion, by marking off along the horizontal axis the numbers 1, 2, 3,..., and along the vertical axis the corresponding hitting probabilities.

If we now put up another target and take another 100 shots, the arrangement of the holes will be different from that of the first target. But the overall number of 10s, 9s, etc., will be about the same, and the ellipse will have a similar outline. Its shape will be different, of course, for different marksmen: for a sharpshooter it will be narrower and for a novice wider. But each marksman will have his characteristic shape (at least for some time) and a skilled coach can tell to which of his trainees it belongs by just glancing at the target.

Even this simple example illustrates that the "laws of chance" are not a mere play on words. Admittedly, each bullet will hit a random point of the target, which we cannot predict. After a large number of shots the hits will form so regular a pattern that we perceive it as being a true fact and we entirely forget about the probability on which it is based.

Electron Diffraction

The example with target shooting resembles experiments in quantum mechanics much more than might appear at first sight. To see that this is so we replace the rifle by an "electron gun", the target by a photographic plate and place a piece of thin metallic foil between them.

"Electron gun" is no figure of speech but a good scientific term; it denotes a device for obtaining a beam of electrons, as in the cathode-ray tube of a TV set. Using diaphragms and focusing lenses we can produce a very narrow pencil beam of electrons, in which all the electrons will travel at the same velocity. We direct this pencil through the metal foil onto the photographic plate and then develop the plate. What pattern will we find on the plate? A point? An ellipse of dispersion as in target shooting? Or something else? The answer is well known — diffraction rings. We can now explain why.

Actually, the electron is not only a particle, it is also a wave. If we haven't yet become accustomed to this fact, we should at least have remembered it. Therefore, electron diffraction should not come as a surprise for us: diffraction always occurs when a wave passes through a substance. Well and good, but a wave of *what* passes through the foil together with the electron?

Waves roll over the surface of the sea; they consist of water. Electromagnetic waves travel through space: they are oscillations of electric and magnetic fields. But what is the wave of an electron if it itself is indivisible and has no internal structure?

Before we try and answer these questions, let us stage the electron beam experiment in a somewhat different manner. We will release electrons one by one (as rifle bullets) and replace the photographic plate after each electron. We will then develop the plates to find a spot on each of them, a hit point made by the electron. (Even if there were no other proof, this piece of evidence alone goes far to convince us that the electron is really a particle.) At first sight, the points on the plates are scattered entirely at random and, of course, none of them hints at a diffraction pattern. But if we put all the plates in a stack and hold them up to the light we will find, much to our surprise the same diffraction rings as before. It follows that the points on the plates are not as random as it might appear at first sight.

This experiment is so simple that some readers may even be offended by its triviality. No wonder it was conducted as recently as 1949; physicists had been so sure of its outcome, although they had deemed it useful and convincing. (The experiment, a technically challenging exercise, was performed by the Soviet scientist Valentin Fabrikant.)

It is not necessary, of course, to use a separate plate for each electron, one target-plate is enough, provided electrons are sent out one by one as before. And as before we cannot predict what point of the plate each successive electron will hit. This is a *chance event*. But if we release a sufficiently large number of electrons, we will obtain a *regular* diffraction pattern.

We have already encountered such phenomena when we discussed coin tossing, die casting, and target shooting. This analogy suggests the natural assumption that electron scattering obeys the laws of probability. Further analysis and acquaintance with the ideas of Born show this guess to be true.

Probability Waves

Max Born taught physics at Göttingen, that famous centre of German learning. He closely followed the developments in atomic theory and was one of the first to put Heisenberg's quantum ideas into a strict mathematical form. In the middle of 1926 he became interested in electron diffraction experiments. After the work of de Broglie this phenomenon as such no longer seemed astonishing to him. A glance at the diffraction pattern enabled him to explain it using the hypothesis of "matter waves" and even to work out the length of those waves. But no explanation was forthcoming for the words "matter waves". Some pulsation of electron-spheres perhaps? Or, the vibration of some kind of ether? Or, the oscillation or something even more exotic? In a word, how material were these "waves of matter"?

In the summer of 1927 Born came to the following conclusion: *"matter waves"* are simply *"probability waves"*. They describe the motion of an individual electron, specifically the probability of its hitting a given point on the photographic plate.

Any new and profound idea has no logical roots, although the loose analogies that have led to it can nearly always be traced. Therefore, instead of trying to prove the validity of Born's ideas (which is impossible) we will try and gain an impression of how natural his hypothesis was. Let us return to the game of coin tossing and recall the reasons why we had to resort to probability:

1. Independence of each successive toss from the previous one.

2. Complete indistinguishability of individual tosses.

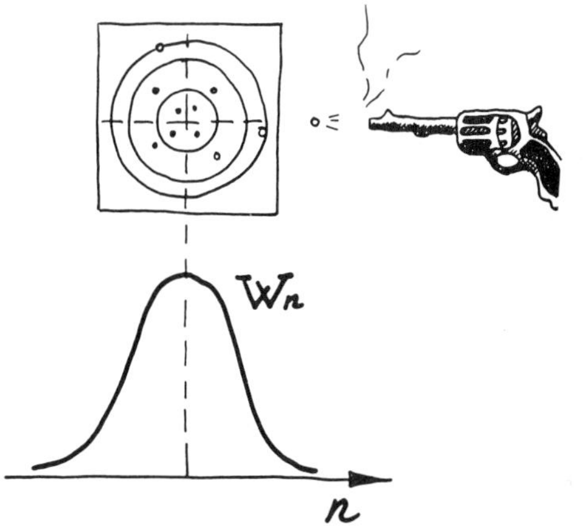

3. Randomness of the outcome of each toss, which is due to the complete lack of knowledge of the initial conditions of each test, i.e., due to the uncertainty of the initial coordinates and momentum of the coin.

All three conditions are met in atomic phenomena, specifically in electron diffraction experiments. In fact:

First, the electrons are scattered one by one and therefore independently of each other.

Second, electrons are so poor in properties (charge, mass, spin — and nothing more) that in quantum mechanics they are indistinguishable, and so are the individual scattering events.

Third, most important, precise values of the coordinates and momenta of electrons cannot be specified in principle, because this is forbidden by Heisenberg's uncertainty relation.

It would then be pointless to seek the trajectory of each electron. Instead we have to learn to compute the probability $\rho(x)$ that electrons will hit a definite spot x on the photographic plate (to use the parlance of physicists, to calculate the distribution function $\rho(x)$).

With the coin it is all very straightforward: it is clear, even without any calculation, that the probability of heads is 1/2. In quantum mechanics things are not that simple. To work out the function $\rho(x)$, which describes the distribution of elec-

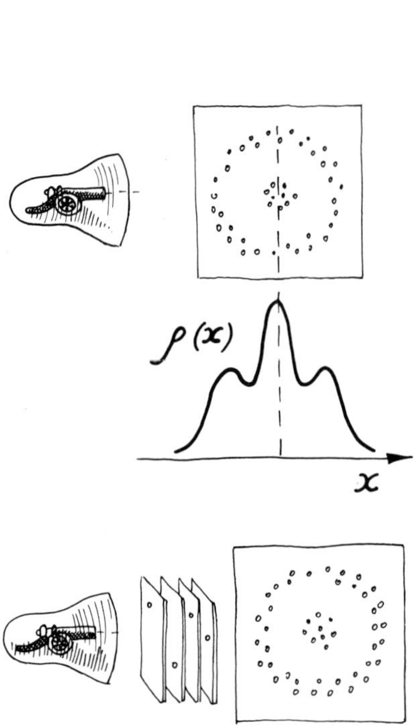

trons over the plate, we have to solve Schrödinger's equation.

Born contended that the probability $\rho(x)$ for an electron to be found at a point x is the square of the wave function $\Psi(x)$, i.e.,

$$\rho(x) = |\Psi(x)|^2.$$

The plot of $\rho(x)$ looks more complicated than the ellipse of dispersion in target shooting. But though we are unable to predict the shape of the ellipse, we can calculate $\rho(x)$ beforehand. Its form is uniquely governed by the laws of quantum mechanics. However bizarre, they exist, unlike the laws of human behaviour upon which the ellipse of dispersion depends.

Electron Waves

When we stand on a beach, we have not the slightest doubt that it is waves and not something else that run up the beach. Nor are we surprised by the hard fact that all waves consist of a vast number of particles, or molecules.

Probability waves are just as real as sea waves. We should not be put off by the fact that these waves are made up of a large number of separate, independent, and random *events*.

Sea water has wave-like and particle-like aspects at the same time. This seems entirely natural to us. And if we are surprised to find the same properties in probability, our bewilderment, is, to say the least, illogical.

When the wind blows at sea, regular rows of waves are formed of a random accumulation of molecules. In much the same way, in our electron beam experiment, the separate random events — electron hits — combine to form a united probability wave that describes the distribution of these hit marks.

To satisfy oneself that sea waves are real, you do not necessarily have to be tossed about by them, although it is desirable to have at least one look at the sea. But to detect probability waves requires sophisticated instrumentation and painstaking experimentation. These experiments are more complicated, of course, than a simple glance at a seascape from a seaside cliff, but this is no good reason for denying the very existence of probability waves.

Voluminous textbook on hydrodynamics tell us that the paths of molecules making up a sea wave are not wave-like trajectories. The molecules move about in circles and ellipses, up and down, and are not involved in the forward movement of the waves. They constitute the wave but do not follow its motion. The shape of this wave is determined by the laws of hydrodynamics.

In much the same manner, the motion of individual electrons in an atom bears no resemblance at all to the vibrations we mentioned previously. But the generally unobservable paths of the electrons belong to a unified observable ensemble, the probability wave. The shape of this wave is governed by quantum laws.

Analogies of this kind could be pursued further but at the moment it is more important to clear up another point: how are we now to understand the phrase "the electron is a wave"? If this is not a material wave but one of probability, it cannot even be detected in experiments with an individual electron.

Sometimes the wave-like nature of quantum-mechanical phenomena is interpreted as being a result of some special kind of interaction of a large number of particles with one another. This interpretation is motivated precisely by the fact that the wave-like and statistical aspects of atomic phenomena cannot in general be detected in one particle experiments. The above reasoning is erroneous, and the error comes from an elementary misunderstanding of the laws of probability: it is possible to compute the wave function $\Psi(x)$ and probability distribution $\rho(x)$ for an *individual* particle. But to *measure* $\rho(x)$ requires *many* identical tests with identical particles. (Born said: "The motion of particles follows the laws of probability, but probability itself propagates in accordance with the laws of causality.")

The preceding examples and reasoning should help us to understand what the electron outside the atom is like and why this particle is also endowed with wave properties. How then can the wave- and particle-like properties be reconciled *within the atom*?

The Atom and Probability

As you may have noticed, so far we have not tried to determine the atom's shape by direct experiments. We computed it using Shroedinger's equation and believed in it because the same equation correctly predicted the finest details of atomic spectra that we can observe. Now this shape of atoms has been generally recognized and earlier in the book we provided some outlines by way of illustration. When we look at those shapes we cannot help visualizing the atomic electron as a sort of charge cloud whose shape depends upon the degree of excitation of the atom. For many reasons the picture is unsatisfactory, however.

In the first place, the electron is nevertheless a particle — just take a look at its track in a Wilson chamber. And if we also remember the photoelectric effect, we can hardly restrain ourselves from jumping to conclusions about the electron's true nature. Now we run to the other extreme and state that the atomic electron is some kind of charge cloud. This interpretation is especially convenient when we try to comprehend the stability of the atom, but it is of no use when we handle the photoelectric effect. Nobody has ever seen a piece of electronic cloud emanate from an atom — it is always a whole electron that is detected. How is it possible that electron clouds of a wide variety of shapes always manage to concentrate in a split second into one and the same indivisible particle? The concept of probability waves gives a clue to this paradox as well.

Let us make a thought experiment to determine the shape of the hydrogen atom. We take our old acquaintance, the electron gun, and aim it now at an individual hydrogen atom, not a foil. What would we observe? Most electrons would pierce the atom, like a shell travelling through a wispy cloud, without being deviated. Sooner or later, one of them would collide with the atom's electron, kick it off, and be deflected in the process. Behind the atom we would now find two electrons instead of one — one from the gun and the other from the atom. Suppose now that we could measure their paths with such precision that we could then reconstruct their point of collision within the atom. Could we, on this evidence, assert that the electron in the atom had been located exactly at that point? No, we could not. We could not even check our supposition since our atom no longer existed — our measurement has destroyed it.

This state of affairs can easily be put right however. All hydrogen atoms are identical and so we can repeat our experiment with any one of them. A second run will disappoint us: we will never find the electron in the atom where we expected to find it. A third, fifth, or tenth trial will only make us more sure that the electron has no definite position within the atom, so that each time we will find it at a new place. But if we take very many atoms and make very many measurements, marking each time the location of the electron within the atom, we will in the end find, to our surprise, that the marks are not arranged in a disorderly fashion but cluster to form the familiar figures, whose two-dimensional pre-images we have computed earlier using Schrödinger's equation.

We already discussed this when we were dealing with electron diffraction. Indeed, we did not know at the time what point on the photographic plate the electron would hit; now we don't know where within the atom we will find the electron. As before, we can only indicate the probability of finding the electron at some definite place within the atom. At one point within the atom this probability is higher, at another it is lower, but the probability distribution as a whole forms a pattern, which we accept as the shape of the atom.

We have no alternative. One can of course counter that this is not an individual atom but some generalized image of an ensemble of atoms.

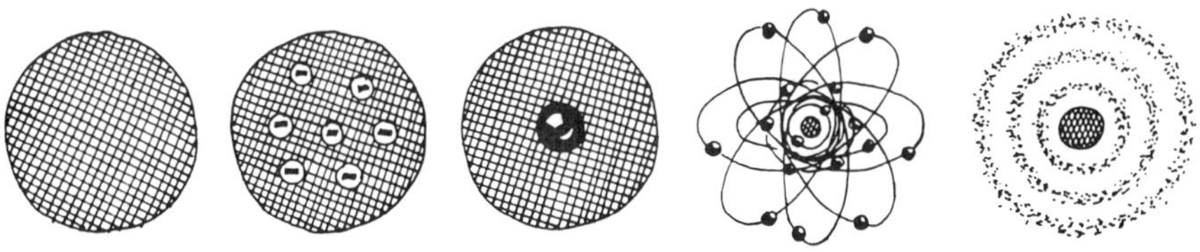

This is but a poor argument, however, since all atoms in the same quantum state are indistinguishable. Therefore, the point-by-point pictures obtained in electron-scattering experiments with *many identical* atoms also determine the shape of a *single* atom.

Here, as elsewhere where chance reigns supreme, we should be aware of the laws of chance. For a single atom the function $\rho(x)$ is but the distribution of the probability of finding the electron at point x within the atom. It is precisely in this sense that we can speak of the "probabilistic shape of a single atom". But the picture is plausible since it is absolutely unique for any collection of identical atoms.

I must confess that I am inclined to give preference to the more intuitively immediate concept of the particulate electron. And so it is with some relief that I, and I hope you, accept the notion that the wave-like behaviour of the electron is probabilistic in nature: this formulation does not evoke in us such instinctive protest as does the rigid statement "the electron is a wave".

At this point we have reached the frontier accessible to those who try to probe into the atom without the aid of formulas and equations. Without resorting to the "mathematical tool kit" of quantum mechanics we cannot predict a single atomic effect, but we can explain certain facts if we exploit the new image of the atom competently and are aware of its origin.

Strange as it may seem, the originator of the antique atomism, Democritus, absolutely denied the role of chance in natural phenomena. What is more, he preached that extreme form of determinism which would later be associated with the name of Laplace. Only Epicurus blunted the edge of Democritus's teaching by allowing atoms to have the ability (he called it *"clinamen — a deviation"*) to vary their paths even under the same forces. (In this postulate, if you wish, you may feel some anticipation of the uncertainty relation and the probabilistic treatment of quantum mechanics.)

Our picture of the atom today is a far cry from Democritus's views. Essentially, only their core has remained. But fruitful delusions are always better than barren infallibility. If it were not for a delusion Columbus would never have discovered America.

Probability and Atomic Spectra

It turned out that probability governs not only the atom's shape, but also all the processes occurring in it. When we deal with an atom we can never be certain where its electron is now, where it will be in the next instant, and what will happen to the atom as a result. The equations of quantum mechanics, however, enable the probabilities of all these processes to be estimated at all times. Probabilistic predictions can then be checked to see that they are correct by conducting a sufficient number of identical tests. But even great minds, such as Rutherford, were not very quick to comprehend this characteristic property of quantum processes.

Rutherford was the first reader of the manuscript of Bohr's paper on atomic structure. When he was returning the paper, Rutherford in his characteristically straightforward and brusque manner, asked Bohr how an electron in the nth orbit knows where it was to jump, to the kth or the ith orbit. In 1913, Bohr had no convincing answer. Now we could answer as follows: an electron knows nothing beforehand, it obeys quantum laws. According to these laws there always exists a definite probability that an electron in the nth state will go over to any other state (for instance, the kth state). As always, the probability W_{nk} of such a transition is a number that depends upon the selection of the pair n and k. By running through all the possible combinations of n and k we will obtain a square table of numbers W_{nk} that represents the internal structure of the atom and is now known as a matrix. Using it we can, for example, explain why in the yellow doublet of the D line of sodium the intensity of the D_2 line is twice that of the D_1 line. And by consistently using the equations of quantum mechanics we can also understand some finer details of the structure of these lines, for example, the laws governing the variation of intensities within them. But all these insights are, clearly, available only to specialists.

Causality and Chance, Probability and Certainty

The probabilistic interpretation of quantum mechanics went against the grain with many physicists who made numerous attempts to return to the previous classical picture. Psychologically, this desire to lean at all costs on old knowledge under new conditions is understandable but in no way justifiable. One is reminded of the old Prussian soldier who returns home, at the turn of the century, after years of service and tries to interpret all the diversity of civilian life from the viewpoint of the military drill regulations. He will undoubtedly be outraged at the disorder he finds in the local dance hall, and it will be difficult to make him understand that the rules on a dance floor are quite different from those of the army drill ground.

Not so long ago not too honest interpreters of quantum mechanics leaned over backwards to reject it on the grounds that it fitted badly with the scheme they themselves had worked out. They were furious with the "free will" that had allegedly been allowed the electron, made fun of the uncertainty relation, and seriously contended that quantum mechanics is a useless science since it only deals with the probabilities of events, not the real events themselves. Those who have had the patience to follow our previous arguments can see how nonsensical these charges were. But even those who treat the theory of the atom with due respect do not always realize how they are to understand the causality of atomic phenomena if each one of them is random, and how reliable the predictions of quantum mechanics can be if they are based on the concept of probability.

The everyday notion of causality — "each event has its own cause" — is in need of no explanation but it is useless in science. Causality in science requires a *law* to guide us through the sequence of events in time. Mathematically, this law takes the firm of a differential equation, which is called an equation of motion. In classical mechanics such equations — Newton's equations of motion — enable us to predict the trajectory of a particle's motion. It is this scheme of the explanation and *determination* of things in nature we have just sketched that has always been accepted as the ideal causal treatment in classical physics. It leaves no room for doubt or misunderstanding; that is why the causality of classical physics came to be known later as *determinism.*

In quantum physics we find no *such* causality. Instead, we find there a peculiar, quantum-mechanical, causality and a law — the Schrödinger equation. This law is even more powerful than Newton's equation, since it identifies and singles out patterns in the chaos of random quantum events. In a sense, it can be likened to a kaleidoscope, in which random combinations of bits of tinted glass produce patterns, which have both meaning and beauty.

Phrases like "statistical causality" and "probabilistic regularity" grate on the unpractised ear because of their seeming incongruity ("Soapy soap", as the jocular phrase has it, sounds bad enough, but "unsoapy soap" is just too much.) The phrases are strange indeed but in quantum physics we are obliged to utilize them when dealing with quantum effects. In actual fact, there is no logical paradox here: the concepts of "chance" and "regularity" are complementary ones. According to Bohr's complementarity principle both are needed, equally and simultaneously, to give rise to a new concept, the concept of "quantum-mechanical causality", which is something more than the simple sum of the concepts "regularity" and "chance". Likewise, a "quantum object" is always something more than just the sum of wave- and particle-like qualities.

The randomness of single quantum events is not a result of the joint impact of unknown causes, but rather a primary elementary law that governs them; this is the starting point of the theory, not a fact to be explained. Probability is a property

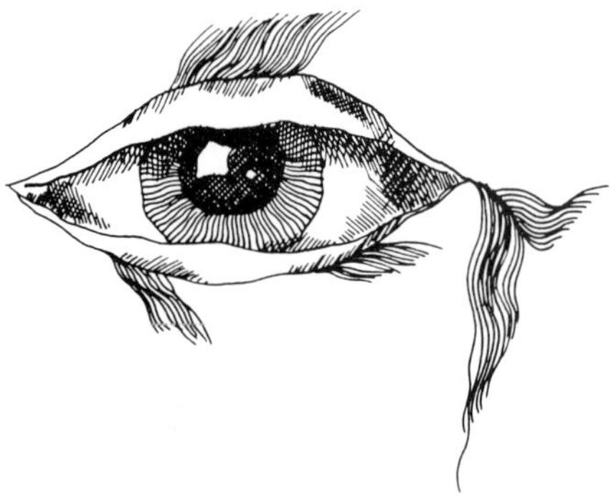

and a category inherent in quantum reality itself, and not just a convenient mathematical trick used to account for experimental evidence.

However elegant and beautiful, these constructs are hard to absorb intuitively and to recognize. As always in quantum physics, the logical hurdles are due to specific features of our language and upbringing. The notions of "regularity" and "chance", "certainty" and "probability" became well established long before the coming of quantum mechanics, and what is normally read into them flouts the wishes of quantum physicists.

The task of probability is one of prediction: What will happen if... "What... if...?" In classical physics two identical tests under identical conditions should always yield the same end result. This is the idea behind classical causality, or determinism. Quantum-mechanical causality is peculiar in that even under invariable conditions it can only give the probability of the outcome of a single test; on the other hand, it can, with absolute certainty, predict the distribution of outcomes for a large number of identical tests. From the quantum-mechanical viewpoint the traditional formulation of the law of causality ("the exact knowledge of the present enables the future to be predicted with confidence") contains a false premise: by the uncertainty relation we *in principle* cannot know the present in all details. But the conclusion remains the same, even if we now understand it in a new fashion.

We can keep on juggling the paradoxes "regular chance" and "certain probability" no end, but this will add nothing to our knowledge of the atom. This is not the point. We must simply grasp that the probabilistic description of the atom does not come from our averaging of as yet unknown atomic phenomena, but is a frontier of principle in today's science. As long as Heisenberg's uncertainty relation remains valid, we cannot refine upon our data on atomic objects indefinitely. Fortunately, we have no need to do so. After all, all bodies in nature consist of a vast number of atoms, and quantum mechanics is able to predict the properties of such systems unequivocally, with no arbitrariness whatsoever.

The probabilistic approach added consistency to quantum mechanics. Probability alone allowed the consistent and ultimate synthesis of complementary pairs of concepts, such as wave and particle, continuity and discreteness, causality and chance, phenomenon and observation. Only then did it become possible, at last, to establish that all these concepts form an indivisible *system* and that each one should be viewed in the context of the others.

Likewise, we cannot, say, explain who Heracles was without mentioning Zeus, Alcmene, Megera, Hydra, Augeas, and others. Only when taken together do they form the unique fabric of an ancient myth.

The answers that quantum mechanics gives to the questions we put to nature depend on which aspect of an atomic phenomenon we want to look into more closely.

In our studies of nature we — consciously or unconsciously — tend to divide it into two parts, the system and the observer. This division is not hard and fast, it depends upon *what* phenomenon we study and *what* we want to learn about it. If under *phenomenon* we understand the motion of a single *particle,* then such an event is *discrete, random* and for the most part unobservable. But if under phenomenon we understand the results of *observation* of the motion of a host of identical quantum objects, then this event is *continuous, regular* and is described by the wave function.

Quantum mechanics is only concerned with the second kind of event. For them it can make reliable and unambiguous predictions, which have not been disproved so far by even a single experiment.

Quantumalia

People, Events, Quanta

Scientific results are unaffected by the psychology and wishes of individual persons; this largely accounts for the power and value of science. But science is a human endeavour, a human enterprise, and so its history is not just the accumulation of fresh facts and the development of physical ideas and mathematical tools — it is also a history of human passions, triumphs, and failures. Discoveries are important, but then so is any detail in the lives of scientists; we always wonder how this or that event, one of those trifles that make up the everyday lives of even the greatest of men, affected the work that made them immortal.

Quantum mechanics is a child of European culture, and its fathers are some of the best representatives of that culture. Einstein, Born, Heisenberg, Ehrenfest, and Laue were excellent musicians; and Planck lectured on the theory of music at university and in his formative years he studied to be a professional pianist. (He was also the mentor of a choir in which the young Otto Hahn sang, who would thirty years later discover the fission of uranium.)

P. Dirac

Heisenberg, Pauli, Laue, and Schrödinger read in ancient languages, de Broglie was a historian by profession, and Schrödinger was fond of philosophy and religion, especially Indian, wrote poetry and later in his life published a book of poems. Even in their scientific correspondence Planck and Sommerfeld used to exchange verses.

The history of the early years of quantum mechanics has preserved for us some lively reminiscences, which help us to re-create that atmosphere of tension and enthusiasm in which people of different nationalities, ages, and temperaments erected the modern edifice of quantum mechanics.

When did it all start? Perhaps on the day when Sommerfeld entered a room, told Heisenberg, then a student, to stop playing chess, and gave him a photographic plate showing the emission spectrum of an atom in a magnetic field. He urged him to find some regularity in the arrangement of the spectral lines. Or maybe three years later, in June 1922, during a prolonged stroll taken by Heisenberg and Bohr, when Bohr at the invitation of Göttingen University was lecturing there on quantum theory. Or at the end of May in 1925, when the assistant professor Heisenberg got hay fever and Born, his mentor at that time, proposed that he take a holiday on the island of Heligoland in the North Sea. There he carried out his famous calculations and experienced a rare elation of which he later related: "The evening came at last when I could plunge into calculating the energies of individual terms in the energy table or, as they say today, in the energy matrix. The excitement that gripped me... prevented me from concentrating, and I began to make mistake after mistake in the calculations. I could only reach the final result at three o'clock in the morning. At first I was frightened... The thought that I had become the holder of all these treasures — the elegant mathematical structures that nature

W. Pauli

had revealed to me — took my breath away. As to getting some sleep, it was absolutely out of the question. It began to grow light. I left the house and walked toward the southern extremity of the island, where a detached rock jutted out into the sea... Without much trouble I climbed the cliff and met sunrise on its summit."

On 5 June, back from his holiday, he wrote to Krönig about his findings. On 24 June he sent a detailed letter to Pauli, and gave a draft of a paper to Born asking him to dispose of it at his discretion. Born approved of his idea, and on 29 July Heisenberg's paper "On the Quantum-Theoretical Interpretation of Kinematical and Mechanical Relations" was submitted to the editorial board of a journal. Heisenberg himself, it seems, took some time to realize the true significance of his work, since on 28 July, when he, at the invitation of "Kapitsa's club", made a report at Cambridge, he chose another topic, "On Therm-Zoology and Zeeman-Botany".

Born continued to think hard about the import of the work of his assistant. "Heisenberg's multiplication rule," he reminisced in his Nobel Prize acceptance lecture, "gave me no peace, and after eight days of intensive thinking and checking there popped up in my memory the algebraic theory taught to me by professor Rosenas at Breslau.... I'll never forget that profound agitation I experienced when I succeeded in concentrating Heisenberg's ideas on quantum conditions in the form of the mysterious equation $pq - qp = h/2\pi i$".

In the world of human thought generally and in physical science particularly, the most fruitful concepts are those to which it is impossible to attach a well-defined meaning.

Hendrick Kramers

A while later en route to Hannover Born confided his troubles with the new calculus to a colleague from Göttingen. As chance would have it in the same railway compartment was the graduate Pascual Jordan, one of the few experts in matrix calculus at that time; he was helping Richard Courant to prepare for print the famous course *Methods of Mathematical Physics* by Courant and Hilbert. At the railway terminal at Hannover Jordan introduced himself to Born and offered his help. The proposal was most welcome, since Pauli refused point blank to cooperate with Born and urged him not to interfere with the course of events, for, he believed sincerely, the new science was *Knabenphysik,* a science for boys (at the time Born was forty-two — too old in Pauli's opinion). Born and Jordan finished their paper by the autumn; soon they were joined by Heisenberg, and the team produced their consistent presentation of matrix mechanics (on 16 November 1925 their paper "On Quantum Mechanics" was submitted to a journal).

A short while earlier, on 7 November of the same year, the journal received a paper by Dirac entitled "The Basic Equations of Quantum Mechanics", in which he presented his mathematical formulation of Heisenberg's ideas. Dirac had received training as an electrical engineer, but during the post-war depression he could not find employment as such and so he decided to continue his education at Cambridge under Fowler, who told him about Heisenberg's paper — in September of 1925 Fowler got the galley proofs from Born.

That same autumn Born went to America and during his stay there in the winter of 1926 he, together with Norbert Wiener, the future father of cybernetics, introduced one of the most important concepts of quantum mechanics — the concept of the operator corresponding to a physical quantity, which, in particular, could also be represented as a matrix, as in Heisenberg's picture.

That same winter Pauli used matrix mechanics to obtain the energy levels of the hydrogen atom and to show that they coincided with the energies of stationary states in the Bohr atom.

A year previously, on 29 November 1924, de Broglie defended his dissertation "Studies in the Theory of Quanta". In 1910 he obtained from the Sorbonne the title of licentiate in history, but under the influence of his brother and impressed by Langevin's lectures on relativity and Poincaré's books *Science and Hypothesis* and *The Value of Science* he, with all the ardour of youth, devoted himself to physics.

A moderately satisfying picture of this world has only been reached at the high price of taking ourselves out of the picture, stepping back into the role of a nonconcerned observer.

Erwin Schrödinger

Louis de Broglie's brother, Maurice, was an authority on the physics of X rays and he was pondering over their nature. He agreed with Sir William L. Bragg (1890-1971), who in 1912 immediately after Laue's discovery and ten years before Compton's experiment, wrote: "The problem now is not to choose between two theories of X rays but rather to find... one theory that would combine the possibilities of both". Louis de Broglie recalled in 1963: "My brother regarded X rays as a sort of combination of wave and particle, but since he wasn't a theoretician, he had no clear-cut idea of that subject... He persistently drew my attention to the importance and unquestionable reality of the dual aspects of wave and particle. Those long conversations helped me to gain a profound understanding of the necessity of an obligatory fusion of the wave and particle points of view."

In his first paper of 1923 Louise de Broglie had already made the assumption that "a beam of electrons passing through a sufficiently narrow hole must also possess the power of interference". At the time, this note passed unnoticed by major experimenters, although the profession already knew about Davisson's experiment as well as the experiments by Karl Ramsauer (1879-1955) and John Townsend (1868-1957), which indicated that electrons of certain energies suffer nearly no scattering when they are passed through gases — a phenomenon similar to coated optics and opposite to the resonance absorption observed in Franck and Hertz's experiment.

Paul Langevin, under whom Louis de Broglie was studying for his doctoral degree, accepted the ideas of his student with reserve, but without hostility. In April 1924 he reported them at the IV Solvay Congress, and in December he sent de Broglie's dissertation to Einstein to be reviewed. Einstein, in turn, earnestly advised Born: "Read it! Although it seems to have been written by a madman, it has been written soundly". Later on he quoted from it with sympathy in his works, and Schrödinger was afterwards grateful to Einstein that he so opportunely "gave him a flick on the nose by pointing out the importance of de Broglie's ideas".

Not everyone gave the idea of matter waves such a warm reception. Planck remembered later that when he heard from Kramers at a seminar about the work of de Broglie he "only shook his head", and Lorentz, who was also present at the seminar, said: "Those young men consider it exceedingly easy to discard old concepts in physics!"

Early in 1925 Max Born discussed his ideas with his close friend and colleague from the University of Göttingen James Franck. The discussion was in the presence of Born's student Walter Elsasser, who proposed on the spot to make an experiment on electron diffraction.

"It's hardly necessary," responded Franck, "Davisson's experiments have already established the presence of the observed effect". (Davisson himself, by the way, did not think so and it is doubtful that he was well acquainted with de Broglie's idea.) After the discussion Elsasser wrote a short note in which he gave an explanation of Davisson's results as well as the Ramsauer-Townsend effect in terms of matter waves.

G. Gamow

Elsasser's note was published in July 1925, before the first paper by Heisenberg had been submitted for publication. It passed practically unnoticed: before long the new matrix mechanics took the fancy of the majority of physicists.

Erwin Schrödinger in 1925 was already thirty-eight and was not one to fall for every fashion and whim. Like Heisenberg he had graduated from a classical gymnasium, where the main subjects were Latin and Greek, and by the cast of his mind he was a poet and thinker. Unfortunately Schrödinger has not left, as Heisenberg did, first-hand reminiscences about the *Sturm und Drang* period of quantum mechanics. One possible reason for this was because he made his most important discoveries in his mature years, when the ardour of youth for action has dissolved in the clear wisdom of knowledge, and when the immediate joy of discovery is subdued by the understanding of the relative value of all that exists.

Later in his life Schrödinger wrote about his first impression of Heisenberg-Born-Jordan's theory: "... I was scared, if not repelled, by the techniques of transcendental algebra, which seemed to me so difficult, and by the absence of any graphic representations". The views of de Broglie were more congenial to him, and now an opportunity presented itself to take a closer look at them: late in 1925 Peter Debye, whose successor he was at the Physics Department of Zürich University, asked him to talk about de Broglie's work to post-graduate students at the famous Zürich Polytechnic. Soon after this Schrödinger published his paper "Quantization as an Eigenvalue Problem", the first of his series of publications. (The paper was submitted for publication on 27 January 1926, at about the same time as Born and Wiener introduced the concept of an operator and Pauli used matrix mechanics to obtain the spectrum of the hydrogen atom.) On 21 June 1926 Schrödinger sent to the journal the sixth paper of the series and by 25 June Born had submitted for publication his report in which he put forward a statistical interpretation of the wave function. This essentially completed the construction of the basics of quantum mechanics.

We all agree that your theory is crazy. What we are not sure of is whether it is crazy enough.

Niels Bohr

Many years later Born said about these papers by Schrödinger: "Whatever is there more outstanding in theoretical physics?" and Max Born added: "The Schrödinger equation enjoys in modern physics the same place as in classical physics do the equations derived by Newton, Lagrange, and Hamilton." But at the time theoreticians accepted wave mechanics with suspicion, because it obviously lacked quantum jumps — which they had become accustomed to only recently and with such difficulty, and which were now considered the principal feature of atomic effects.

In June 1926 Heisenberg went to Munich to visit his parents and "... gave way to absolute despair" after having heard at a seminar the report by Schrödinger and his interpretation of quantum mechanics. He wrote to Pauli: "The more I reflect on the physical side of Schrödinger's theory the more dreadful it seems to me."

On the other hand, experimental physicists (Wilhelm Wien and others), who called Heisenberg's theory "atomistics" (i.e., mystics of the atom), welcomed Schrödinger's theory with enthusiasm. (Wien perhaps recalled Heisenberg's ignorance of the resolving power of a microscope in his oral examination.)

Arguments about wave mechanics went on for hours. They reached a climax on September 1926, when Schrödinger came to Copenhagen at Bohr's invitation. Schrödinger got so tired of discussions that he fell ill and spent several days in Bohr's home with Bohr nearly always at his bedside. From time to time, Niels Bohr would raise his finger with a characteristic gesture and utter: "But Schrödinger, you must nevertheless agree..." Once, almost in total despair, Schrödinger exclaimed: "If we are going to stick to those damned quantum jumps, then I regret that I ever had anything to do with the quantum theory." Bohr replied: "But the rest of us are much obliged to you for that."

Nothing is more real than nothing.

Samuel Beckett

With the passage of time, the positions of the advocates of matrix and wave mechanics drew nearer to each other. Schrödinger himself proved that the theories were mathematically equivalent in March 1926, and independently the same conclusion was reached by Carl Eckart in the United States, and Cornelius Lanczos and Wolfgang Pauli in Germany.

In August 1926 Davisson came from America to a congress of the British Association for the Advancement of Science, where he discussed his latest experiments on electron reflection from crystal surfaces with Born, Hartree, and Franck. The European colleagues supplied Davisson with Schrödinger's papers, which he studied diligently on his return journey across the ocean. Within a year while continuing his experimental work with Germer, he showed that electron waves do exist, and half a year previously, in May 1927, G.P. Thomson had discovered the diffraction of electrons, and so wave mechanics acquired a solid experimental base.

The words or the language, as they are written or spoken, do not seem to play any role in my mechanism of thought. The psychical entities which seem to serve as elements in thought are certain signs and more or less clear images which can be "voluntarily" reproduced and combined.

Albert Einstein

Experiments on electron diffraction, which became known in the summer of 1926, greatly strengthened belief in the theories of de Broglie and Schrödinger. Bit by bit physicists came to understand and became reconciled to the idea that the wave-particle dualism is a well established fact, not just an elegant hypothesis. Now scientists endeavoured to comprehend the consequences to which this fact leads and what restrictions it imposes on their views on quantum processes. Here they ran into dozens of paradoxes, whose meaning they often could not grasp.

In the autumn of 1926 Heisenberg lived in the garret of the Physics Institute in Copenhagen. In the evenings Bohr would climb the stairs to his room and their discussions would last well into the night. "Sometimes they would end in complete despair, caused by the incomprehensibility of the quantum theory, in Bohr's apartment over a glass of port", Heisenberg recalled. "Once, after such a discussion, deeply perplexed, I went downstairs and outside to take a walk in the fresh air of the Felled-Park behind the Institute and to calm down before going to bed. During this stroll under the starry night sky an idea flashed through my mind: Why not postulate that nature permits the existence of only such experimental situations in which... it is impossible to determine the position and velocity of a particle simultaneously?" This idea anticipated the future uncertainty relation.

M. Born

It was perhaps to reduce the tension of those days that Niels Bohr went on vacation to Norway in February 1927. Left to himself, Heisenberg continued to think intently. Specifically, he was preoccupied with a question put to him long ago by his classmate Burkhard Drude, the son of the well-known physicist Paul Drude: "Why can't the orbit of an electron in an atom be observed using rays with a very short wavelength, for instance, gamma rays?" Consideration of this experiment quite soon led him to the uncertainty relation. (It is quite probable that Heisenberg now recalled with gratitude the severe examiner Willy Wien, who was on the point of turning him down because Heisenberg was ignorant of the resolving power of the microscope. As Heisenberg confessed later he was conscientious enough to revise this section of optics after the exam, which he only passed thanks to the intercession of Sommerfeld. And that knowledge came in handy now.)

A few days later Bohr returned from his holiday to put forward the completed idea of complementarity, which he had thoroughly thought over in Norway. After several more weeks of heated discussion, with the participation of Oscar Klein, everybody came to the conclusion that the uncertainty relation was a special case of the complementarity principle, one that could be represented quantitatively in the language of mathematics. On 23 March 1927 Heisenberg submitted for publication his paper entitled "On the Graphic Contents of Quantum-Theoretical Kinematics and Mechanics" with comments by Bohr.

By that time quantum mechanics was already being studied in many scientific centres, notably, of course, in Göttingen and Copenhagen. In the winter semester of 1926/1927 David Hilbert at Göttingen University twice a week read lectures on mathematical methods in quantum mechanics (the course was published by the spring of 1927). His assistant was the twenty-three-year-old Hungarian-born scientist John von Neumann (the future designer of computers, originator of game theory, and one of the greatest mathematicians of the twentieth century), who within two years would succeed in adding mathematical rigour and conceptual independence to the quantum theory.

Since the day that Heisenberg published his first paper the mathematical tools of the new mechanics had continuously been improved, and its interpretation painstakingly supplemented and refined. Within two years, by the autumn of 1927, more than 200 papers had been published, most of which have not yet become outdated. On 16 September 1927 at Como in Italy at the International Congress to commemorate the centennial of Alessandro Volta, Niels Bohr made a report entitled "The Quantum Postulate and the Newest Development of the Atomic Theory". It contained a consistent exposition of the system of concepts of the new quantum physics and introduced the term "complementarity". Several weeks later, at the end of October 1927, in Brussels, at the V Solvay Congress Planck, Einstein, Lorentz, Bohr, de Broglie, Born, Schrö dinger, and the younger physicists Heisenberg, Pauli, Dirac, and Kramers gathered together. Here that view of quantum mechanics and that system of concepts, which later came to be known as the "Copenhagen interpretation", became firmly established. The disputes at the congress turned out to be an acid test for all aspects of quantum mechanics. The new branch of physics passed the test with flying colours and has not undergone any major changes since.

The most beautiful experience we can have is the mysterious. It is the fundamental emotion that stands at the cradle of true art and true science.

Albert Einstein

Those years in Bohr's institute at Copenhagen saw not only the birth of the science of the atom but also the establishment of an international family of young physicists. Among them were: Kramers, Ehrenfest, and Rosenfeld of the Netherlands; Klein of Sweden; Dirac of England; Heisenberg of Germany; Brillouin of France; Pauli of Austria; Nishina of Japan; Uhlenbeck of the USA.; and Gamow and Landau of Russia.

This alliance of scientists, unprecedented in the history of science, was noted for its uncompromising striving for truth, sincere admiration for the grandeur of the problem they were to solve, and an ineradicable sense of humour, which harmonized so well with the general spirit of intellectual chivalry. "There are things so serious that they can only be spoken of jokingly," was a favourite saying of Niels Bohr, who became a teacher and father confessor for the small community.

The man with a new idea is a crank until that idea succeeds.

Mark Twain

They all had the spark of cosmic feeling, which is a mark of really great men. They preserved that feeling of eternity through all the civil disturbances that were to befall them. Many years later they would find themselves scattered all over the world by political storms: Heisenberg would become head of the German "Uranium Project", Nishina would become head of the Japanese uranium programme, Bohr himself would escape from the Nazis and go to Los Alamos to work in the American secret nuclear centre, and Goudsmit would become leader of the Alsos Mission, whose purpose would be to find out how far Heisenberg had progressed in his German atomic bomb project.

Very few of those pioneers are still alive now; they are the last of the Mohicans of a whole epoch in physics, comparable only to those of Galileo and Newton.

Tot, Egyptian

Chapter Eleven

What Is an Atom? What Is Quantum Mechanics? Physical Reality

To formulate and refine concepts is a pursuit that is hard and not always safe. At one time Socrates paid with his life for his insistent attempts to grasp the essence of the basic moral and ethical concepts, such as good and ill, truth and delusion, law and justice... Socrates lived in the antique Greece of the golden age. Being a true sage, he spent his days in the sun-flooded squares of Athens and tested his compatriots with questions such as: "Tell me, oh most learned Hippias, what is the beautiful?" The learned opponent plunged into his discourse with ardour, but quite soon he found himself hemmed in by examples: he was more or less on safe ground when he described a beautiful woman, a beautiful pot of porridge, or a beautiful horse, but to say what is the beautiful as such was always beyond him.

The drama of this typical mental situation has always been understood. It was understood and taken for granted: "Truth lies beyond the confines of consciousness and so it cannot be expressed in words," was said in ancient India. When we seek an answer to the question "What is the atom?" we invariably run into similar difficulties. Just like ancient philosophers, we cannot overcome them with words. Specific examples tell us that the atom is neither the spectral lines it emits, nor the infinite variety of crystals made up of atoms, nor the heat of red-hot iron, nor the electrons ejected by atoms.

Like Socrates's opponents, we here have to recognize that the atom is something not definable in itself, a certain general cause of quantum phenomena, all of which are needed in some degree or other to define it. When we observe red-hot iron and spectral lines, crystals and electrolysis, electrons in a Crookes's tube and the scattering of alpha particles, we touch upon various facets of the atom. Can we now give meaningful answers to the two basic questions we posed at the outset?

What is an Atom? What is Quantum Mechanics?

We began our story of quantum mechanics with the definition

Quantum mechanics is the science of the structure and properties of quantum objects and phenomena.

We discarded it at once because it was obviously useless until we could define the concept of the "quantum object" itself. We set out to analyse experiments that revealed the properties of the atom and other quantum effects and to examine the relationships that enable us to predict the results of these experiments.

As we progressed along this path we were amazed to find that all the formulas that describe the behaviour of quantum objects invariably contain Planck's constant h. And now when a physicist sees an equation containing the quantum of action h, he accepts this as a tell-tale sign that he is dealing with an equation of quantum mechanics.

It follows that quantum mechanics could be defined as a system of equations that include Planck's constant h. This definition, however, would only satisfy our desire for unambiguity and formal rigour; otherwise it is of no use — the definition of a science should concern the object being studied, not just the method used to this end.

We could define the atom as a physical object, whose wave- and particle-like properties are of equal significance for its complete description. We are safe in saying, however, that this approach, too, fails to take into account all the properties of a quantum object, although it establishes the duality inherent in it.

After so many attempts to answer the question, we could simply say, "The atom is the sum total of our present knowledge about it." We must admit, of course, that this again is no definition but rather a plausible excuse to evade one.

What words could define the *concept* of an "atom" concisely and unambiguously? We have seen time and again that no single word of our speech is capable of accommodating all the diversity and complexity of this concept. Then we turned to the equations of quantum mechanics and used formulas, without words and strict definitions, to construct an *image* of the atom for our private use. In so doing, we deliberately followed the principles of quantum physics.

So one principle recommends us, if possible, to avoid talking about phenomena in themselves without any reference to the method of observing them. The two concepts of a "phenomenon" and its "observation" exist independently only in our minds and then perhaps only vaguely. For a physicist these concepts are two aspects of the same physical reality, which he studies and whose objective existence is for him an article of faith. These concepts are incompatible: without observation we know nothing at all about a phenomenon. The complexity of their unity and interaction does not allow us to comprehend the essence of a phenomenon by itself, but together they help us to reveal the *relations* between phenomena.

We can represent these relations by formulas or words. But the words alone will remain empty unless the formulas are given alongside them. And the formulas are still-born until we find a way to explain what they actually mean. For an explanation of "phenomenon-observation" to be complete we require a concord between concepts and formulas. Only then can we form a satisfactory image of a physical phenomenon for ourselves.

At this stage the chain of cognition of the new physics (p. 115) is modified once again. It becomes more involved and assumes the form

$$\begin{pmatrix} \text{phenomenon} \\ \\ \text{observation} \end{pmatrix} \rightarrow \begin{pmatrix} \text{concept} \\ \\ \text{formula} \end{pmatrix} \rightarrow \text{image}\,.$$

In all our attempts to define the concept of the "atom" we were unconsciously seeking this scheme.

Nowadays young physicists are at first taught formulas. This is perhaps reasonable: when tackling a foreign language it is better to learn to speak it from the very beginning without bothering to find out why a word is written the way it is and not some other way. Students first learn the words that are to be pronounced to make contact between people possible at all.

Formulas do not have precise verbal equivalents, however. The teaching of modern physics therefore consists in couching uncommon things in common idiom, but each time from a slightly different angle. The aim is to translate new concepts from the sphere of the logical and conscious into the sphere of the intuitive and subconscious, a vital condition for any creative activity.

This training philosophy imperceptibly deforms (or reforms) in young physicists the world of their images and concepts and even their system of associations. Physicists are jarred by the immaculately sterile verbal structures of most popular-science books; they infallibly detect an almost imperceptible foreign accent. It is frequently impossible idiomatically to convey the meaning of a foreign phrase without initially destroying its original structure. The language used by physicists is often English, German, Russian, or whatever only by name and some of the words used. In actual fact, it is a special language whose vocabulary and grammatical structures drive editors to despair. But any attempt to "polish" a seemingly ungainly physical phrase to make it fit the standards of literary language does to it what even the best of translations does to foreign poetry.

Now we have two uncomfortable physical truths:

Quantum mechanics is a system of formulas concepts and images that enables us to explain and predict the observable properties of quantum objects.

A quantum object is a physical reality whose fundamental principle is duality and whose properties can be described in terms of quantum mechanics.

The above two definitions, when placed side by side, seem to flout common sense. Actually, their exact meaning is only beyond comprehension if they are approached individually; when taken together they become meaningful. Of course, to understand and visualize all the diversity and unity of the formal definitions of quantum phenomena alone will not suffice: a knowledge of their origins and evolution is also needed. That is why we have dwelt so long on the experiments that would later give rise to the concept of a "quantum object". Taken alone, isolated from these experiments, this concept tells us nothing; it only consolidates in the language of formal logic that intuitive image which is gradually taking shape in our mind, for the most part not of our own will. Our last definition of quantum mechanics coincides nearly word for word with the one we provided at the very beginning of the book. And if now its import for you is absolutely different, the trouble you have taken to read thus far in the book has not been in vain.

We could finish our story about quantum mechanics at this point if it were not for one important circumstance. When we said "the atom is a physical reality" we unintentionally touched upon an extensive boundary region between physics and philosophy.

"Physical reality" is the ultimate category to which we come inevitably when we undertake any serious attempt to explain something in physics. A universal notion is so vast and all-embracing that we cannot define it in terms of physics alone. And so we have to appeal for help to philosophy, a field of learning concerned with the notion of *objective reality*.

It is well known that objective reality is everything there is, or has ever been, independently of our consciousness. But for science this definition is not concrete enough, since nothing arises out of it, it only calls upon us to believe in the objective nature of the knowable world. But this is something all scientists believe in anyway; otherwise they would hardly devote their lives to the knowing of this reality. Opinions differ only as to the *nature* of physical reality, its plausibility, and its uniqueness. A great many physicists admit that physical reality is the part of objective reality that we get to know through our experience and our consciousness, i.e., it is all the facts and numbers that we obtain using our perceptions, instruments, and speculation.

Opinions are vacillating. Why are we so sure that the picture of physical reality thus obtained is true? Or, to put it more mildly (who really knows "what truth is"), why are we so sure that this picture is the only possible one?

Physical Reality

Enter any physical laboratory, look around and try to tell what phenomenon of nature is being investigated there. You will see stacks of hardware and a tangle of wires amidst which you will not make out either the subject of investigation or even the investigators themselves. If you are then told, say, that "we are studying the splitting of spectral lines in a magnetic field", you may feel mildly interested but you may also feel that you are having your leg pulled.

Even if you are given a photographic plate showing some thin black lines, you will hardly imagine atoms from whose depths (as the physicists will assure you) light was emitted that was subsequently transformed by the spectroscope and left the lines on the plate. These explanations will not seem very convincing to a non-physicist. It is more or less clear to him how a motor mechanic tells what is wrong with an engine by listening to its noise, or how a doctor diagnoses a disease on the basis of the patient's complaints. Our layman knows that it is always possible to take the engine apart without changing the components. If the worst comes to the worst, the medical diagnosis can always be checked by a post-mortem. In both cases we know all the parts of which the whole is comprised. Even if you are not a watch-maker, you can get an idea of the workings of your watch after you have taken it apart, and you can say why the visible motions of its hands are so unlike the usually invisible motions of its springs and cogs.

With atoms things are not that simple. We observe the external manifestations of their properties: spectra, the colour of bodies, their specific heat and crystalline structure, but we cannot remove the "back of the watch case" to see what is "actually" inside the atom. From an assembly of facts, concepts, and formulas, we have formed a certain image of the atom. But since there is no independent method in existence to check this image, a natural question arises, "Could we not think up a different image of the atom that would still lead to the same observed effects?"

This is no idle question; it has occupied the minds of nearly all major physicists. Sceptical common sense will put the matter somewhat differently: "Everything you have thought up is wrong, actually everything is different!" This objection is hard to refute because the concept "actually" has never been defined. In the everyday world, all that "actually" exists is what we can check with our five senses, or what we can verify using the extensions of our senses, i.e., instruments.

Even the last statement raised objections at first. The contemporaries of Galileo contended that his discoveries of sun spots and the moons of Jupiter were not actually discoveries at all, but came from imperfections of the telescope he used.

Since the time of Galileo we have become less suspicious of the readings of instruments, but there still remains some freedom for the interpretation of these readings. Now "actually" means "How unequivocal is the interpretation of experiments concerned with phenomena inaccessible to direct sense perception?" The common sense of a person, even one concerned with science, should tell him that such an interpretation is ambiguous. A fleeting visit to a physics laboratory may even strengthen this *a priori* conviction. But physicists themselves know that the facts and conceptions of their science permit free interpretation only as they are discovered and established. Once incorporated into the general system of physical knowledge and adjusted to it, they are almost impossible to change unless we venture beyond the scope of their applicability. (Just try and drop a phrase from a poem, even though it all seems to be just "poetical invention".)

As we extend and refine our system of scientific knowledge we are compelled to depart farther and farther from direct perceptions and from the concepts that are based on them. This process of abstraction is irreversible, but this should not distress us; we may well pride ourselves that our reason is capable of understanding even what we cannot imagine. The abstraction of scientific ideas is just as great a necessity as the invention of alphabets in place of ancient picture scripts and hieroglyphics. Not a single letter in the word "rhinoceros" reminds us of the animal but the whole word infallibly brings to mind the image. Everyone realizes that our modern culture is inconceivable without printing. But comparatively few people realize that the growth of science is impossible without further abstraction of scientific concepts. In other words, abstract science, just like music, requires profound understanding, not justification. Only abstraction can help us to grasp the bizarre quantum reality, although this reality is of a different kind from that of tangible and visible stones and trees.

But even this "abstract reality" man endeavours to make *visualizable*, i.e., to reduce it to a few proven images. This striving is deeply ingrained in man, and so physicists have with time developed their own peculiar system of images. This system almost certainly corresponds to nothing real in nature and defies any verbal descriptions. It is still useful to physicists, seeking relations between phenomena at moments of the highest mental intensity.

The cognition chains that we drew previously, from phenomena through concepts and formulas to images, are no more than skeleton schemes, which only give one a faint idea of the most complex processes that go on in the mind of a scientist when he tries to discern in a seeming chaos of facts some simple connexions, to define them in words, and to fit them within the general jig-saw picture of nature.

A single word does not constitute a language — you need a set of words and rules of grammar to arrange the words. Likewise, a single scientific fact, however important it may seem, does not mean anything by itself if we do not know its place within the entire body of knowledge. Only when it has some interpretation can we attach some meaning and significance to it.

Recall the story of the *D* line of sodium. It was first observed by Fraunhofer. But could he possibly suspect that he had the key to all of quantum mechanics in his hands? He perceived that the line is split into two components. But could he possibly know that the effect was due to electron spin. Electron, quantum mechanics, spin — in Fraunhofer's day there were no such concepts. But without them the *D* line of sodium was simply a curious fact without any consequences. Only after the experiments of Crookes, Rutherford, and Thomson and after the system of ideas and formulas called quantum mechanics had been found did it become clear that the *D* line was one of those facts destined to revolutionize our views of nature.

It is common knowledge that concepts grow out of new facts, but few are aware of the extent to which the meaning of new facts depends upon the concepts employed to interpret them. It is only through a good theory that we can appreciate the harmony of the subatomic world: a description of the experimental arrangement alone would be dreadfully boring and featureless. Theory makes a picture of nature not only coherent but also aesthetically appealling. "Only ideas make an experimentalist into a physicist, a chronologist into a historian, a script scholar into a philologist," wrote Planck.

Theory is an intuitive penetration to the core of observable things. It makes it possible to describe those properties of phenomena that lie on the yonder side of our consciousness and sense experience, and explain using these properties the apparent complexity of phenomena by their latent simplicity. This novel thinking, exercised by such geniuses as Dalton and Bohr, created modern atomism.

The interlacing facts, concepts, equations, and images of science are very hard, or perhaps even impossible, to untangle. All attempts of this kind inevitably run into the sacramental question "Which came first, the chicken or the egg?" Nobody will ever find out the first scientific fact and the first scientific concept that started modern science. Therefore, more often than not, instead of the "explanation of nature" natural scientists speak about the "description of nature".

"We now realize, better than previous natural science did, that there is no reliable starting point, from which paths originate to all the domains of our cognition, but that all cognition is obliged, to a certain extent, to hover above the fathomless pit. We always have to start somewhere in the middle and, in discussing reality, apply concepts that only gradually acquire a definite meaning owing to their application..." These words of Heisenberg's will appeal to any physicist. Einstein would often repeat that "the only mystery of the world is its cognizability".

Physical reality is a very profound conception and, like all profound notions in our language, it is not restricted to one meaning. This is a primary entity and it cannot be logically defined strictly enough through simpler concepts. We must simply accept it after giving it the meaning deduced from our whole previous life and the knowledge we have accumulated. Evidently, this meaning will change as science advances, and so will the meaning of the concept of an "atom".

Science has changed the understanding of reality beyond recognition, and the reality of man in the twentieth century is as far from that on the ancient Greeks as the atom of today is from that of Democritus. The decisive touches to the new picture of physical reality were made by quantum mechanics. This is perhaps the main reason behind the popular desire to perceive "what quantum mechanics is". This endeavour is, as a rule, more deep-rooted than a quite natural professional interest. A knowledge of quantum mechanics equips a person with much more than special skills enabling him to design a laser or a nuclear reactor. Exposure to quantum mechanics is a kind of emotional process that causes one to retrace its whole history. An illogical process, it is strictly individual and leaves ineradicable traces in the mind. This abstract knowledge, once acquired, irreversibly influences the entire life of a person. It influences his attitude towards physics, towards other sciences and even his moral criteria. It is quite probable that a person is influenced in the same manner when he learns to appreciate music.

The previous chapters of the book have only acquainted you with the first notes of quantum mechanics and, at most, several sonorous chords. Only a musician, of course, can appreciate the depth of a musical conception, and only a physicist can derive an aesthetic satisfaction from the beauty of formulas. Those of you who are looking forward to devoting your lives to science may understand this some day. If, nevertheless, you have, without going into the details of its "laws of harmony", perceived the beauty of the "melody" in quantum mechanics, the task of our narrative has been accomplished.

Quantumalia

In Search of the Last Concepts

Uncertainty and expectation are the joys of life.

William Congreve

At the 1927 Solvay Congress, the one at which Bohr in his report presented quantum mechanics as a new comprehensive theory of atomic phenomena, Lorentz said: "For me the electron is a particle that at each given instant of time is located at a certain point of space; and if I imagine that this particle at the next moment is at another point, then I will have to visualize its trajectory as a line in space... I would like to retain this previous scientific ideal — to describe all things occurring in the world using clear images."

Lorentz, an outstanding Dutch physicist at the turn of century, expressed the common frame of mind of the scientists of his day. As such this thinking is quite easy to understand — any new theory inevitably must overcome established stereotypes of thought. On the other hand, it is surprising that so many outstanding physicists, even the founders of quantum mechanics, put in question its basics and completeness. Among them were Planck, Einstein, Schrödinger, de Broglie, Laue, and Landé... Over the years they were becoming ever more entrenched in their doubts despite all the spectacular triumphs of quantum mechanics. (This was perhaps one reason why Max Born was not awarded a Nobel Prize until 1954 — twenty-eight years after his famous work on the statistical interpretation of the wave function.)

In the summer of 1926 Einstein wrote to Besseau: "My attitude to quantum mechanics is a mixture of admiration and distrust." Already in the following year, at the Solvay Congress, his attitude had become quite well established and over the years it became ever more irreconcilable.

On 31 May 1928 he wrote to Schrödinger: "The Heisenberg-Bohr philosophy (or religion?) of reassurance is so subtly engineered that for just so long it supplies a believer with a soft pillow from which he is not so easily aroused. Let him sleep..."

On 7 November 1944 he wrote to Bohr: "The grand initial triumph of the quantum theory cannot make me believe in the game of dice that lies behind it all." (Till his last days Einstein used to reiterate *Gott würfelt nicht!* — "God does not play dice".)

Three years before his death, on 12 December 1951, he wrote to Besseau: "All of these fifty years of incessant pondering brought me not an inch nearer an answer to the question of what light quanta are. These days any youngster imagines that he knows this. But he is quite mistaken..."

Einstein was not alone in his doubts.

"Quantum mechanics cannot be regarded as totally complete," wrote Planck in 1941.

Words are like leaves;
And where they most abound,
Much fruit of sense
Beneath is rarely found.

Alexander Pope

Laue, as recently as the early 1930s, thought of Bohr's interpretation of the fundamental principles of quantum mechanics as a "bad palliative", and in April 1950 he wrote to Einstein: "You and Schrödinger are the only ones among famous contemporaries who in this respect are my fellow believers." Schrödinger agreed with him and when told about the successes of quantum mechanics and its universal recognition, would retort: "Since when is the correctness of a scientific conclusion decided by a majority of voices?" In this vein he wrote to Born in 1950.

Remarkably enough, all of these statements contain some reference to faith. Einstein and Schrödinger, Planck and Laue — they all recognized the might of quantum mechanics but *did not believe* in its completeness, although all their attempts to *prove* that it is incomplete and inconsistent were failures.

Their stand took much courage: the Copenhagen interpretation fairly quickly became a dogma and any criticisms of it might cost a physicist his professional reputation. And yet disputes on quantum physics are still going on and there are even several specialized journals devoted exclusively to the interpretation of quantum mechanics.

Embittered and irreconcilable fights, these disputes sometimes resemble feuds between sects within one religion. No one among the disputants puts in question the existence of the god of quantum mechanics, but each prays to his own god and only to his own. And, as always in religious disputes, logic is of no use, because the other side is deaf to it: there is a primary emotional barrier, an article of faith, which baffles all the impeccable proofs of the other side without ever allowing them to sink to the level of consciousness.

The doubts of physicists about the fundamental points of quantum mechanics did not exactly strengthen confidence in it by the general public. But the task of any true scientist is not to establish by hook or by crook his views and authority by fair means or foul, but to quest for truth and to accept it, even if it is in sharp contrast with his *a priori* attitudes.

When you have no basis for argument, abuse the plaintiff.

Cicero

What then lies behind this interminable dispute? It is akin to attempts to seek out the principal truth and the last concept from which all the rest follow. However diverse are the doubts of the opponents of the orthodox theory and however subtle are the paradoxes they discuss, their objections come down to the negation of the probabilistic interpretation of quantum mechanics and the concept of the "state of a physical system" adopted in it.

It is well known, for instance, that each radioactive element has its characteristic property — its half-life, i.e., the time during which

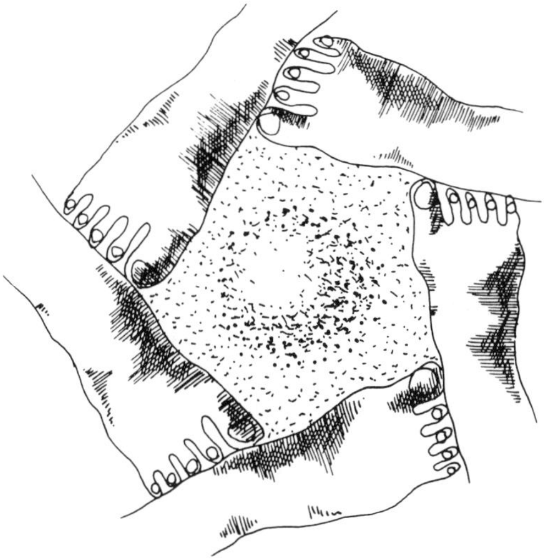

The more success the quantum theory has, the sillier it looks.
Albert Einstein

a half of the available nuclei decay. This fact is accepted by all — particularly since it is easily verified. But the advocates of today's quantum mechanics hold that the half-life at the same time characterizes each individual nucleus, just as 1/2 is the probability that an individual tossing of a coin will give heads. The opponents disagree and appeal to common sense: the time each nucleus decays, although random, is still quite definite, individual and different from the average lifetime of the nucleus. And so this lifetime is no more than a convenient fiction, one that has nothing to do with physical reality. They are answered by the argument that quantum mechanics forbids using concepts that correspond to unobservable properties, such as the lifetime of an individual nucleus. But the opponents turn this explanation down: they regard it as going against their common sense.

One of the most frequently debated effects is the diffraction of electrons fired through two closely spaced slits. Both sides agree that the trace on the photographic plate can only be produced by an electron particle. But then — to be logical — it will have to pass only through one of the slits, i.e., it is impossible to get an interference pattern, since it results from a wave passing through the two slits at the same time. The "traditionalists" in quantum mechanics recall the wave-particle duality and complementary types of devices. They view the two slits as a device that brings out the wave-like aspect of the electron, and the photographic plate as a device that brings out its particle aspect.

M. von Laue

The other school is not happy with this explanation, since quantum theory in principle does not enable one to observe how the transition from the particle picture to the wave picture takes place. When told that this is a purely statistical process, governed by the laws of probability, they answer in the words of Sir Karl Popper, who wrote that Heisenberg tries "to give a causal explanation of the impossibility of causal explanations". They lean on the authority of Laue (who did not accept Heisenberg's uncertainty principle for "it sets a limit to the search for more profound causes") and the views of Einstein, who always insisted that probability is just our way of representing experimental data, and not an inherent property of quantum systems.

The many opponents of quantum theory cannot yet reconcile themselves with the fact that within the framework of quantum mechanics all questions about the true characteristics of individual quantum objects and unobserved effects are strictly forbidden. To overcome this constraint many attempts have been made to introduce into the theory the so-called *hidden parameters,* which ostensibly can describe in detail the "true" properties of objects, obliterated subsequently by our averaging over the parameters introduced. (Such a possibility as well as the term "hidden parameters" were discussed by Born as far back as 1926.) All these attempts, however, have remained futile so far and contributed nothing to quantum theory.

Fanaticism consists in redoubling your efforts when you have forgotten your aim.

George Samtayana

Disputes about the interpretation of quantum mechanics rage on and on: the pride of man and his faith in the omnipotence of reason have difficulty in accepting the frontiers of knowledge discovered by man himself. But not everyone takes these disputes seriously.

Characteristically, none of the opponents denies the fact that the conclusions of quantum mechanics are constructive and true within the region of the applicability of the science. Bohr was aware of that weakness in the dissidents' stand and with his famous mild humour he liked to tell the story of his neighbour at Tisville. The neighbour had a horseshoe nailed on the door. Somebody asked him once whether he really believed that it brought happiness to a home. "No, of course not," answered the neighbour, "but it is said to help even those who don't believe in it."

Well, "man does not live by bread alone" — and as long as disinterested doubts are alive this dispute cannot be considered finished. Admittedly, it will never change the planks of existing theory but will perhaps make easier the search for new paths and the understanding of newly discovered phenomena.

3. Results

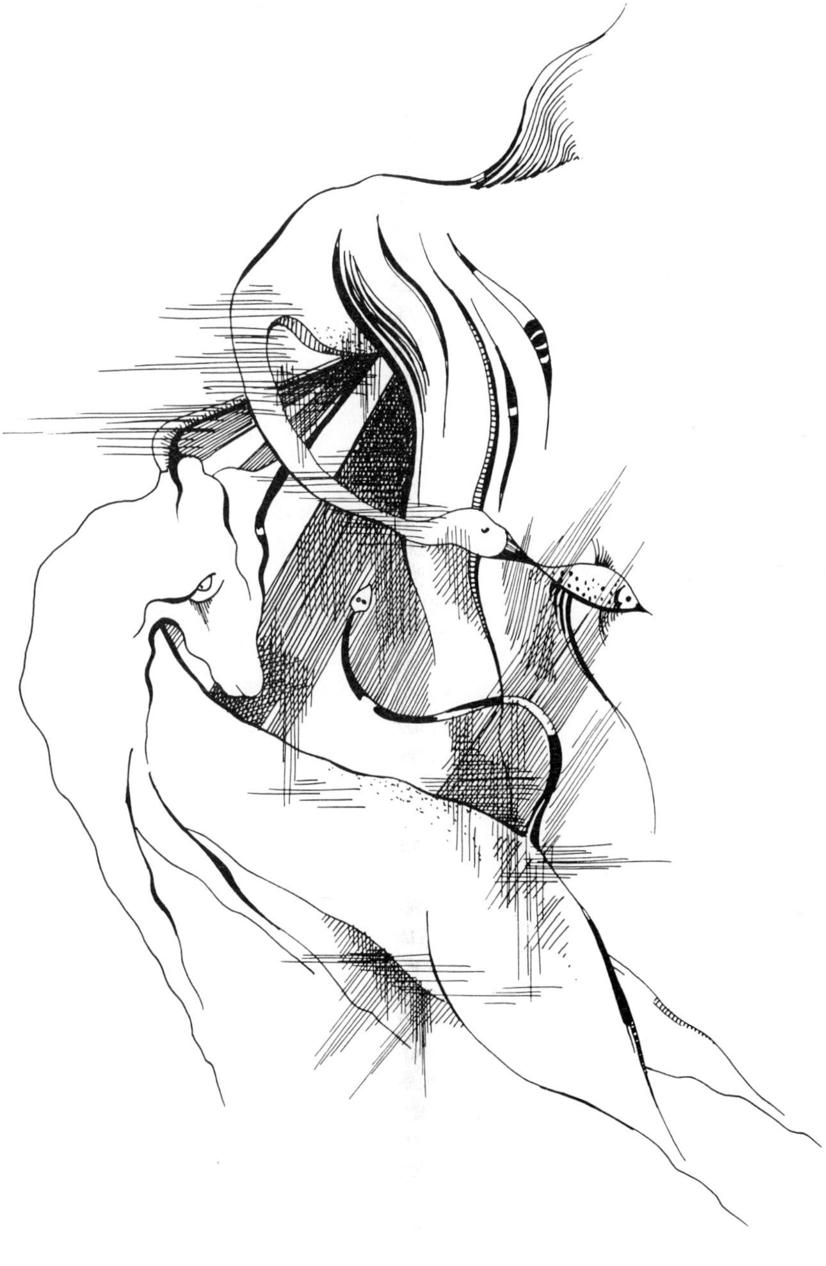

They who know the truth are not equal to those who love it, and they who love it are not equal to those who find pleasure in it.

Chinese proverb

Reality provides us with facts so romantic that imagination itself could add nothing to them.

Jules Verne

Kui, Chinese

Chapter Twelve

Wilhelm Konrad Röntgen • Antoine Henri Becquerel
Pierre and Marie Curie
Ernest Rutherford and Frederick Soddy
The Energy of Radium

Without fire there would be no man, man who has split the atom, landed on the Moon and computed the path of the Earth among the stars. In a century or two the fire, which has warmed humanity for millenia, will go out: fossil fuels on Earth are running out, a fact most people have not yet fully comprehended.

Our hope in the foreseeable future is atomic, or rather nuclear, power. Less than five decades have passed since the time man mastered the energy of the nucleus — we are all contemporaries of that breakthrough. Considering that human life is so short, it is for us a rare and unique occurrence because, since the time fire was first tamed human history has seen nothing of more importance and significance than the discovery of "atomic fire". Therefore, the mastering of atomic energy is not just another episode in the long line of scientific discoveries, but rather a revolution in the development of our civilization.

Whether we like it or not, fear it or not — the fate of mankind lies with the disintegrating nucleus: either we will put it to work or it will put us to death — no other alternative. This dilemma is quite real and all too important for its solution to be entrusted to narrow specialists only — be it atomic scientists, generals, or politicians. And so *each* human being, just as he understands the nature of fire and lightning, should have a proper understanding of his *own* of the principles of the physical processes on which his life is now essentially dependent.

Wilhelm Konrad Röntgen

Traditionally the coming of the atomic era is associated with the thunder of the first atomic explosion. Well, the blare of a brass band is loud and impressive, but music started not with that band; it started with a simple tune produced with a single string.

An amazing photographic picture still exists: in a room with a large window, at the wall at the right stands a laboratory desk with devices; at the wall on the left is a high cabinet with chemicals; the room is flooded with sunlight from the window, and the window looks out on a yard leading to an old park. This is the laboratory of Wilhelm Konrad Röntgen (1845-1923) at the University of Würzburg. On Christmas Eve of 1895 (just imagine — no electricity yet in today's sense of the word; no radio or cars!) in this very room for the first time man looked into the deepest recesses of matter, into the atom (the nucleus is still unheard of, the electron would be discovered in two years' time, the concepts of the "quantum" would make its appearance five years later).

When you look at this peaceful picture you can hardly believe that in a bare fifty years the impartial logic of scientific research would relentlessly lead humanity from this room to the Alamogordo nuclear testing site and the ruins of Hiroshima and Nagasaki. Nor could this be known to the Royal Swedish Academy of Sciences, but still they made Röntgen in 1901 the first Nobel Prize laureate. The future proved that the decision of the Academy was right: the work of Röntgen heralded an era of brilliant discoveries, which Rutherford dubbed the heroic period in the history of physics.

By the time Röntgen made his discovery he was fifty. He was leading the equable life of a German professor and was noted for his austere judgements and independent views. He was a pupil of Rudolf Clausius, and also of the eminent German experimental physicist August Adolph Kundt; among the students of the latter were the Russian physicists Pyotr Nikolaevich Lebedev and Boris Borisovich Golitsyn. By 1895 Röntgen had already published fifty papers, and his experimental skills had enjoyed wide recognition among his colleagues.

By the time of Röntgen the famous tube of Geissler (or Plücker, Hittorf, Goldstein, or Crookes) had been known for more than forty years and was employed by the outstanding experimentalists of the nineteenth century, and still the nature of cathode rays remained unclear. The Crookes tube could be encountered in nearly any laboratory, and each researcher introduced into it some modifications in order that he might test his conjecture or hypothesis. Quite recently, in the spring of 1895, Jean Perrin focused cathode rays within a "Faraday cylinder" and proved that they are negatively charged; in that same year von Lenard let the rays come out of the tube and measured their free path in air — another step, it seemed, and the nature of cathode rays would be unravelled. Their enigma was the talk of the town in those days and Röntgen too was not indifferent to it. He decided to repeat some of von Lenard's experiments.

Like hundreds of experimenters before him, Röntgen could observe a beautiful yellow-green glow at the point where the cathode rays struck the wall of the tube; he watched the spot shift in a magnetic field, and so on. Things remained unchanged till that memorable evening of the 8th of November 1895, when Röntgen noticed, much to his surprise, the glow of a piece of paper covered with a layer of fluorescent barium salt. The paper was located far away from a Crookes's tube that was in operation; moreover, at that time the tube was covered with an opaque cardboard casing. Röntgen did not leave this chance observation unheeded. A mature and experienced researcher, he immediately understood the importance of his discovery. Five weeks of back-breaking labour followed, during which time he had his food brought into the laboratory and even slept there.

By the end of December Röntgen already knew all the major characteristics of the novel kind of rays he had discovered (he called them X rays), including their value for medicine. (Later, when asked by a reporter "What were you thinking when you saw the glow of the fluorescent screen?" he would answer in the roughish manner characteristic of him, "I was investigating, not thinking"!) The day of the 22nd of December 1895 may be regarded as the birthday of fluorography; the picture of Mrs Röntgen's left hand taken on that day would be included in all later books on X rays. On 28 December Röntgen reported on his findings to the German Physical Society and sent a paper describing the properties of X rays to a scientific journal (it was published by 6 January 1896). Furthermore, following the practice of the day he wrote a letter to the French Academy of Sciences.

Röntgen's discovery produced quite a stir both among the men of science and among the general public. Suffice it to say that within a few weeks Röntgen's paper was issued five times in booklet form and translated into several languages. The year 1896 alone saw the publication of more than 1000 scientific papers and about 50 books devoted to X rays. The medical profession started to use X rays within a few weeks. Immediately thereafter F rays, N rays, Blondlot rays, and so on were "discovered".

Newspapers were quick to pick up and spread the sensation, and before long one British firm was already advertising underwear protecting against X rays, and the Senate of one of the states in America was considering a draft bill to prohibit the use of X rays in opera glasses. Already in mid-January Röntgen himself was invited to Kaiser's court to demonstrate his rays. Public demonstrations of the new phenomenon were performed all over the place, people thronged to them as to circuses, the sight of human bones on the screen causing fainting and hysteria. In May 1896 in New York the celebrated inventor Edison constructed a demonstration apparatus that enabled each member of the Public to see the shadow cast by the bones of his hand. (The demonstrations ended tragically: Edison's demonstrator died of heavy burns. It is quite probable that Röntgen's own work with X rays was not without cost, because a quarter of a century later he died of cancer.)

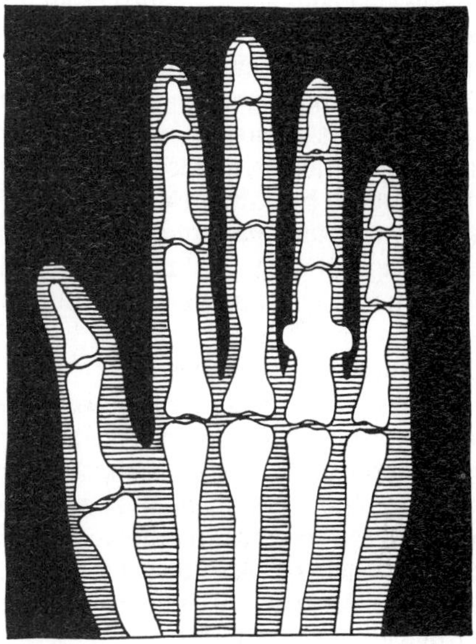

Arguments flared up about priority. An aggressive claimant to the discovery of X rays was von Lenard (he regarded Röntgen as his enemy to his last day). Then an X ray picture was found that had been obtained in the United States five years before Röntgen; Crookes was also remembered — he complained that photographic plates darkened when he left them near an operating tube. But, as always in disputes like these, the claimants forgot that what they had spoken and thought at the time looked quite different after Röntgen had made his discovery.

Antoine Henri Becquerel

The news of the discovery of these mysterious X rays, which pass freely through all objects and even enable one to observe things within the human body, spread like wildfire. By 20 January 1896 Henri Poincaré read Röntgen's letter at a regular session of the French Academy of Sciences, which for once was quite well attended. A member of the Academy, Henri Becquerel (1852-1908), was also present; he sat in the front row and listened attentively to the report. After the letter had been read he asked Poincaré about the latter's views on the nature of X rays. Poincaré replied that, in all probability, they emerged from within the bright fluorescent spot that is formed where the cathode rays strike the glass wall of a Crookes's tube.

Luminescence (and its special case, fluorescence) occurs widely in nature. Examples are the glow of a TV screen or rotten wood, the shimmering of glow-warms on a southern night, and an aurora borealis in the polar night. Interest in luminescence had arisen naturally at least 300 years before Röntgen's work.

For Becquerel all things concerned with fluorescence were a family business: this effect was studied by his father Alexandre Edmond Becquerel (1820-1891) and grandfather Antoine César Becquerel (1788-1878), and Henri maintained his family tradition, having devoted his life to fluorescence. (Later he would carry on another family tradition: in 1908, shortly before his death, he would be elected president of the Academie des Sciences.)

In the Museum National d'Histoire Naturelle where the Becquerels held professorships was a beautiful collection of fluorescent minerals gathered throughout the three generations of the Becquerels, and so on the morrow after the meeting of the Academy Henri Becquerel embarked on his research effort. (To this end he needed photographic plates, whose preparation at the time required much skill. Here, too, he succeeded thanks to the family tradition: his father had long been interested in applications of photography to astronomical studies. Specifically, he was the first to produce colour photographs of the solar spectrum.) Becquerel's reasoning was simple and consistent: X rays issue from the area of a glass vacuum tube struck by a beam of cathode rays; it is not unlikely that they are somehow related to fluorescence; accordingly, it makes sense to see whether or not X rays are given out by fluorescent materials.

His experimental arrangement was exceedingly simple as well. You take a photographic plate wrapped in opaque black paper and place a mineral on it, then expose this to sunlight, and a short while later develop the plate. If X rays are really produced in the process of fluorescence of the mineral, the plate on development will show the silhouette of the mineral sample — the black paper will not stop the X rays.

For some reason from his rich collections of minerals Becquerel took for his experiment a fairly rare salt of uranium (perhaps because his father devoted many years to the study of that salt, or perhaps because about fifteen years before he had prepared the sample himself). Nature lived up to his expectations: he exposed his experimental arrangement to sunlight on a balcony for several hours, and the plate, upon development, showed outlines of the uranium salt crystals.

On 24 February Becquerel reported at the Academy on the results of his first trials, which seem to have supported Poincaré's hypothesis. The report evoked much sympathy and interest and it was decided that at the next meeting, on 2 March, Becquerel would inform the Academy about his new findings. As luck would have it, on 26 February the weather broke up and for several days the Paris skies were overcast. Becquerel was vexed but there was nothing to be done, and the wrapped plates with uranium salt crystals on them, which he had prepared for the experiments, lay for three days in the drawer of his desk.

On Sunday, 1 March 1896, Becquerel came to the laboratory. The weather in Paris was still inclement. On the morrow he had to make his report at the Academy, but there was decidedly nothing to be reported. For this reason or otherwise, he decided to develop at least those photographic plates that were in his desk, awaiting experiments. Imagine his feelings when he found clear outlines of the crystals, much sharper than when he had exposed the crystal for hours to sunshine? Most certainly those feelings were utter surprise and the fever of a discoverer.

Each major discovery is made up of important, often tiresome trifles, and only after much preliminary work has been done in the dark and all the details have been systematically examined, can a happy flare of chance momentarily illuminate the contours of the whole. But it takes a true naturalist to unravel within that split second nature's design. The details of the picture that unexpectedly occurred to him can then be filled in easily, a task within the competence of any sufficiently skilled investigator.

Some sticklers among science historians would later unearth the fact that thirty years before Becquerel two researchers, N. de Sent-Victor and L. Arnodon had observed the darkening of photographic plates exposed to uranium salts. But neither they nor three other contemporaries of Becquerel, who hastily cried out that they had observed X rays emitted by fluorescent minerals, became pioneers of the new natural phenomenon. What prompted Becquerel to develop the unexposed photographic plates? The reasons may be different, insignificant, or purely accidental; they might have occurred to any researcher. What was not accidental, however, was that Becquerel made the correct inferences from the stunning and unclear facts. He was absolutely consciously in search of his discovery and, most important, he was prepared for it.

On 2 March 1896 at the Academy, Becquerel made a short report on his experimental findings, and within another half year he had formed a really compete picture of the phenomenon. Becquerel found that:

- photographic plates are affected in the manner just described only by those minerals in his collection that contain uranium;
- the effect does not depend on the species of mineral, only on its uranium content;
- the effect is in no way associated with luminescence; pure metallic uranium, which does not possess this property, exerts the same (or even stronger) action on a plate as its fluorescent salts.

On 23 March 1896 Becquerel reappeared before the Academy with his progress report. His new findings suggested that uranium emits some unknown rays (they were quickly christened as "uranium" or "Becquerel" rays), which like the X rays of Röntgen affect a photographic plate and ionize air. So was discovered the fascinating phenomenon, which Marie Sklodowska Curie in 1898 would call *radioactivity*.

Pierre and Marie Curie

The life of Marie Sklodowska Curie (1867-1934) has been described in many superb books. Even during her lifetime she was elected an honorary member of 106 academies and scientific societies, and recent polls indicated that even now she still remains the most respected woman in the world. Her life was not rich in spectacular external manifestations — she was simple and austere, like a pure harmonic tone, and she was totally, every inch of her, devoted to the service of science. For Marie Curie science was not a means and even not an end in itself, it was rather a way of life. She often said that for her to come into science was like taking the veil.

Marie Sklodowska was twenty-four years of age when she came from Poland to Paris and entered the Sorbonne, one of the few places in the world at that time where a woman could receive formal scientific education. Here she met Pierre Curie (1859-1906), a man of rare talents, integrity and nobility, with whom she threw in her lot and made her main scientific discoveries. She gave birth to two daughters. One of them, Irene, would carry on the cause of her parents' life.

In the spring of 1896 Marie Curie was looking for her thesis subject ("Choosing the topic of one's first scientific study is like first love; it is for life," she would say half jokingly). She became intrigued by the first reports about Becquerel's findings, especially by the ability of "uranium rays" to ionize air. That was no idle interest: shortly before that her husband Pierre Curie and his brother Jacques had invented a handy and sensitive electrometer that relied on the piezoelectric effect (discovered by them). An electrometer was much simpler in use than photographic plates; besides it not only detected the presence of a radiation but also measured its intensity with reasonable accuracy. It was precisely that quantitative approach to radioactivity that enabled Marie Curie to go further than others.

Marie Curie's first aspirations were not that ambitious: she wanted to find answers to simple questions: "Is it uranium alone that emits the new rays? And, if so, what makes it an element apart from others?" By that time uranium had already been known for more than a hundred years and had distinguished itself with nothing — it was just another metal; at the time it was rarely used, and then mainly as a stain for glass and ceramics of yellow-greenish colour. Marie Curie used the electrometer painstakingly to test for radioactivity essentially all of the then known elements (more than eighty) and soon found that among them only thorium had that property, to an even greater extent than uranium. (At the same time and independently this fact was established by the German chemist Gerhard Carl Schmidt (1865-1949).) This was an exceedingly important result because it immediately removed the question of the exceptional position of uranium: if two radioactive elements can exist, why couldn't there be more of them?

After a short break in work (she was expecting), and two months after she had given birth to Irène, in December 1897, Marie returned to work. Among the many chemical substances and minerals, her attention was drawn to the mineral pitchblende from Czechia, where uranium was extracted at that time. Pitchblende appeared to be four times as radioactive as the uranium it contained. This was quite an unexpected development since chemical analysis showed that the mineral contained no thorium. Then Marie Curie hypothesized — an audacious hypothesis — that pitchblende contains some hitherto unknown radioactive element in amounts that are not amenable to chemical detection. For this to be the case, its activity should be thousands of times higher than that of uranium, which made up about 30 per cent by weight of the ore.

The first entry in Marie's laboratory log appeared on 16 December 1897. In March 1898 Pierre joined her in the work. Already on 12 May 1898 they were pretty confident that they had discovered a new element, which would later be called "radium", meaning "ray". In July of the same year they found in the ore tailings another radioactive element, which they called polonium in honour of Marie's native country. At last, on 25 December 1898 they announced their discoveries at the Academie des Sciences. By that time the Curies could already demonstrate to the audience a radium preparation, which was 900 times more active than uranium.

From then on Marie Curie concentrated on one task: to isolate pure radium. But how? Without a decent laboratory, without assistants, without ore (which was fairly expensive)? But where there's a will there's a way, especially if it is linked to a readiness to persevere.

With the assistance of Eduard Süss, the then president of the Austrian academy of sciences, the Austrian government agreed to present the Curies with a tonne of uranium pitchblende. They found a suitable barn to store the ore, and also another 10 tonnes acquired for them by the multimillionaire Baron de Rothschild. For Marie began years of strenuous, monotonous, and exhausting work: day in, day out, year in, year out to dissolve, evaporate, and again dissolve. She had processed 11 tonnes of ore by hand and carried out several thousand crystallizations alone. Later Marie Curie remembered: "...The discovery of radium was made under miserable conditions: a barn where this happened now became a legend. But this romantic detail was no advantage: it had devoured our strength and impeded the coming of the discovery..."

It was dirty and heavy work, rain or shine, in an old barn, without any protective measures: a radioactive radiation counter even now clicks menacingly when a page from the laboratory log of those years is brought up to it. "We were aware of the fact that our health was deteriorating, that we were subjecting it to heavy trials. As it happens with all who know the price of life together we were sometimes overcome by fears of the irretrievable. Then some feeling, maybe simply courage, would lead Pierre invariably to the same conclusion: even though we may seem to be heartless beings, we still have to work," wrote Marie Curie many years later.

Both of them passed away prematurely: Pierre in 1906 was killed by a dray on a Paris street. Marie died in 1934 of leukeamia caused by radioactive radiation.

By 1902 Marie Curie had isolated from a tonne of ore several tenths of a gramme of the concentrated radium preparation; in another three years she had 0.4 gramme of pure radium chloride, and only in 1910, twelve years since the beginning of her work, was her dream fulfilled — at long last, she saw a silvery drop of pure metallic radium weighing just 0.0085 gramme. But the drop was 3 million times more radioactive than the same drop of uranium.

The scientific feat of labour of Pierre and Marie Curie was recognized throughout the world even in their lifetimes. In 1903 they, together with Becquerel, shared the Nobel Prize for Physics; in 1911, when Pierre was no more, the Royal Swedish Academy of Sciences awarded Marie Curie a second Nobel Prize (now for chemistry), an honour bestowed upon only three researchers throughout the entire history of Nobel Prizes.

Ernest Rutherford and Frederick Soddy

The picture given on the next page first appeared in 1903 in the doctoral thesis of Marie Curie. Now it is provided in all school textbooks throughout the world and every schoolboy can explain that radioactive materials emit three types of rays, which Rutherford called alpha (α), beta (β), and gamma (γ) rays.

The emission of beta rays was discovered back in 1898 by Becquerel; it was later shown that their properties coincide with those of cathode rays, i.e., they constitute a beam of fast electrons (the electron was discovered in the previous year, a most timely development). Two years later the French scientist Paul Villard (1860-1934) established that another component of "uranium rays" are gamma rays, whose properties turned out to be similar to the X rays of Röntgen. In January 1899 Rutherford found the third component — alpha rays — whose nature was unknown at the time. Nowadays even schoolchildren know that alpha particles are "simply" nuclei of helium, but proving this in those days, even for such figures as Rutherford, Soddy, and Ramsay, took more than five years of strenuous work.

What made the task so complex? Above all, such a concept as the "nucleus of the atom" was unknown at the time; it would come eleven years later. And although the electron had been known for two years already, the very existence of atoms had not been strictly proved — Perrin's experiments would be staged nine years later. Some reverberations from those days we feel even now when we attempt to present the history of discoveries in the field of radioactivity. Indeed, as we are now well aware, all of these phenomena come under the heading of nuclear physics, but we have to narrate them taking especial care to avoid using the word "nucleus".

When radium was discovered Ernest Rutherford was a doctoral student of J.J. Thomson at the Cavendish Laboratory. After having learned about Becquerel's and Curie's discoveries he gave up his studies of the ionization of gases and by the autumn of 1898 had already accomplished a major work on the radioactivity of uranium. Before long he moved to Canada and was appointed professor of physics at McGill University, Montreal, where he, with his characteristic energy and imagination, set about a comprehensive investigation into the new phenomenon. Unlike Marie Curie, who concentrated on the chemical isolation of radium in pure form, Rutherford was above all interested in the physical nature of radioactivity.

What is the essence of radioactivity? What are its workings? What is its true cause? These were some of the questions that most intrigued Rutherford.

He started by probing into the behaviour of alpha particles.

Within three years Rutherford was almost certain that alpha particles were just doubly ionized helium atoms. He used a simple and well-known fact: all the uranium and thorium compounds were found to contain large amounts of helium. So, by roasting 1 gramme of thorium one can obtain about 10 cubic centimetres of helium, one hundred times the volume of the source thorium. Helium, which had been found on Earth seven years previously (in thorium minerals, by the way), was by 1902 well enough studied, and so it was known to belong to the group of noble gases and not to enter into any chemical reaction. Therefore, chemistry could not account for the presence of that amount of helium in thorium.

Consequently, helium is formed from radioactive elements. But what then happens to the elements themselves?

Soon after coming to Canada Rutherford made another important observation: thorium compounds emanate some unknown radioactive gas, which he named "emanation". (A year later Pierre and Marie Curie also observed emanation of radium.) Further experiments indicated that the gas was very heavy and that it lost its radioactivity quite swiftly — every four days it was halved.

A new puzzle — by that time radioactivity came to be viewed as an unchangeable characteristic of elements, just like the atomic weight.

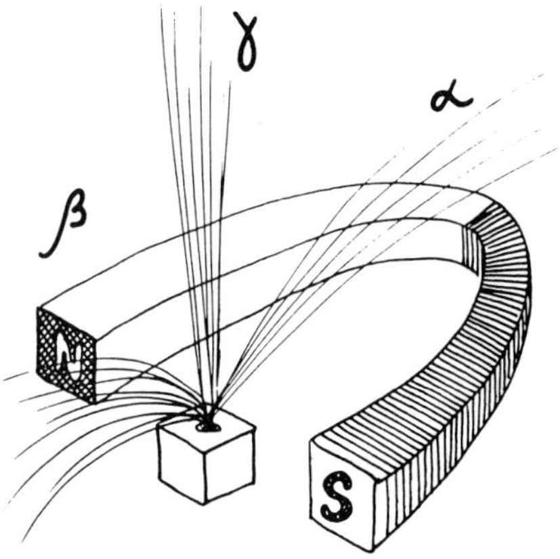

What is wrong with radium? And whence come the two gases — helium and emanation of radium?

In the autumn of 1900 Rutherford was joined by a young gifted chemist, Frederick Soddy (1877-1956). They soon proved that emanation of radium is an inert gas, now known as radon, whose properties are similar to those of all noble gases — helium, neon, argon, krypton, and xenon. (They were discovered by Ramsay together with Rayleigh. Ramsay also surmised in 1901 that in the periodic table they form a special group, the 8th, with zero valence.)

In two years of inspired work (they were young at the time, one was aged twenty-nine, the other twenty-three!) Rutherford and Soddy came to a shocking conclusion:

Radioactivity is nothing but the decomposition of an atom into a charged particle (it appears for us as a radioactive emission) and an atom of another element chemically different from the original one.

The resultant atom may also be radioactive and undergo a further decay. This conjecture, now known as the hypothesis of radioactive decay, appeared to be surprising and audacious. Judge for yourselves — as recently as the end of the century it was accepted (if only by a majority!) that atoms do exist. But all refused to admit the possibility that they could be transmuted, and even more so transmuted spontaneously. This conviction was in line with the age-old tradition of atom-

ists — from Democritus to Newton and Maxwell. Therefore the disintegration hypothesis was a revolution in fundamental views on the structure of matter. Further, all this put one in mind of medieval alchemy with its attempts at the transmutation of elements, and to return to what in an enlightened age would be sheer folly. We can thus understand the fears of the physicists at Montreal that if the crazy idea were published it might do harm to the prestige of a young university.

Nevertheless, what was done was done, what was said was said.

The hypothesis of radioactive decay, like any good hypothesis, had consequences that allowed for experimental verification. The first of these was performed by Soddy and Ramsay in the summer of 1903, soon after Soddy's return from Montreal to London. The idea of their experiment was exceedingly simple: they collected in a thin glass tube the emanation from the 50 milligrams of radium bromide they possessed, and passed through it an electric discharge to observe the characteristic spectrum of the new element. With the passage of time, however, the spectrum weakened, and instead showed more clearly the spectrum of helium — the emanation decayed into helium and radium *A*. The participants and witnesses of that experiment, even years later, could not conceal their emotions when relating the story — so astonished had they been. They had seen one element turn into another — for a physicist and a chemist it was as if a zoologist saw a cat turn into a dog.

A short while later Rutherford would give emanation its present name radon (Rn), and Ramsay would demonstrate prodigous experimental skills — with as little as 0.1 cubic millimetre of radon available he would measure its atomic weight to be 222. The atomic weight of radium (226) was measured by Marie Curie in 1902 and three years previously Ramsay found the atomic weight of helium to be 4. Now the hypothesis of the radioactive decay of radium by the scheme

$$\mathrm{Ra} \rightarrow \mathrm{Rn} + \mathrm{He}$$

could be tested not only qualitatively, by observing spectra, but quantitatively as well. In fact, the atomic weight of radium 226 = 222 + 4 appeared to be equal to the sum of those of radon and helium. The hypothesis of radioactive decay could now be promoted to the status of scientific truth without reservations.

It was found quite soon that radioactive elements decay at a definite rate, which is as inherent a characteristic of a radioelement as its atomic weight. In 1900 Rutherford suggested that this rate be characterized by the radioactive half-life of an element $T_{1/2}$, i.e., by the time in which a half of the initial number of atoms of the radioactive element would disintegrate. For example, the half-life for radon is 3.82 days, for plutonium 138 days, for radium 1600 years, for uranium 4.5 billion years.

Lastly, the hypothesis of radioactive decay explained why radium is always encountered together with uranium — it seems to be the product of its decay. Further, if radium is a decay product of uranium, radium gives rise to radon, and the latter disintegrates further, then there must exist whole *radioactive families* with the first element (radioactive) and the last one (stable). The decade that followed was devoted to the search for these radioactive families, disentangling the sequence of radioactive decays in them, measuring the disintegration rates, and so forth. By 1913 the work was as good as done.

The summarizing article by Rutherford and Soddy was entitled "The Cause and Nature of Radioactivity". Following its publication it would be safe to say that the nature of radioactivity was well established. But the reason why atoms of radioactive materials explode spontaneously would only be clarified a quarter of a century later, after quantum mechanics had been constructed. Another decade would have to pass before the nature of X rays would be absolutely clear, those X rays whose discovery heralded the beginning of radioactivity research.

The Energy of Radium

Soon after they started their research work, Pierre and Marie Curie noticed that vials with radium concentrates glow in the dark with a soft bluish light (radium owes its name to this radiation). "Here's the light of the future!" said Pierre to his friends, without in the least suspecting how right he was. Even then it was understood that the observed glow is due to the fluorescence caused by radium's radiation in the glass. But unlike common fluorescence, which decays fast after the material has been irradiated, the radium preparations kept on glowing without apparent weakening for years. What is more, radium compounds were always slightly warmer than the surrounding objects. This all suggested that radium continuously emits energy.

Early in 1903 Pierre Curie and Albert Labourd measured the amount of heat given off; it appeared that 1 gramme of radium liberates 100 calories in an hour (according to the latest data, 135 calories), i.e., enough to bring 2 grammes of water to the boil or to melt 1 gramme of ice. Rutherford, who measured the ionization of gases by radioactivity, reached the same conclusion, i.e., that radium emits energy constantly. This energy is very large; it can be readily worked out that annually 1 gramme of radium releases more than 1 million calories, the heat produced by burning 170 grammes of coal; and the total decomposition of 1 gramme of radium liberates the enormous energy of 4 billion calories, a heat yielded by 0.5 tonne of coal.

Where does this huge amount of energy come from? For a quarter of a century scientists were racking their brains over this question, all in vain. Opinions were divided: Crookes, Pierre and Marie Curie, and some other scientists, were inclined to believe that radium atoms work as energy transformers, i.e., they at first absorb the energy of waves of unknown nature that permeate the entire universe, like the ether, and then re-emit the accumulated energy. Others likened the ejection of alpha particles to the process of evaporation of molecules. But then the particles should have different energies, and William Bragg in 1904 definitely established that this is not so: all the alpha particles coming from radium have the same energy. Rutherford insisted on the subatomic origin of radium's energy and was shown to be absolutely right by later research.

Quantumalia

In the fields of observation, chance favours only the prepared mind.

Louis Pasteur

X ray Waves

The diffraction and interference of X rays was discovered in 1912 at Munich, and it was no accident. At the time the director of the Physics Institute at Munich was Röntgen, the professor of theoretical physics there was Sommerfeld, and Laue worked under him as a *privatdozent* . They were all interested in X rays and believed in their wave nature, although in different ways. Röntgen himself thought that he had discovered longitudinal oscillations of the ether,

W. Röntgen

There are two ways to slide easily through life: to believe everything or to doubt everything; both ways save us from thinking.

Alfred Korzybski

M. Curie

P. Curie

similar to sound. Sommerfeld contended that X rays are excited when electrons are decelerated sharply, and he even estimated their wavelength from that model.

In February 1912 Sommerfeld's assistant Peter Paul Ewald (1888-1985) was handling the problem of the scattering of light waves from a spatial lattice. He approached Laue for help. Laue did not know the answer, but while discussing the problem it occurred to him to pass X rays through a crystal — not a novel idea, since many, including Röntgen himself, had undertaken similar experiments. But Laue did not simply suppose, he *predicted.* He hypothesized that if X rays are very short waves, and crystals are really ordered arrays of atoms with spacings comparable with the wavelength of X rays, then if we pass the rays through the crystal, they will have to diffract and interfere. These two suppositions, however, were not that obvious at the time and many (including Planck and Sommerfeld) regarded them as an "ingenious but still fantastic combination of ideas".

Young physicists, who learned about that assumption of Laue at Café Lutz where they gathered on Wednesdays, were more optimistic. One of them, Sommerfeld's research assistant Walter Friedrich (1883-1968), rushed to test Laue's hypothesis. Sommerfeld was at first sceptical of his enthusiasm (he entrusted him with other work), but Friedrich devoted his evenings to this problem and, before long, in April 1912, he was joined by Röntgen's doctoral student Paul Knipping (1883-1935). Together, they fairly quickly uncovered the phenomenon, which was predicted by Laue and which for nearly two decades had evaded the attention of other investigators. Within two weeks from the time that the first photographic evidence became available Laue completed his theoretical treatment of the findings and on 8 June 1912 he reported the results to the German Physical Society. On that day he demonstrated the famous Laue diffraction patterns, which are still invariably provided in all texts on nuclear physics.

The Society reacted immediately and vehemently. Einstein wrote: "This is the most surprising of all that I have ever seen". William and Laurence Bragg immediately constructed their crystal diffraction spectrometer to measure the wavelengths of X rays (which was at once used by Henry Moseley in his famous work) and the Royal Swedish Academy in only a year and a half awarded Laue a Nobel Prize, an exceptionally rare occurrence.

The reaction of his contemporaries may now seem to us to be a bit effusive (remember: Planck received his Nobel Prize in 1918, Einstein only in 1921), but this is perhaps because Laue's discovery very quickly passed into the ranks of the "obvious". Even Laue himself, at a later date, believed his idea to be so self-evident that he "could never understand the stir it produced in the world of specialists". We should not, however, overlook the fact that this "obviousness" is of the same class as "Columbus's egg" or the astronomical discoveries of Galileo. Hundreds of people before Galileo had held in their hands a spyglass but it had never occurred to them to direct it at the skies. Likewise, in many laboratories at the University of Munich, which for many years had been a recognized centre of crystallographic studies, one could see models of crystals, but common things are less likely to attract fresh attention — it takes an element of genius.

A. Becquerel

There are other factors as well that made Laue's findings extremely popular. In the first place, the discovery was unbelievably timely to supply decisive ammunition to the proponents of the atomic hypothesis — this was the time when Perrin had just concluded his experiments with emulsions, Rutherford had come up with his planetary model of the atom, and Wilson had constructed his famous chamber to observe the motion of charged particles. This avalanche of discoveries made Wilhelm Ostwald write in 1913, "Atoms became visible!"

Finally, Laue's findings at last explained the nature of Röntgen rays, and this alone made him worthy of a prominent place in the history of physics. (At Laue's anniversary in 1939 Planck said that the year 1879 was for science a special year — that year saw the birth of Einstein, Hahn, Laue, and, several months before them, of the inquisitive girl Lise Meitner.)

Ganesa, Indian

Chapter Thirteen

The Chemistry of Radioelements
Isotopes • Uranium Family • Stable Isotopes
Radioactive Decay Energy • Nuclear Binding Energy

The idea that history is a limitless advancement of humanity seems to have originated at the turn of the century. It was at that time that nearly all the inventions that characterize our life today were made: practical electricity, telephone and radio, gramophone and cinema, motor car and aeroplane. World wars, depressions, and revolutions lay ahead, and hardly anybody gave any thought to the fact that the Earth's resources were limited and soon would be depleted.

That time was marked by a tide of popular interest in exact knowledge, and for many people radium became a symbol of the scientific achievements of modern times: people admired radium, people feared radium, and people pinned their hopes on radium. Radioactivity was studied throughout the world by an international community of brilliant scientists: Pierre and Marie Curie in France; Rutherford in Canada; Soddy, Crookes, and Ramsay in England; Bragg in Australia; Fajans, Hahn, and Meitner in Germany; Schweidler in Austria; Boltwood in the United States; Strömholm and Swedberg in Sweden; Antonov and Petrov in Russia. The noisy progress changed but little the traditional silence of scientific laboratories, and scientists themselves hardly suspected that their work might affect the destiny of civilization. Traditionally, they were in quest for truth and had not the foggiest idea of how it would be employed — for good or evil.

The Chemistry of Radioelements

Within fifteen years of the discovery of radioactivity and 10 years of the revelation of its nature, about 30 radioactive elements had been investigated. Names for them were invented in haste, such as uranium X_1, uranium X_2, radium A, B, C,..., G, and so on. Scientists at first did not concern themselves with the relation of the radioactive elements to common chemical elements. By 1913 much evidence had been accumulated:

These elements all decay within periods ranging from millionths of a second to billions of years and in the process they emit alpha particles (helium nuclei), beta particles (fast electrons), and gamma-rays (X rays with exceedingly short wavelengths).

When an element decays and emits alpha or beta particles, its chemical properties are changed.

The energy of radioactive radiations is millions of times higher than the energy of chemical reactions.

The process of radioactive decay cannot be retarded or enhanced: heat and cold, pressure and chemical reactions, electrical and magnetic fields affect it not in the least.

There remained three fundamental questions:

What is the chemical nature of radioelements?
Whence does their energy come?
What is the reason for their disintegration?

The last two questions would be answered by relativity and quantum mechanics. The answer to the first question was soon found by radiochemists themselves.

The main snag with radiochemists was that radioelements, as a rule, could not be isolated in amounts sufficient to carry out standard chemical analysis. Therefore, they were identified by the characteristic radiation they emitted, and distinguished by their inherent half-lives. Gradually, notions became interchanged: they began to use the name radioelement not for a substance with a set of characteristic chemical properties, but rather for an element with a certain half-life. This was a good example of a change in the psychology of researchers — a short while ago they had accepted the hypothesis of radioactive decay as the negation of the very idea of the immutability of chemical elements; now they established the presence of some radioelement in a mixture by using precisely the fact of the decay of its atoms! Whatever the ingenuity and usefulness of this method the final judgement concerning the chemical nature of a radioelement must lie with chemistry. It was by this rule that Marie Curie was guided when she sought to isolate chemically pure radium. When she had achieved her end, she thereby proved that radioactivity in no way changes the chemical properties of the element and, in turn, is independent of them.

The radiochemists of the day were also plagued by the problem that in certain cases radioelements that differed widely in half-life and radiation type could never be separated from one another using the then known chemical means. For example, radium *D* would not separate from lead, radium *C* from bismuth, and "ionium" from thorium. What is more, the radioelements became too numerous, so that it seemed clearly impossible to accommodate thirty elements with various half-lives in the twelve boxes of the periodic table that had remained vacant at that time.

In 1913 several researchers independently came close to the answer. Much more was then known about the atom than at the turn of the century. So in 1909 Perrin reported his early findings; and in that same year Millikan measured the electronic charge; in 1911 Rutherford proved that atomic nuclei exist; the next year Laue published his experimental results. The vintage of 1913 was rich: Bohr's model of the atom, Moseley's work, and van den Broek's hypothesis that the charge on a nucleus equals (in terms of elementary charges) its serial number in the periodic table. This all undoubtedly helped Frederick Soddy to formulate the concept of the *isotope* and to crown thereby the many years of hard labour of radiochemists.

Isotopes

Imagine that we have a handful of coins of various values. To sort them out into stacks according to their value is a piece of cake. And nobody seems to care that, strictly speaking, not all the coins in a stack are identical; they have minted on them the different years of their issue. But if we are only interested in the coins' purchasing power, this detail might as well be overlooked.

Radioactive elements had to be treated in the same manner: chemists had to sort them out according to their chemical properties, whatever their half-lives, and to put them in the appropriate place in the periodic table; some of them had to be placed in the same cell of the table. (As an aside, "isotope" is the Greek for "occupying the same place".)

For us now this idea may appear trivial. All and sundry know that the atom is a system composed of a nucleus and electrons. The nucleus is characterized by two quantities, the charge and the mass. The nuclear charge equals the number of the atomic electrons and is totally responsible for the element's chemical behaviour. The nuclear mass determines the element's atomic weight and half-life; on the other hand, it has no influence on

the chemical properties and, generally speaking, it need not be the same for all the nuclei of the same element. One should not be surprised, therefore, to find in one cell of the periodic table several atoms with varying atomic weights and — it is a must — with the same charge. This blend of isotopes does not lend itself to separation by chemical techniques, although from the viewpoint of a radiochemist it is a mixture of different radioelements, each of which has a specific half-life.

Returning to our analogy with coins, chemists can be likened to men in the street, who are only interested in the value of a coin. On the other hand, radiochemists are rather like numismatists, who normally do not care about the value of coins but attach great importance to the year of minting and other curious details.

The deep meaning of the isotopic hypothesis would only become clear in 1932, after the neutron had been discovered and the proton-neutron model of the nucleus had been well established. But some consequences of the hypothesis could be derived at once. Now, for an element to be completely defined researchers had to give two characteristics, the charge and the mass of its nucleus. They are generally indicated to the left of the element's symbol. For example: the symbol

$$^{226}_{88}\mathrm{Ra}$$

means a radium isotope with the nuclear charge 88 and mass 226; the symbol $^{2}_{1}\mathrm{H}$, the heavy hydrogen isotope, deuterium; and $^{4}_{2}\mathrm{He}^{++}$, the doubly ionized atom of helium with a mass of four, or simply an alpha particle.

The radioactive decay of radium to radon and helium can now be represented by the scheme

$$^{226}_{88}\mathrm{Ra} \xrightarrow[\text{(1600 years)}]{\alpha} {}^{222}_{86}\mathrm{Rn} + {}^{4}_{2}\mathrm{He}\,,$$

which looks very much like an equation of conventional chemical reactions. As a matter of fact, this is a reaction, but a *nuclear* one (at last we can make use of this term!). In an alpha-decay reaction the mass of the initial nucleus is reduced by four units, and the charge by two, and so the chemical properties of the resultant atom are absolutely different from the initial one, since in the periodic table the resultant atom shifts by two cells to the left.

In a beta decay, according to the knowledge of the time, the nucleus somehow emits an electron with the result that its charge increases by a unit (in the table our element shifts to the right by one cell), and its mass remains virtually unchanged. For instance, precisely by this scheme polonium, discovered by Marie and Pierre Curie, is formed from radioactive bismuth:

$$^{210}_{83}\mathrm{Bi} \xrightarrow[\text{(5 days)}]{\beta} {}^{210}_{84}\mathrm{Po} + e\,.$$

These simple rules, known as the *Soddy-Fajans displacement law,* allowed the radiochemists to disentangle at last the chains of radioactive transmutations and to assign each radioelement its place in the periodic table. And it turned out that they all settled down peacefully in eight cells, from lead to uranium. It also turned out that all the radioelements known at the time could be grouped into three radioactive families, those of

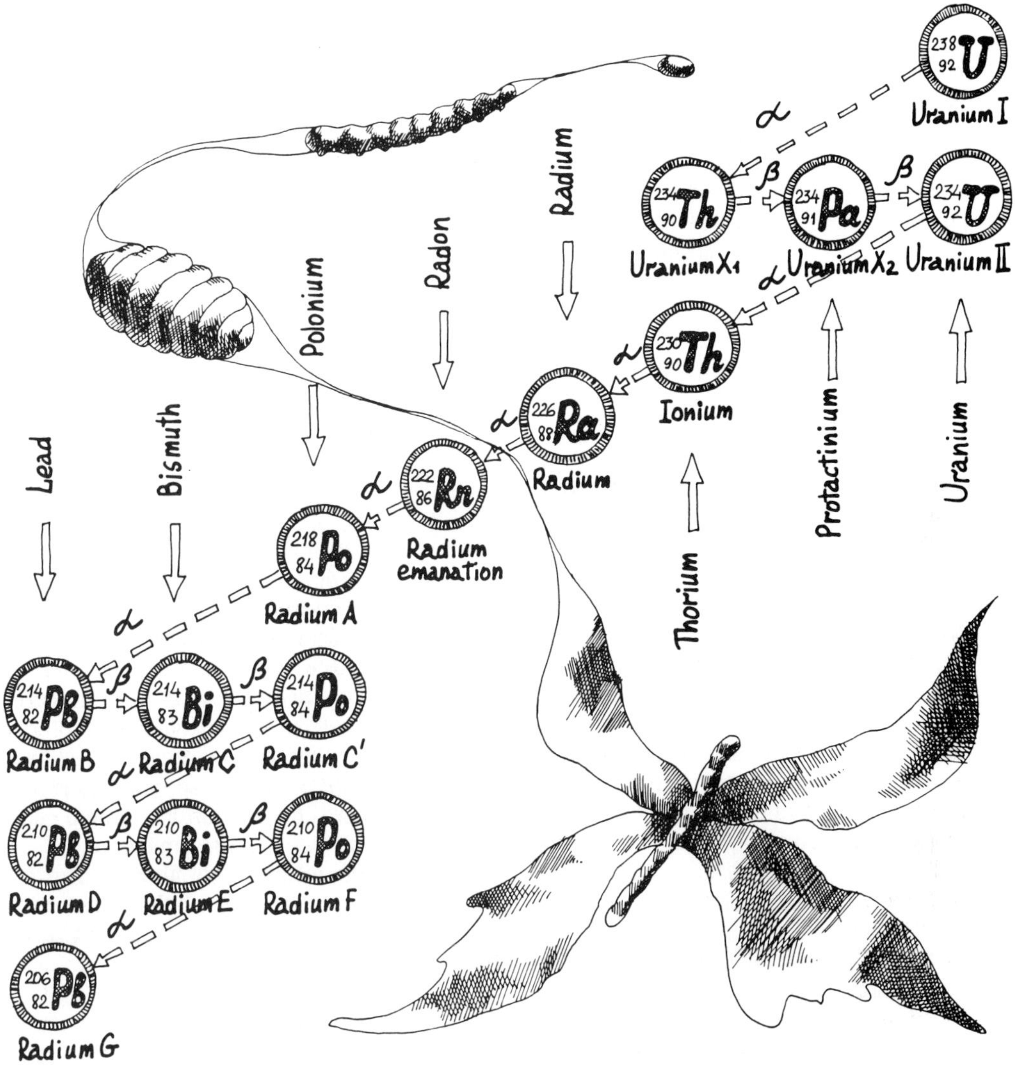

uranium, thorium, and actinium, so that each element in them is formed from the previous one with the emission of an alpha or beta particle.

Uranium Family

The uranium family is presented in the accompanying figure. Before you take a look at it just imagine how much effort has gone into it, how many great lives and happy discoveries have been associated with it! To draw it required long years of back-breaking toil by physicists and radiochemists, thousands of experimental runs, innumerable sedimentations, extractions, and analyses.

The arrival of that scheme gave a moment's respite to all radiochemists, like the one you experience at a saddle point between two mountains after a long arduous ascent, when you see your earlier mistakes and can map out your further route. Among other things, all the secrets of the mysterious elements, which for so many years had been the nightmare of chemists, were now unravelled. In the figure the earlier names of those elements are given under their true names: uranium X_1 (discovered by Crookes in 1900), uranium X_2 (Fajans and Hering, 1913), uranium II (Geiger and Nuttall, 1912), ionium (Boltwood, 1907), radiums *A*, *B*, *C*,.., *G* (Rutherford, 1904). Many of them appear as our old acquaintances, the chemical elements thorium, bismuth, lead, and so on.

In the chain of transmutations in the scheme we find two uraniums, two bismuths, three poloniums, and three leads, the last lead, $^{206}_{82}$Pb, being a stable one. Notice the spread of the half-lives; from 4.5 billion years for $^{238}_{92}$U to 1.6×10^{-4} seconds for $^{214}_{84}$Po. The clues to that diversity would only be found fifteen years later, after quantum mechanics had been created.

Also, it became clear that any uraniferous mineral should contain all the products of uranium disintegration, that is, all of the fifteen members of the radioactive family. It is hardly surprising then that the Curies discovered radium (Ra) and polonium (Po) in pitchblende. Further, we can easily see that when they observed alpha decay from the bismuth fraction derived from pitchblende, this was the radiation of at least two polonium isotopes with atomic weights of 214 and 210.

All the elements in the family are at equilibrium, that is, the number of atoms of each element formed in a second equals the number of atoms of the element that decay in a second. It is intuitively clear that the shorter the half-life of an element the fewer atoms of it are present in the mixture. This is supported by accurate calculations, which also show that for any two elements of the family their concentration ratio is equal to their half-life ratio. For instance, the ratio of the concentrations of radium and uranium is 1.6×10^3 years/4.5×10^9 years $= 0.36 \times 10^{-6}$, that is 1 tonne of uranium must contain as little as 0.36 grammes of radium — precisely the amount isolated by Marie Curie.

When and how uranium formed will be discussed later in the chapter. But once it has been formed, it lives and dies according to laws we are now well aware of. In a sense uranium is even more fascinating than radium. Its half-life is an unbelievable 4.5 billion years. A witness of the birth of our planet, it produces radium, polonium, and a dozen of other radioelements, which would have long disappeared on Earth if it were not for uranium. Each second it produces more and more of these elements again and again, just like the ancient god of heavens, Uranus, who threw out of his maw the Titans and Cyclopes.

Between $^{238}_{92}$U and $^{206}_{82}$Pb lie eight alpha decays and six beta decays with various rates and various energies of emitted particles, each beta-decay being accompanied by an emission of a gamma quantum. It is this mixture that Becquerel observed in his first experiment. Even now it is surprising that this mixture could be sorted out into its component parts. Leeuwenhoek, perhaps, experienced a similar feeling when he first looked at a drop of water through his microscope.

The scheme brings to mind the idea of the evolution of elements, and this is no random thought. From his early years Rutherford was influenced by the ideas of the eminent English astrophysicist Lockyer, who was an ardent proponent of the "inorganic evolution" of elements, i.e., of their transmutation into one another in stars. No wonder, perhaps, that it was Rutherford who became an author of the hypothesis of radioactive decay — the idea of the decay of radioactive elements was for him not so absurd as for others. Later in his life Rutherford would consciously seek the artificial "transmutation of elements" and in 1919, twenty years after he had started his radioactivity studies, he would achieve his goal.

Stable Isotopes

In the scheme of decay for $^{238}_{92}$U the last element is the stable lead isotope $^{206}_{82}$Pb. It must have an atomic weight of 206, otherwise the entire scheme is nothing more than just a beautiful mental picture. At first the inference baffled chemists: they were aware that the atomic weight of naturally occurring lead is 207.2. But soon the US scientist Theodore William Richards (1868-1928) determined the atomic weight of lead isolated from uranium minerals to be 206. Soon Soddy established that the lead formed as a result of the decomposition of thorium has an atomic weight of 208, in perfect agreement with predictions for the thorium family. This strongly suggested that naturally occurring lead is a mixture of stable isotopes with integral atomic weights. What is more, isotopy is not an exclusive privilege of radioactive elements, and *all* chemical elements are a mixture of isotopes with different integral atomic weights.

Some supporting evidence for this hypothesis was obtained by J.J. Thomson in 1913 in cooperation with Francis William Aston (1877-1945). They photographed neon-ion beams using Thomson's famous "method of parabolas" and noticed on the photographs along with neon-20 some traces of neon-22. To be on the safe side, Aston undertook the first ever attempt to separate the neon isotopes, and not without success. But science at that time was in general not yet up to the task, and Aston constructed a *mass-spectrograph* that measured masses of isotopes to within 0.1 per cent, without having to isolate them first from the natural mixture. (Thirty years later the method of gaseous diffusion, employed for the first time by Aston to separate neon isotopes, would be applied to uranium isotopes. In today's nuclear industry this is one of the most valuable processes.)

The First World War was a major set-back for Aston's work. Only in 1919 could he complete his device and embark on a systematic research effort. Quite soon, by the end of 1920, Aston had already studied the isotopic composition of 19 elements and in nine of them found isotopes. At about that time the US scientist Arthur Jeffrey Dempster (1886-1950) designed another version of the mass-spectrograph, and together with his colleagues established that the majority of naturally occurring elements (83 out of 92 known by 1940) consisted of a mixture of 281 isotopes (we know 287 stable isotopes, 210 of which were discovered by Aston). In 1931 Aston discovered the uranium isotope $^{238}_{92}U$, and in 1935 Dempster found the uranium isotope $^{235}_{92}U$ with an atomic weight of 235, which had a great future in store for it. Hydrogen was found to have two stable isotopes, neon three, iron four, and mercury seven. The largest number of stable isotopes, ten, was found in tin.

With the discovery of isotopes scientists began to distinguish "simple elements" from "mixed elements" (although the terms are incongruous). A simple element is a combination of atoms with the same mass and charge (like gold, for example, which consists of the only stable isotope $^{197}_{79}Au$). A mixed element is a natural mixture of simple elements, which were once formed in cosmological transformations.

For chemistry "simple" or "mixed" is the same, for it cannot distinguish between them even using its most delicate analytical methods. It is all the more beyond the capabilities of man, with his imperfect sensory organs. But at times this minor distinction makes all the difference. So the survivors of the Hiroshima disaster have remembered for life the difference between the harmless uranium isotope $^{238}_{92}U$ and the isotope $^{235}_{92}U$, of which the first atomic bomb was made.

Radioactive Decay Energy

In modern science frontal assaults on fundamental problems, with the help only of logic and speculation, are rarely a success. As a rule what is required is a comprehensive and, above all, *quantitative* probing into the phenomenon at hand and establishing its relations with other natural phenomena. This was also the case with radioactive decay. An answer was found only when physicists learned how to measure accurately the masses of isotopes and, in addition, recalled Einstein's formula $E = mc^2$.

The early mass-spectrographs were accurate to within 0.1 per cent at best, although even that poor precision was enough to establish that all isotopes, without exception, have atomic weights expressed by whole numbers. Aston at once set out to construct a new device, ten times as precise, which would enable him to measure isotope masses to 0.01 per cent. (In 1937 he would improve the accuracy up to 0.001 per cent.) In 1925 he reported the results of his fresh measurements. It appeared that the masses of all the isotopes are indeed close to whole numbers, although there was still a slight difference.

Improved measurements made it necessary to distinguish the *atomic weight A*, i.e., the atom's mass in appropriate units, from the *mass number N*, which is the atomic weight rounded up to the nearest integer. Previously, the convention was that the unit of atomic weight was the mass of the hydrogen atom (not without the influence of Prout's hypothesis, it seems). Now it became inconvenient, since the atomic weight and the mass number of nearly all the elements appeared to vary widely (widely from the new viewpoint, of course).

It was at first decided to take as the unit one

sixteenth of the mass of the oxygen isotope ^{16}O. This reference had long been in use, up to 1961, when it was decided to switch over to ^{12}C. In other words, scientists agreed that the atomic weight of the isotope ^{12}C is exactly its mass number: $A_c = N_c = 12.00000$. This *atomic mass unit* (amu) is currently in use throughout the world, and its value in grammes is now known with great accuracy:

$$1\,\text{amu} = \frac{1}{N_A} = 1.66054 \times 10^{-24}\ \text{gramme},$$

where $N_A = 6.022136 \times 10^{23}$ is our old acquaintance Avogadro's number.

Since Aston's times measurement technology has been improved at least a hundred-fold, and now we know the masses of isotopes with a relative accuracy of 10^{-7}, or to 0.00001 per cent. For example, in terms of ^{12}C units the atomic weight of hydrogen ^1_1H is 1.0078250; the mass of the heavier hydrogen isotope, deuterium ^2_1H is 2.0141018, and the mass of helium ^4_2He is 4.0026033. The electron mass in these units is $m_e = 0.00054858$ amu, and it too should be taken into account since with such accuracies the nucleus and the atom differ noticeably in mass.

The maximum difference between A and N for all isotopes is not larger than several thousandths of the isotope's mass. However minute at first sight, these differences are highly significant. Suffice it to say that an atomic power station derives its energy exactly from those differences, converted into energy by Einstein's formula

$$E = mc^2.$$

Early in the century and much later on this formula was the subject of long and bitter strife. Without bothering ourselves with some gnosiological niceties we can explain the idea behind it as follows: a body of mass m has stored in it the energy $E = mc^2$, where $c = 3 \times 10^{10}$ centimetres per second is the velocity of light. This energy is enormous — 1 gramme of matter contains $E = 1 \times (3 \times 10^{10})^2 = 9 \times 10^{20}$ ergs $= 9 \times 10^{13}$ joules, an amount of energy contained in 3000 tonnes of first-grade coal (a kilometre-long train). If just one thousandth of a gramme disappears, turning into energy, the energy released will be that obtained in burning 3 tonnes of coal. Just compare 1 milligramme and 3 tonnes — 3 billion times as much — this gives us an appreciation of the scale of nuclear energy, the energy of the future.

To make our future reasoning clearer, we will now perform some simple calculations. The exercise is as simple as taking the reading of your electricity meter. Nevertheless, it gives us some insight into one of the most guarded mysteries of matter.

In nuclear physics energy is generally measured in electronvolts, or the derivative units megaelectron volts (MeV). One electronvolt is the energy acquired by an electron accelerated through a potential difference of 1 volt. This unit of energy is related to the conventional units (erg, joule and calorie) by

$$\begin{aligned}1\,\text{MeV} &= 1.602 \times 10^{-6}\,\text{erg}\\ &= 1.602 \times 10^{-13}\,\text{joule}\\ &= 3.829 \times 10^{-14}\,\text{calorie}.\end{aligned}$$

Using the formula $E = mc^2$ we can now readily find that 1 amu harbours an energy of 931.5 MeV, since

$$1\,\text{amu} = 1.66054 \times 10^{-24}\,\text{gramme},$$

$$\begin{aligned}E\,(\text{amu}) &= (1.66054 \times 10^{-24}) \times (2.9979 \times 10^{10})^2\\ &= 1.492{,}4 \times 10^{-3}\,\text{ergs} = 931.5\,\text{MeV}.\end{aligned}$$

(We have here used the accurate value of the velocity of light $c = 2.9979 \times 10^{10}$ centimetres per second.)

Compare this energy with that released in the burning of one carbon atom. It is well known that 1 gramme-atom of any substance contains the same number of atoms, namely Avogadro's number $N_A = 6.02 \times 10^{23}$. Since the atomic weight of carbon is exactly $A_C = 12.00000$, then 1 gramme of coal contains

$$\frac{N_A}{A_C} = 0.5 \times 10^{23}\,\text{atoms}.$$

When 1 gramme of coal is burned away completely, the heat given off is 7,800 calories, or 33,000 joules, or, per atom,

$$\frac{3.3 \times 10^4}{0.5 \times 10^{23}} = 6.7 \times 10^{-19} \text{joules} = 4.2\,\text{eV}.$$

On the other hand, the energy stored in one carbon nucleus is $12 \times 931.5 = 1.1 \times 10^{10}$ eV. Therefore, if we succeed in putting to use one 1000th of the energy stored in the nucleus, even then we will get 3 million times more energy than from just burning coal.

The kinetic energy of atoms at room temperature is 0.04 eV, and their velocity is about 1 kilometre per second. The energy of alpha particles given off in the decay of radium is 4.8 MeV, i.e., 100 million times more, and their velocity is 15,000 kilometres per second, i.e., just twenty times less than the velocity of light. The alpha particle takes this enormous energy from the radium nucleus. Given the accurate atomic weights of the elements, we can readily work it out.

The respective atomic weights are

$$A_{Ra} = 226.02544,\ A_{Rn} = 222.01761,$$

$$A_{He} = 4.0026033.$$

Therefore, in the decay

$$^{226}_{88}Ra \rightarrow\ ^{224}_{86}Rn +\ ^{4}_{2}He$$

the mass of the system is reduced by

$$\Delta m = A_{Ra} - (A_{Rn} + A_{He}) = 0.00523\,\text{amu},$$

and the corresponding energy released is

$$E = 0.00523 \times 931.5 = 4.88\,\text{MeV}.$$

The radon nucleus carries away 2 per cent of this energy, the balance by the alpha particle, which is in agreement with experiment. In one second 1 gramme of radium emits 3.7×10^{10} alpha particles. The atoms decay at a rate of $3.7 \times 10^{10} \times 3600 = 1.33 \times 10^{14}$ atoms per hour with the total energy output of $E = 1.3 \times 10^{14} \times 4.8 = 6.4 \times 10^{14}$ MeV = 24 calories. This is about one fourth of what Curie and Laborde obtained and one sixth of the later result of 135 calories. An examination of the decay scheme reveals the reason. The fact is that pure radium can only be isolated with enormous difficulties; it always contains some products of its decay. In several days a radium preparation reaches radioactive equilibrium between all the decay products of $^{226}{}_{88}Ra$ down to $^{210}{}_{82}Pb$ (which lives on average for 19.4 years). At equilibrium for each species of nuclei the number of nuclei produced equals the number of nuclei that decay; therefore each decay of $^{226}{}_{88}Ra$ entails the decay of all the other members of the family down the chain. The total energy given off in the process is 28 MeV, which is 5.8 times more than the energy of a singly radium decay. Consequently, in several days after preparation 1 gramme of radium must release $24 \times 5.8 = 140$ calories of heat per hour, in satisfactory agreement with experimental measurements.

We have taken this trouble with the calculation in order to dispel all doubts that after Aston's work there were any secrets left concerning the source of energy in radioactive decay. Since the turn of the century scientists had been seeking an answer; now they had it.

Einstein derived his formula $E = mc^2$ in 1905, as a simple consequence of his theory of relativity. The result was so outlandish that the physics community did not take it seriously, and for nearly two decades it only served as a subject of reflection for philosophers and a target for witticisms of comedians in variety shows.

It is easy to see the reason behind that attitude to this formula. It seemed impossible to test it. In fact, when we burn 1 gramme of coal we obtain on average $Q = 7000$ calories, or 3×10^{11} ergs. This implies that the total mass of the system is only reduced by the amount

$$\Delta m = \frac{Q}{c^2} = \frac{3 \times 10^{11}}{(3 \times 10^{10})^2} = 3 \times 10^{-10} \text{ gramme}.$$

By way of comparison, the best analytical balance has as accuracy of only 10^{-8} gramme... (Consequently, the law of mass conservation can no longer be regarded as valid, although hardly anybody will rebuke Lavoisier and Lomonosov for the categorical nature of their original formulation.)

When Einstein came up with his formula he was fully aware of the difficulties of testing it and already at that time he pointed to radioactive transmutations as one of the ways of checking the theory. But it was not until 1913 that that idea was recalled by Paul Langevin (1872-1946) in France and J.J. Thomson in England. And, once Einstein's formula had been recalled, deducing its consequences was a straightforward exercise.

On 6 August 1945 the world would once and for all believe in Einstein's formula. At 8.16 a.m. during a millionth of a second only 0.7 gramme of an atomic bomb's 20 kilogrammes would vanish. The energy stored within that tiny mass would be sufficient to destroy the town of Hiroshima and take the lives of 70,000 people.

Nuclear Binding Energy

When Aston published his measurements and conclusion that all isotopes have whole atomic masses, Prout's hypothesis was unearthed at once. It stated that all the elements in nature are constructed by way of consecutive condensation of hydrogen atoms. True, universal acceptance of the hypothesis was hindered by the fact that the nuclear charge and the mass number are unequal. But immediately in many research centres — Rutherford in England, Harkins in the United States, and Masson in Australia — the assumption was made that atomic nuclei are made up of protons and some other, neutral, particles that are in turn some exceedingly compact system composed of a proton and an electron. For that system Rutherford immediately coined a name, the *neutron.* A bit prematurely, because the true neutron would only be discovered 12 years later. However, there remained one more question: What forces hold the protons in place within the nucleus? By that time it was already known that the scale of nuclear dimensions is less than 10^{-12} centimetre, and the forces of electric repulsion are enormous at such distances.

To gain some appreciation of the forces, suppose that we succeeded in separating electrons from protons in 1 gramme of hydrogen and separate them by a distance of 1 kilometre. Even in that case they will be attracted with an incredible force of 6×10^5 tonnes. The average separation between the electron and the proton in the hydrogen atom is 0.5×10^{-8} centimetre and the pull on them is enormous indeed. If we replace the electron by a proton, the resultant repulsion forces will be just as large. And if we take into account the fact that to form a complex nucleus protons must be brought together to within 10^{-12} centimetre, it will become clear that here some special *nuclear forces* are involved.

In 1915, before Aston's precision measurements, the US physicist William Harkins (1873-1951) assumed that a source of such forces might be precisely the energy stored in the nucleus. He predicted that the helium atom must be lighter than the four hydrogen atoms from which it is made up, according to Prout's hypothesis. He maintained further that it is this mass difference $\Delta m = 4m_{\mathrm{H}} - m_{\mathrm{He}}$, which he dubbed the *mass defect,* that makes the helium atom so stable, and the energy $\Delta E = \Delta m \times c^2$, which corresponds to it, retains the protons in the nucleus, however strong the electrical repulsive forces on them are. The repulsion is determined by the well-known Coulomb law

$$E_{\mathrm{c}} = \frac{e^2}{a}.$$

Knowing that the charge on the proton is $e = 4.8 \times 10^{-10}$ esu, and the average separation between protons in the helium nucleus is $a \approx 2 \times 10^{-13}$ centimetre, we find

$$E_{\mathrm{c}} \approx \frac{(4.8 \times 10^{-10})^2}{2 \times 10^{-13}} = 1.1 \times 10^{-6}\,\text{ergs} = 0.7\,\text{MeV}.$$

This is very large, although much less than the energy of nuclear attraction. It is well known now that the nucleus of any atom is made up of *nucleons,* i.e., of protons and neutrons, two particles that differ slightly in mass:

$$m_p = 1.007276\ \text{amu},$$

$$m_n = 1.008665\ \text{amu}.$$

When two protons are combined with two neutrons the result is a helium nucleus (alpha particle) with mass $m_\alpha = 4.001506$ amu, the mass defect of the helium nucleus is

$$\Delta m = 2m_p + 2m_n - m_\alpha = 0.030377\,\text{amu},$$

and the corresponding binding energy is

$$E = \Delta m \times c^2 = 0.030377 \times 931.5 = 28.3\,\text{MeV},$$

which is forty times the energy of the electrostatic repulsion of the protons in the nucleus.

Lastly, we can also introduce some average characteristic of the integrity of the nucleus, which is called the binding energy of a nucleon in the nucleus E_b; it is the total binding energy divided by the number of nucleons in the nucleus. For instance, for helium $E_b = 7.1$ MeV. For heavier nuclei the binding energy per nucleon goes up at first with mass (i.e., nuclei become stronger), passes through a maximum at $E_b = 8.5$ MeV in the middle of the periodic table for elements around tin and then falls off in a monotonic manner down to $E_b = 7.6$ MeV for the uranium nucleus. (Recall for comparison that the energy of the chemical bond between two hydrogen atoms in a molecule is 4.5 eV, or a million times less, and that to evaporate one water molecule, i.e., to overcome the attraction between molecules, takes as little energy as 0.1 eV.)

As you go through this chapter you may well think that nuclear physics is an extremely simple science. To reveal the sources of decay energy and to understand the causes of stability of the majority of nuclei it is really sufficient to know Einstein's formula $E = mc^2$, the masses of the isotopes, and the four arithmetic operations. But simple calculations provide no answer to the question of *why nuclei of radioactive elements decay.* We know that if we want to knock at least one nucleon out of a nucleus of uranium, we will need the energy $\Delta E_b = 7.6$ MeV; to knock out an alpha particle we have to remove four nucleons! What then makes alpha particles leave atoms of uranium, radium, and other radioelements, and even carry away energies of the order of megaelectronvolts?

The answer came only in 1928, three years after quantum mechanics had made its appearance and thirty-two years after radioactivity had been discovered.

Quantumalia

We are not to imagine or suppose, but to discover, what nature does or may be made to do.

Francis Bacon

Science is to see what everyone else has not seen and think what no one else has thought.

Albert Szent-Gyorgyi

Uranium

In 1789, the year of the French Revolution, the German chemist and natural philosopher Martin Heinrich Klaproth (1743-1817) for the first time isolated uranium oxide, UO_2. Only a half-century later, in 1841, the French scientist Eugène Péligo (1811-1890) separated some pure uranium. It appeared to be a heavy steel-grey metal with a density of 19.04 grammes per cubic centimetre and a melting point of 1132°C. It has the appearance of silver, the density of platinum, and the chemical properties of tungsten. At first it was assigned an atomic weight of 120, but in 1874 Mendeleev corrected it to 240. It is well known now that natural uranium is a mixture of two isotopes: 99.28 per cent of uranium-238 and 0.72 per cent of uranium-235.

Uranium is fairly abundant on Earth: each gramme of the terrestrial crust contains on average 3×10^{-6} grammes of uranium, i.e., more than lead, silver, and mercury. In granites there is even more uranium: 25 grammes per tonne. About 200 compounds and minerals of uranium are known. Of these, the most remarkable is uranium fluoride, UF_6, a colourless crystalline material, which at 56.5°C turns into a poisonous gas. This is the only known gaseous compound of uranium; and, but for it, it would be much more difficult to separate uranium isotopes.

The uranium isotopes have exceedingly long half-lives: 7.1×10^8 years for $^{255}_{92}U$ and 4.5×10^9 years for $^{238}_{92}U$. In addition to these two isotopes 12 more isotopes are currently known, the most short-lived of them being $^{237}_{92}U$ with a half-life of 1.3 minutes.

Earth and Radium

F. Aston

It has long been known that as you descend into a mine the ambient temperature grows by about 3°C for each 100 metres. This fact was explained in a quite natural manner. Once the Earth was a hot ball, and it has been cooling down ever since. That is why it is hotter inside than outside. But when in the mid-nineteenth century William Kelvin estimated the cooling time, it turned out to be strikingly short. Less than 100 million years.

This result amazed Charles Darwin, since for species to evolve takes enormous times, and on a cool earth at that. (He even made comments to this effect in the second edition of his famous work *On the Origin of Species.*) Geologists too disagreed: to account for the facts of geology they needed at least a ten-fold increase in time. This dispute between physicists on the one side and biologists and geologists on the other had lasted for a fairly long time and was abandoned by tacit accord as leading nowhere.

The discovery of radioactivity made it possible to return to the issue, now in a new context. It was noted that if each gramme of terrestrial matter contains at least 10^{-13} gramme of radium, this quantity is quite sufficient to maintain the inner temperature of the Earth at a constant level due to heat input from radioactive decay. Further estimates have shown that each gramme of the terrestrial crust contains 10^{-6} grammes of uranium, and hence 3×10^{-13} gramme of radium, i.e., even more than needed. And so geologists are now inclined to believe that the Earth is not cooling down at all but, on the contrary, is being heated up from within by the energy of the decay of radioactive materials. (One of the first to come to this conclusion in 1910 was the Russian scientist Alexey Petrovich Sokolov (1854-1928).) Owing to this decay the total heat flux at the Earth's surface is 3×10^{13} watts, which is three times the total capacity of the world's power industry.

Change is one thing, progress is another. Change is scientific, progress is ethical; change is indubitable, whereas progress is a matter of controversy.

Bertrand Russell

As to the real age of the Earth, it can be estimated by determining the relative concentration of lead in uranium ore. In the oldest uranium ores about a fifth of the uranium has decayed to lead, which suggests that the rocks date back to over 1 billion years.

Knights of the Fifth Decimal Place

Four things come not back — the spoken word, the sped arrow, the past life, and the neglected opportunity.

Arabian proverb

"Clock, balance and scale are symbols of progress", wrote James Maxwell more than one hundred years ago. Throughout the book we have repeatedly stressed the importance of precision measurements in physics and the role they play in establishing new laws of nature. This work seems mundane and does not catch the imagination of the young, but it is the bread and butter of physics without which exact sciences are unthinkable. Michelson said: "In our time new laws of nature can only be uncovered in the fifth decimal place." He himself was a true enthusiast of precision measurements. Suffice it to recall his measurements of the diameter of the star Betelgeuse, the development of the optical metre reference standard and the famous Michelson-Morley experiment that proved the absence of the ether wind (1907 Nobel Prize).

The discovery of the noble gases began with the revelation of a difference between two numbers in just the third decimal place. In 1892 John William Rayleigh (1842-1919) found that 1 litre of nitrogen derived from the air weighs 1.2521 grammes, and 1 litre of nitrogen produced from a chemical compound weighs 1.2505 grammes. Quite soon he realised that this tiny difference is caused by the presence in atmospheric nitrogen of the noble gases, which in combination are heavier than nitrogen. Later on he, together with William Ramsay (1852-1916), isolated from this minute fraction nearly the entire 8th group of the periodic table (1904 Nobel Prize).

A theory has only the alternatives of being right or wrong. A model has the third possibility: it may be right, but irrelevant.

Manfred Eigen

Without Ångström's spectroscopic measurements there would be neither Balmer's formula nor Bohr's atom. Precision measurements of the wavelengths of spectral lines led Willis Eugene Lamb (b. 1913) to the discovery of vacuum polarization (1955 Nobel Prize).

Among those "knights of the fifth decimal place", to use Rayleigh's words, was Francis William Aston (1922 Nobel Prize). He devoted a quarter of a century to continual improvement of his mass-spectrograph. His careful measurements of isotopic masses laid the foundation to many discoveries: they provided indications as to the primary sources of the energy of radioactive decay, solar and stellar radiations, explained why nuclei are so stable, and, immediately after Hahn and Strassmann's discovery, enabled the fission energy of uranium nuclei to be worked out.

Mermaid, European

Chapter Fourteen

Probing into the Nucleus • The Neutron
Artificial Radioactivity • Slow Neutrons
Nuclear Fission

For many scientists World War I was an unexpected and terrible ordeal. From the many triumphs of positive science and high-minded ideas about universal progress Europe suddenly found itself at the level of the worst instances of medieval barbarism. In those years scientific thought in laboratories was barely alive: Marie Curie with her daughter Irène was busy with X ray installations at hospitals; Louis de Broglie served as a signalman; Max Born and Max Laue fought on the other side. Francis Aston worked at an aircraft factory and Henry Moseley died in action. For the first time scientists then understood that science is not always a blessing and that the knowledge they produce can at all times be turned against humanity. They could see, for instance, soldiers in the trenches killed by toxic gases invented by chemists.

At that time it also became clear to many scientists that science was no longer an ivory tower endeavour and that it would now have to live in the limelight, under inspection by reporters and businessmen, generals and politicians. Even at that time scientists with a keen perception of moral values tried to separate the instinct for knowing from the fears of its uncontrolled implications. Thirty years later the dilemma would turn into a tragedy for scientists and many of them would be willing to declare, together with Otto Hahn, after Hiroshima and Nagasaki: “I had nothing to do with it!”

Probing into the Nucleus

After the war scientists returned to their interrupted studies. The year 1919 will go down in the history of science: in 1919 Ernest Rutherford for the first time carried out the artificial transmutation of elements. At the time the very possibility of such transformations no longer seemed absurd. Radioactivity provides many examples of transmutation. Nonetheless, it was found that heat and cold, electrical and magnetic fields, pressure and chemical reactions changed the process of radioactive decay not a bit. There was something magnificent in that indifference with which nature flouted all human attempts at interference in the course of natural processes. One can understand therefore the interest and excitement with which the scientific community accepted Rutherford's experiments.

In 1919 Ernest Rutherford was forty-eight, he was a Nobel Prize laureate, the director of the world-famous Cavendish laboratory, a member of nearly all the academies of the world, a recognized authority in atomic and nuclear physics, he had been knighted by the king of England for his scientific achievements, he was surrounded by a cohort of disciples, many of whom would later win Nobel Prizes themselves. But just as twenty years ago, when he had been a young scientific worker, he liked to pore over a microscope and experiment with alpha particles.

This time he continued the pre-war measurements started by his assistant Marsden and found that when alpha particles pass through air some new particles emerge whose free path was much longer than that of the initial alpha particles. Quite soon Rutherford found that the secondary particles were protons produced when alpha particles collide with nitrogen nuclei. But how? Rutherford recognized two possibilities: either an alpha particle knocks a proton out of the nitrogen nucleus and turns it into a carbon nucleus by the scheme

$$\alpha + {}^{14}_{7}\mathrm{N} \rightarrow \alpha + {}^{13}_{6}\mathrm{C} + p\,,$$

or the alpha particle gets caught in the nitrogen nucleus, thus turning it into an oxygen nucleus,

$$\alpha + {}^{14}_{7}\mathrm{N} \rightarrow {}^{17}_{8}\mathrm{O} + p\,.$$

Six years later Rutherford's colleague Patrick Maynard Stuart Blackett (1897-1974) observed this *nuclear reaction* in a Wilson chamber and proved that it is the second scheme that is true. This implied that man could now change what, to use Newton's words, "God himself created on the first day of Creation". The age-old dreams of alchemists and the youthful hopes of Rutherford himself came true. *The Newer Alchemy* — that is what he later called his book on nuclear transformations and to his last days he would be fascinated by the world that had opened up before him.

In the subsequent four years Rutherford, together with James Chadwick (1891-1974), established that when bombarded by alpha particles at least a dozen other elements, as far as sodium, enter into nuclear reactions. But this exhausted the possibilities for alpha particles. The charge of a potassium nucleus is 19 and of the alpha particle 2; the energy of the alpha particle was insufficient to overcome the repulsion of nuclei with charges more than 20. The proton charge is half that of the alpha particle, and so it is preferable as a missile to bombard nuclei than are alpha particles. But how could high-energy protons be obtained? There are no naturally occurring radioactive elements that emit protons.

It was then that the idea of the proton *accelerator* was voiced for the first time. It was made real nearly a decade later. In 1931, nearly concurrently, Robert van de Graaf (1901-1967) suggested his electrostatic generator, Ernest Orlando Lawrence (1901-1958) invented the cyclotron, and John Douglas Cockroft (1897-1967) and Ernest Thomas Sinton Walton (b. 1903) constructed the cascade generator to accelerate protons.

Back in 1932 Cockroft and Walton observed in Rutherford's laboratory the first proton-induced nuclear reaction. When they bombarded a lithium target with 0.2-MeV protons they found that about one proton per billion split a lithium nucleus into two alpha particles, which flew apart in opposite directions each with the enormous energy of 8.5 MeV. In symbols

$$p + {}^{7}_{3}\mathrm{Li} \rightarrow {}^{4}_{2}\mathrm{He} + {}^{4}_{2}\mathrm{He} + 17\,\mathrm{MeV}\,.$$

This nuclear reaction became as famous as Rutherford's first reaction that turned nitrogen into oxygen. If we compare the energies before and after the reaction (0.2 MeV and 17 MeV), we may be led to question the energy conservation law. That is, of course, if we do not take into consideration Einstein's formula. In actual fact, the validity of the formula $E = mc^2$ was proved precisely in this reaction. To see that this is so, we may compare the masses of the particles involved before and after the reaction:

Before	*After*
$m_p = 1.007276$ amu	$m_{He} = 4.001506$ amu
$m_{Li} = 7.014359$ amu	$m_{He} = 4.001506$ amu
Total: 8.021635 amu	Total: 8.003012 amu

The mass defect is $\Delta m = 0.018623$ amu.
The energy liberated is $E = \Delta m \times 931. = 17.3$ MeV.

You would, I think, agree that this simple exercise was worthwhile, because it demonstrates the validity of one of the most fundamental laws of nature.

The Neutron

The neutron is a key to the stores of intranuclear energy. We now know much about it: it has no charge, it has mass $m_n = 1.008665$ amu, which is slightly — about two electron masses — more than the proton mass; its spin is like that of the proton, and all nuclei are close packages of a mixture of protons and neutrons. The hypothesis that nuclei have proton-neutron structure was put forward immediately after the discovery of the neutron by several scientists nearly simultaneously: the Soviet physicist Dmitry Dmitrievich Ivanenko (b. 1904), Werner Heisenberg, and the talented short-lived Italian scientist Ettore Majorana (1906-1936). It has never since been put in question. In 1932 Harold Urey (1893-1981) discovered the heavy isotope of hydrogen, *the deuteron,* whose nuclei represent a bound state of a proton and a neutron.

In a free state a neutron disintegrates fairly quickly, with a half-life of 10.7 minutes, into a proton, an electron, and an electron antineutrino by the scheme

$$n \rightarrow p + e + \tilde{\nu}_e .$$

Let me explain briefly what an anineutrino is. The neutrino is a neutral, massless particle with spin 1/2; it was introduced, originally quite formally, by Pauli to maintain conservation of energy and spin in β-decay. The antineutrino is the antiparticle of the neutrino just as, as we shall see below, the positive electron or positron is the antiparticle of the electron. Indeed all elementary particles have their antiparticles, which are in some sense their opposites.

A nuclear neutron is bound by strong nuclear forces and, as a rule, is stable. Every now and then it decays by the conventional scheme with the result that the proton remains in the nucleus and the electron and the antineutrino are emitted.* It is these electrons that we observe as beta rays of radioactive elements. Nuclear forces change the behaviour of neutrons dramatically and, depending on the nucleus species, its beta-decay period may vary widely: from hundredths of a second to billions of years.

It is unclear why the neutron was discovered so late. After all, Rutherford and Harkins had predicted it as early as 1920, and to discover it nothing was needed save for commonplace alpha particles. Nevertheless, it was not until ten years later that scientists got on its track.

In 1930 Planck's pupil Walter Bothe (1897-1957), together with Becker, irradiated beryllium with alpha particles, in keeping with the tradition of nuclear reaction studies started by Rutherford in 1919. But it was not protons that they observed but some other radiation, a radiation that passed through a lead layer 2.5 centimetres thick. They decided that it was hard gamma-radiation from the excited beryllium atom, and were satisfied with the assumption.

Two years later Frédéric Joliot (1900-1958) and Irène Curie (1897-1950) became interested in the nature of the new radiation. They directed it at a paraffin target and observed protons that left the target with high energy. This led them to conclude that they had discovered "a new way of interaction of radiation with matter" whereby the "gamma-quanta" of Bothe and Becker bombarding paraffin knock out not only protons from the hydrogen atoms but also some carbon atoms.

In the autumn of 1920 Rutherford invited James Chadwick (1891-1974) to work with him on the artificial transmutation of elements. They counted scintillations in the dark, and during those long and tedious sessions Rutherford would initiate Chadwick into his ideas concerning the neutron and its possible role in nuclear architecture. Later on Chadwick even made several attempts to discover the neutron. They appeared to be unsuccessful but not wasted; indeed, they paved the way for the discovery. When he learned about the experiments by the Joliot-Curies, he understood within a month that Bothe and Becker had observed the nuclear reaction of the transmutation of beryllium into carbon with the emission of a neutron, i.e.,

$$\alpha + {}^{9}_{4}\mathrm{Be} \rightarrow {}^{12}_{6}\mathrm{C} + n\,,$$

and the Joliot-Curies had observed just the recoil of protons hit by neutrons, similar to that observed when two billiard balls collide.

Amazingly simple, isn't it? For this elegant discovery Chadwick three years later, in 1935, would be awarded a Nobel Prize. But still, why not the Joliot-Curies or Bothe? Why had such a simple idea never occurred to them?

With such situations, of which there have been many in the history of science, one should be cautious. To give a clear formulation of a fundamental discovery, one must, as a rule, contradict some generally accepted views, and this formulation leaves no room for an honourable retreat in case of failure. Such a venture, you will agree, is fraught with professional risks for any scientist. And the more venerable the scientist, the more dangerous for him is a mistake of such a kind. This is perhaps one of the reasons why it is generally the young who make truly revolutionary discoveries, although it is the older generation that normally formulate a problem and indicate ways of solving it. (Enrico Fermi liked to say that problems are solved by post-graduate students and that it is the task of their supervisors to formulate them.) In the history of physics such examples are legion. Among them is the theory of relativity, Bohr's atom, matrix mechanics, the electron spin, and so forth. Just like any empirical factor, this rule is not without exceptions. So, Röntgen, Planck, Rutherford, Schrödinger, Born, and Chadwick himself, made the discoveries of their lifetime at a mature age.

Artificial Radioactivity

When they learned about Chadwick's discovery, Irène and Frédéric Joliot-Curie were a little vexed. But they were true scientists and were guided in their careers by the joy of the scientific quest, not the sting of false pride. With fresh zeal they carried on their research. Their efforts were soon crowned with success, they discovered *artificial radioactivity.*

Early in 1934 they irradiated aluminium with alpha particles and, as before, observed a radiation of high penetrating power. It was clear to them now that they observed the reaction

$$\alpha + {}^{27}_{13}\mathrm{Al} \rightarrow {}^{30}_{15}\mathrm{P} + n\,,$$

or, in words, when an alpha particle is captured by an aluminium nucleus, a neutron is emitted and the nucleus of a phosphorus isotope is formed. But then again something strange occurred: the irradiated aluminium gave off not only neutrons but also *positrons,* i.e., particles with the mass of the electron and positive charge. They were predicted by Dirac in 1928, but few believed in them until Carl David Anderson (b. 1905) discovered them in 1932 in cosmic rays, nearly contemporaneously with the discovery of the neutron.

Positron emission could be accounted for by assuming, say, that a proton may beta-decay into a neutron, a positron, and a neutrino:

$$p \rightarrow n + e^{+} + \nu_e\,.$$

This reaction is analogous to the neutron beta-decay reaction, but proceeds in the opposite direction. Such an assumption may appear to be illegitimate, since hydrogen nuclei are known to be stable. If they were not, hydrogen would have vanished on Earth long ago. However, to be stable a proton must be free, and whether or not it will be stable when included in the composition of any nucleus, there is no saying. We only know that the nuclear neutron differs from the free neutron in disintegration rate and that it can even become stable in the process.

In other words, positron radiation, although an unusual fact in itself, was accounted for somehow. Another aspect was unclear: when the source of alpha particles was removed the neutron flux stopped at once whereas positrons were still observed. It obeyed the well-known law of radioactive decay with a half-life of 2.5 minutes. Why? Examination of the reaction scheme leaves no other choice: positrons must be emitted by the nuclei of the resultant phosphorus isotope by the scheme

$$^{30}_{15}P \rightarrow ^{30}_{14}Si + e^{+} + \nu_e .$$

This radioactive decay yields a rare silicon isotope, which in natural silicon amounts to only 3 per cent. But the phosphorus from which it is formed is non-existent in nature. This new, artificial isotope would be called by the Joliot-Curies *radiophosphorus* after they had proved that chemically it is identical with the element phosphorus.

Physicists may write schemes of nuclear reactions and give speculative arguments but chemists will only believe them when they are given *chemical* proofs of the formation of new elements under nuclear transformations. Easier said than done! A routine chemical analysis require a microparticle of matter of at least 10^{-8} gramme, in other words about 10^{14} atoms. But when aluminium is bombarded with alpha particles for a reasonable time, there may form 10^{6} atoms, at best. If, however, phosphorus atoms are not stable, but radioactive, we can still obtain chemical evidence for it using a so-called carrier reaction. The idea is simple and ingenious. If chemists suspect that there is some radiophosphorus in a solution, they bring a Geiger counter up to it, which detects even the smallest admixtures of radiophosphorus. Then they add a large amount of conventional phosphorus to the solution and precipitate it using an adequate reagent, the two phosphoruses being inseparable in the precipitate. Now, the counter is brought up first to the solution and then to the precipitate, and if the counter clicks near the precipitate and is silent near the solution, this implies that initially the solution contained radiophosphorus, which later went completely into the precipitate. In this way the first chemical proof of the artificial transmutation of elements in the course of nuclear reactions was obtained.

Radiophosphorus $^{30}_{15}P$ was the first man-made radioactive isotope. A multitude of them would be obtained thereafter, more than 1000. Only eleven years would elapse before the first atomic bomb would explode in the Alamogordo desert. It contained several kilogrammes of the plutonium isotope $^{239}_{94}Pu$, an isotope that a bare five years previously had been nonexistent in nature.

The discovery of the Joliot-Curies was marked with a Nobel Prize in 1935 — so important it seemed to the scientific community. Marie Curie did not live to see this, she died in the autumn of 1934. Before she passed away, however, she touched with her radium-burned fingers a test tube containing some radiophosphorus and heard the counter click. Frédérick and Irène were to suffer the same fate, for they died of the consequences of radioactive irradiation. But before that they had to live through a Nazi invasion and, still later, constructed the first French atomic reactor.

Slow Neutrons

The great Italian physicist Enrico Fermi (1901-1954) occupies an exceptional place in the history of atomic energy. He is customarily referred to as a theoretician, but his Nobel Prize was awarded to him for his work on experimental physics and the National academy in Rome awarded him a medal for his contribution to chemistry. Any single scientific discipline seems to have been too narrow for him; he was a *natural scientist* in the widest sense of the word. This universality — an exceedingly rare quality in the twentieth century — appeared to be a godsend in a man destined to solve a host of problems in atomic energy, where each step was a step into the unknown.

Fermi was one of the first to understand that the neutron was an ideal means of probing into the mechanism of nuclear reactions and the simplest path to obtaining new radioisotopes. The main feature of the neutron is its electrical neutrality, which enables it to penetrate easily into nuclei of even the heaviest elements.

In the summer of 1934 a group of young Italian physicists, the youngest of whom was Bruno Pontecorvo (twenty years) and the oldest Fermi (thirty-three years), were absorbed in experiments. Their devices were simple, just a radon-beryllium source of neutrons and Geiger counters. The task was formulated clearly: to irradiate various elements with neutrons and to see what would occur as a result. Their scope was wide — the entire periodic table, and their enthusiasm and energies were unlimited — a dozen articles were later required to describe the properties of the several dozen new radioisotopes they had obtained.

When a neutron is absorbed by a nucleus $^{N}_{Z}X$ with a charge Z and a mass number N, it turns into the isotope $^{N+1}_{Z}X$, which in turn tends to get rid of the superfluous neutron. The simplest path is to turn a neutron into a proton with the emission of an electron and an antineutrino. Such a *beta-*

decay of the nucleus yields a nucleus of a new element Y with nuclear charge $Z + 1$ and mass number $N + 1$, that is we have the following nuclear reactions

$$n + {}^{N}_{Z}X \to {}^{N+1}_{Z}X \xrightarrow{\beta} {}^{N+1}_{Z+1}Y + e^- + \tilde{\nu}_e \, .$$

Although it is formally similar, this process of production of artificial isotopes differs dramatically from the Joliot-Curie scheme, which can be represented in the following form:

$$\alpha + {}^{N}_{Z}X \to {}^{N+3}_{Z+2}A + n$$
$$\hookrightarrow {}^{N+3}_{Z+1}Y + e^+ + \nu_e \, .$$

In both cases we start with an element ${}^{N}_{Z}X$ with nuclear charge Z and end up with the element Y with nuclear charge $Z + 1$, but the resultant isotopes are different: in Fermi's experiments, with mass number $N + 1$, in Joliot-Curie's experiments, with $N + 3$. Also if with the Joliot-Curies a target, irradiated with alpha particles, emitted positrons, with Fermi the same target, irradiated with neutrons, emitted electrons.

For instance, the result of aluminium irradiation was with the Joliot-Curies

$$\alpha + {}^{27}_{13}\mathrm{Al} \to {}^{30}_{15}\mathrm{P} + n$$
$$\xrightarrow[2.5\ \text{min}]{} {}^{30}_{14}\mathrm{Si} + e^+ + \nu_e \, ,$$

and with Fermi:

$$n + {}^{27}_{13}\mathrm{Al} \to {}^{28}_{13}\mathrm{Al} \xrightarrow[2.3\ \text{min}]{} {}^{28}_{14}\mathrm{Si} + e + \tilde{\nu}_e \, .$$

(The terrestrial crust contains 28 per cent of silicon isotopes; among these ${}^{28}_{14}\mathrm{Si}$ accounts for 92.2 per cent, ${}^{29}_{14}\mathrm{Si}$ for 4.7 per cent, and ${}^{30}_{14}\mathrm{Si}$ for 3.1 per cent.)

Quite clearly, Fermi's technique of producing new isotopes is preferable, since it is simpler and more universal. Within a short period of time Fermi's "boys" had irradiated sixty-eight elements and in 47 of them they observed artificial radioactivity, that is to say they synthesized several dozen new isotopes.

But their main discovery was not this. On 22 October 1934 they found, much to their surprise, that atoms are hundreds of times more efficient in capturing neutrons when the reaction proceeds under water (conveniently, in the institute's yard there was a pool with goldfish.) Their amazement lasted two hours, the time it took Fermi to sketch with his characteristic elegance the outlines of a new physical phenomenon. His explanation was simple. Water molecules consist of oxygen and hydrogen and the neutron mass is essentially equal to that of the proton. Therefore, when a neutron hits hydrogen nuclei it is slowed down quickly, dozens of times faster than when it collides with heavier nuclei. Only after the neutron has been moderated can it enter into a nuclear reaction.

Surprise is generally the consequence of a collision of unexpected facts with established patterns of thinking. Over the years physicists had got used to the idea that the nucleus, although intangible, is quite a solid entity and to change it requires a missile accelerated as fast as possible. The missile may be an alpha particle, proton, or other particles. It was for this purpose that accelerators had been invented. And with neutrons they observed the opposite picture: the slower a neutron moved the more willingly was it absorbed by nuclei. The reason is two-fold. First, it is attracted, not repelled, by nuclei; second, it obeys the laws of quantum mechanics.

The discovery of the moderation of neutrons with its accompanying effects was not nearly as spectacular as the discovery of the neutron itself or artificial radioactivity, but it was this seemingly modest effect that was destined to have a great fu-

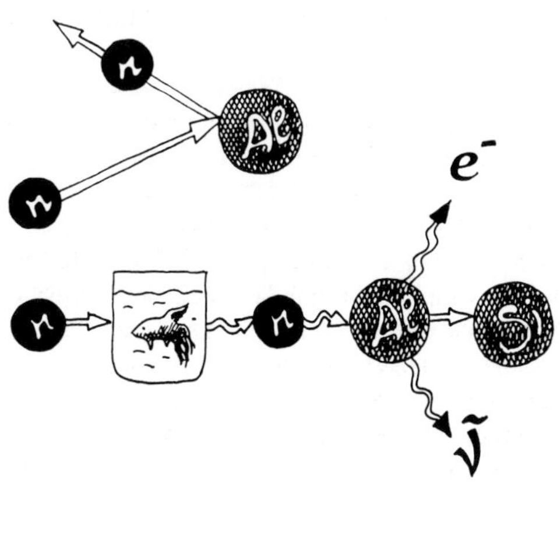

ture. Without it we would not be able to start a nuclear reactor. The participants in that historic experiment were quick to perceive its significance. On that same evening of 22 October, 1934, they met at Fermi's home and worked well into the night on an article entitled "The influence of Hydrogen-Containing Substances on Neutron-Induced Radioactivity". In 1938 for the "disclosure of artificial radioactive elements produced by neutron irradiation" Enrico Fermi was awarded the Nobel Prize for Physics.

Nuclear Fission

In that remarkable summer of 1934 "papa Fermi" (as infallible in things scientific as the Pope in things religious) with his "boys" irradiated many elements with slow neutrons. Among them was uranium. Just like the majority of other elements when exposed to neutrons it turned beta-active, i.e., it gave off electrons. That was striking. At that time uranium occupied the last place in Mendeleev's table. The charge of its nucleus is 92; therefore, if a uranium nucleus captures a neutron and then emits an electron, its charge will increase by 1, and uranium will turn into the *first transuranic* element by the scheme

$$n + {}^{238}_{92}\mathrm{U} \rightarrow {}^{239}_{92}\mathrm{U} \rightarrow {}^{239}_{93}\mathrm{X} + e + \tilde{\nu}_e .$$

From Fermi's experiments this conclusion followed in such a natural manner that even without its detailed checking it became a scientific sensation and hit the front pages. In that chorus some critical voices were drowned; they were only remembered several years later. A German radiochemist team, husband and wife the Noddacks, who discovered rhenium, assumed as far back as 1934 that Fermi had observed not transuranic elements but fragments of uranium fission. But the majority of radiochemists were not yet prepared for such radical inferences, and so they set out to seek "transuranic" elements.

The German radiochemist Otto Hahn (1879-1968), a pupil of Ramsay and Rutherford, had been disentangling for years the chains of radioactive transmutations. He discovered protactinium and nuclear isomerism. In 1937, together with Lise Meitner (1878-1968) and Fritz Strassmann (1902-1980), he decided to repeat Fermi's experiments on the neutron irradiation of uranium. As he did so he observed conventional beta activity with a half-life of twenty-three minutes. Since there was no separating that activity from uranium, they agreed with Fermi that that was really the beta-decay of ${}^{239}_{92}\mathrm{U}$ yielding a new unknown element ${}^{239}_{93}X$. Unlike Fermi, however, they were professional chemists and before publishing their discovery they sought chemical proofs for the new element. A fresh series of experiments ensued. But in the summer of 1938 it was discontinued, for Lise Meitner had to escape from the Nazis and seek refuge in Sweden.

In the autumn of 1938 Hahn and Strassmann resumed their experiments, again using the technique of carrier reactions. They would irradiate some uranium with neutrons and dissolve it with some barium salt added to the solution, and then they would precipitate barium. It turned out that in addition to barium the sediment contained some beta-active substance. Hahn and Strassmann decided that that was radium-231, which might form from ${}^{239}_{92}\mathrm{U}$ through two successive alpha-decays and which chemically behaved very much like barium. But something prevented them from immediately reporting their conclusion. What was more, in similar experiments Irène Joliot-Curie and Pavle Savic (b. 1909) had observed a "lanthanum-like substance". In the long run, after no end of rechecking, Hahn and Strassmann made sure that it was possible to separate their beta activity from radium but not from barium. And for that result Otto Hahn, a radiochemist with thirty years of experience, could vouch. Still, some doubts lingered, and in their paper Hahn and Strassmann confessed: "As chemists we should replace the symbols Ra, Ac and Th in our previous scheme by Ba, La and Ce. As 'nuclear chemists', who are in a sense close to physics, we still cannot bring ourselves to make that step, which would be at odds with all the previous conceptions of nuclear physics."

It is difficult for us now to understand their bewilderment, since from our school years we know that the uranium nucleus splits and we find nothing strange about this. Let us try and look at the situation through the eyes of the trail-blazers and, if not to understand, at least to sense the reasons behind their doubts. They were above all chemists, and for them a chemical element was some exceedingly stable entity, unaffected by hot and cold, dissolutions, crystallizations, and vehe-

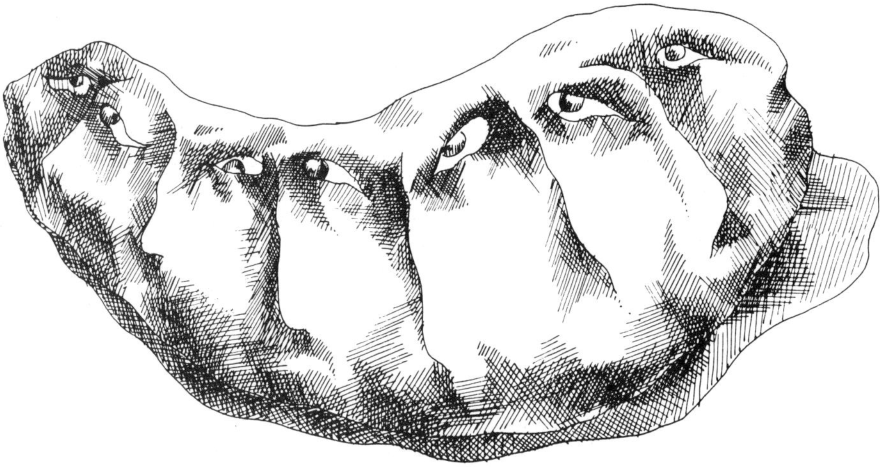

ment chemical reactions. It was just recently that, reluctantly, they had accepted the fact that sometimes, in the process of radioactive decay of nuclei, one element may turn into another. But the most which could result was having the element shifted by one or two cells in the periodic table. But the atomic number of barium is 56, nearly half that of uranium! And if they were to believe that barium did form out of uranium, this would immediately suggest that elements could be shifted at will throughout the periodic table. No chemist could accept that.

On 22 December 1938 Hahn and Strassmann submitted for publication a paper with their findings. Shortly before that Hahn wrote a letter to Lise Meitner and confided his doubts to her. They were linked together by thirty years of friendship, work, and joint discoveries. She received the letter just before Christmas Eve at an hotel in the small town of Kungälv near Göteborg. She was visited there by her nephew Otto Frisch (1904-1979), also a physicist, who had emigrated to Copenhagen, where he was working at Niels Bohr's institute. Together they fairly quickly understood that Hahn and Strassmann had observed the disintegration of the uranium nucleus on the capture of a neutron (shortly after that, on the suggestion of the biologist William Arnold they introduced the now common term *nuclear fission* — in analogy to cell division, just as a quarter of a century previously Rutherford had introduced the concept of the "atomic nucleus" in analogy with the cell nucleus). But what was important was that they were quick to grasp that this fission must liberate enormous energy. After Aston's work it was known that the binding energy per nucleon in the uranium nucleus was 7.6 MeV, and for the elements in the middle of the periodic table it was much larger — 8.5 MeV. The more the binding energy of a nucleus the stronger the nucleus is and the larger its mass defect is, i.e., the difference between the nuclear mass and the total mass of its component nucleons, and hence the higher the energy released as the nucleus is formed. Therefore, when a nucleus of uranium-235 splits, the energy released will be

$$\Delta E = (8.5 - 7.6) \times 235 \approx 200\,\text{MeV}.$$

This is nearly eight times as much as is released in the entire chain of radioactive decay, from uranium to lead. Now we know already that this energy is enormous, and we can understand the excitement Frisch and Meitner felt when for the first time they had sketched their elementary calculation on a scrap of paper. They found that the fission of nuclei in 1 gramme of uranium gives off an energy of 8×10^{10} joules, equivalent to the heat stored in 3 tonnes of coal.

Back in Copenhagen Frisch caught Niels Bohr at the gangway of a ship about to carry him away for several months to America. He informed Bohr about his latest estimates and immediately set out to prepare an experiment to check the uranium fission hypothesis.

In the afternoon of 13 January Frisch began his experiment, and by 6 a.m. of 14 January he had made sure that the uranium-fission hypothesis was true. On 16 January he sent off to *Nature* a paper describing his findings.

From that moment on events were racing at a heady speed, and time was counted not by years and months but rather by weeks and days. This is another story, however, and if we want fully to grasp the significance of these developments, we will have first to get acquainted with certain of the main principles of quantum mechanics.

Quantumalia

Letters about Fission

O. Hahn

Every creative act involves... a new innocence of perception, liberated from the cataract of accepted belief.

Arthur Koestler

Monday night, on 19 December 1938, two days before sending the paper to the journal, Otto Hahn wrote to Meitner: "All day the indefatigable Strassmann and I, with some support from Lieber and Bonne, have been working with uranium products. It is 11 o'clock in the evening now; at a quarter past 11 Strassmann promised to be back, so that I may go home. There is something in those 'radium isotopes' and this 'something' is so rare that we for the time being inform only you... They are separable from *all* elements except for barium... Although we should not exclude any accidental concurrence of circumstances, we nevertheless are more and more inclined to make the terrible conclusion: our radium isotopes behave not as radium, but as barium... We agreed with Strassmann that for the time being we will only initiate you into it. Perhaps, you will be able to come up with some fantastic explanation. We, of course, know that there is no way for a decay into barium to occur, but we want to check additionally whether or not the actinium isotope that emerges from 'radium' has the properties of lanthanum, not actinium... It is unbelievable that we have been in error for so long..."

21 December 1938, Meitner to Hahn: "Your results with radium are shocking. A process due to slow neutrons that leads to barium!... To recognize such an unusual disintegration, it seems to me, is yet difficult, but we have witnessed so many surprises in nuclear physics that there is hardly anything about which one can say directly: this is impossible..."

21 December 1938, Hahn to Meitner: "Since yesterday we have been gleaning our evidence for barium-radium... From it we, as 'chemists' should make the conclusion that the three isotopes we have studied are not radium but, chemically, barium. The actinium that emerges from the isotopes is not actinium, but rather radiating lanthanum!"

On 22 December 1938 Hahn and Strassmann's paper arrived at the journal. They wrote in it: "We should discuss some new studies whose results, because of their strange nature, we provide not without hesitation... We come to the conclusion that our 'radium isotopes' show the properties of barium. As chemists we, strictly speaking, should say that the new substance is not radium but barium;

L. Meitner

Any idea foreign to our way of looking at things and feeling them always seems incongruous to us.

Helvetius

other elements are out of the question."

28 December, Hahn to Meitner: "I am eager to tell you something more about my barium conjectures: perhaps Otto-Robert is currently with you at Kungälv and you can discuss this... Here is my new conjecture: if it were possible for uranium-239 to disintegrate into barium and masurium, 138 + 101 would give 239!... Is this possible energetically? I do not know it, I only know that our radium has the properties of barium..."

29 December 1938, Meitner to Hahn: "Your results with radium-barium are extremely interesting. Otto-Robert and I have already racked our brains."

1 January 1939, Meitner to Hahn, after she had read Hahn and Strassmann's paper: "Perhaps it is energetically possible for a heavy nucleus to disintegrate."

3 January 1939, Meitner to Hahn: "Now I am almost sure that you have indeed discovered decay into barium, and I consider this result beautiful indeed. I cordially congratulate you and Strassmann..."

5 January 1939, Hahn to Meitner: "Today I am no more sure, even afraid for barium; perhaps it is still barium? I cannot make myself believe this."

On 6 January 1939 an issue of *Naturwissenschaften* came out with Hahn and Strassmann's paper "On Proof of the Existence and the Properties of Alkaline-Earth Metals Produced by Irradiation of Uranium by Neutrons".

10 January 1939, Frisch to Hahn: "I have already accumulated so many arguments against the transuraniums that it would be difficult for me to agree with their resurrection."

On 16 January 1939 *Nature* received two papers: "Physical Evidence for the Fission of Heavy Nuclei under Neutron Bombardment" by Otto Frisch and "Disintegration of Uranium by Neutrons: A New Type of Nuclear Reaction" by Meitner and Frisch.

22 January 1939, Hahn to Meitner (after he had received the manuscripts of Frisch and Meitner's papers): "...it appears that all our arduous experiments are no longer necessary after the convincing experiment by Otto-Robert..."

25 January 1939, Meitner to Hahn: "Why 'no longer necessary'? Without your beautiful result on barium instead of radium we would never have come to that..."

26 January 1939, Meitner to Hahn: "All you have done lately seems to me to be just fantastic. A good half of the periodic table is encountered among these uranium fragments..."

The stubborn unwillingness of Hahn to recognize the unquestionable facts obtained by him had psychological reasons in addition to the objective ones. In the first place, it was not easy to recognize that the three years of frenzied toil in search of transuranic elements (twenty-four hours a day spent at counters) were all in vain. Furthermore, by publicly recognizing it he would willy-nilly place Lise Meitner in an uncomfortable position, since it would disprove the results of their joint work, through experiments conducted without her. Lastly, it took courage to recognize that Irène Joliot-Curie with her "lanthanum-like substance", known in Berlin as "curiosium", was right. And yet scientific honesty won out.

Centaur, Greek

Chapter Fifteen

Tunnel Effect • Effective Cross-sections of Reactions Neutron Cross-sections Nuclear Fission

A book written using pictography elicits in the majority of people only a superficial interest or slight perplexity: is it really impossible to do without these mysterious characters? And why do the Japanese and Chinese still use pictographs? But for a professional orientalist, and even more so for a Japanese or a Chinese, hieroglyphics are not just a means of information transfer, they are the mainstay of the history and culture of a people. In some incomprehensible way they embody not only appropriate concepts but also the history of their development and even the attitude of people to them. So there are bad and happy pictographs, and there exist entire pictures consisting of a single pictograph. Clearly, such a picture only allows an approximate verbal description.

The nature of the "atomic fire" can only be explained in the language of quantum ideas. In order that their meanings may be stipulated unequivocally physicists have devised a system of symbols, which is in a way akin to pictographs. Some of these pictographs we know already. For example when we see the symbol for the wave function, Ψ, we visualize a multitude of images and associations, from a shapeless wave-particle to strict rows of formulas that represent it. Now we will get acquainted with at least three other "quantum pictographs": *tunnel effect, reaction cross-section, quantum resonance*. Without them our further discussion would for the most part be useless, if not without interest altogether.

Tunnel Effect

Radioactivity was discovered by Henri Becquerel in 1896. Six years later, in 1902, Rutherford and Soddy gave its explanation: the spontaneous decay of nuclei accompanied by a massive release of energy. As to the true source of this energy the first conjectures were made in 1913, but it was not until ten years later, after the work of Aston, that the hypothesis of the intranuclear origin of the energy of emitted alpha particles was firmly established. By that time Rutherford had already accomplished the artificial transmutation of elements, and the world had got accustomed by slow degrees to the idea that the nucleus, like the atom, has a complex structure. And although the internal structure of the nucleus still continued to be largely unknown, no one doubted any longer that alpha particles fly *out of* nuclei. But this knowledge, or rather belief, clarified almost nothing, however. It was still unclear:

Why are alpha particles ejected by the nucleus? (They are bound there so strongly!)

Why are alpha particles monoenergetic?

What affects the half-lives of nuclei and why are they so different?

What dictates the time and place of a decay event?

In 1928, within three years of the creation of quantum mechanics and thirty-two years of the discovery of radioactivity, these questions were answered nearly simultaneously by the Russian physicist Georgy Antonovich Gamow (1904-1968) and the American scientists Ronald Wilfrid Gurney (1899-1953) and Edward Condon (1902-1974). Their idea was remarkably simple and daring: they supposed that the motion of an alpha particle in the nucleus, just like the motion of an electron in the atom, is described by the Schrödinger equation

$$\frac{d^2\Psi}{dx^2} + \frac{2m}{\hbar^2}\Big(E - V(x)\Big)\Psi = 0 .$$

Much in this equation we know already: $\hbar$ is Planck's constant divided by 2π, m is the mass of the alpha particle, E is its energy, $\Psi = \Psi(x)$ is the wave function that describes the motion of the alpha particle in a potential field $V(x)$ at a distance x from the nucleus's centre. It must be confessed that even now, with all the advances of nuclear physics, the true nature of the nuclear forces, and hence the exact form of $V(x)$, is unknown. What we do know is that nuclear forces are attractive, and extremely strong: they are dozens of times stronger than the Coulomb repulsive forces between an alpha particle and the nucleus. Their range, however, is only 10^{-13}- 10^{-12} centimetre, i.e., tens and hundreds of times shorter than atomic scales.

The general form of $V(x)$ is shown in the figure. Outside the nucleus an alpha particle is repelled by the Coulomb field $V(x) = 2Ze^2/x$ of the

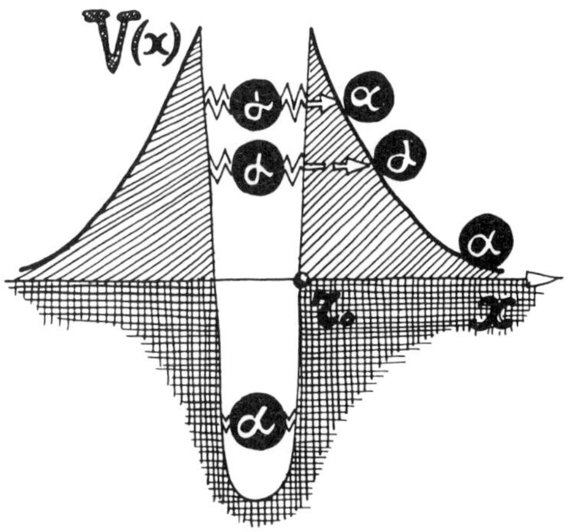

nucleus Z. At the boundary of the nucleus, at $x = r_0$, repulsion changes into attraction, and the alpha particles move about in a narrow deep *potential well*, being separated from the external world by a *potential barrier.* It appears that even this knowledge of the potential $V(x)$ is sufficient for us to understand the main feature of the alpha-decay of nuclei.

Take another look at the figure. It looks like a cross-section through a volcano, doesn't it? Well, even the very phenomenon of radioactivity is somehow reminiscent of a volcanic eruption: the alpha particles within the nucleus can be compared to magma that seethes in the mouth of the volcano and overflows in eruptions. Unlike magma, however, alpha particles obey quantum laws, and so their energy E is "quantized", i.e., it can only take on a discrete set of values E_1, E_2, ..., E_n. Therefore, within the "quantum volcano" they can only be located at certain heights. If an alpha particle moves about below the horizon line ($E_n < 0$), then no spontaneous emission of the particle is possible, the nucleus is stable, the volcano is dead. If $E_n > 0$, i.e., the energy of the particle is higher than when it is at rest remote from the nucleus, then an alpha decay will be at a premium in terms of energy and, as we shall now see, inevitable.

Deep within the volcano, magma seethes constantly but the volcano erupts only when magma ascends to the brim of the crater and overflows. Unlike magma (in conformity with the laws of quantum mechanics) an alpha particle can break through the "potential barrier" and escape out of the "potential well" even if its energy is insufficient to overcome it. This inherently quantum phenomenon is known as the *tunnel effect,* and we shall now see why. Just imagine that through the side of the volcanic cone a tunnel is made through which the magma now can pour out long before a regular eruption occurs.

Beyond the potential barrier the alpha particle is strongly pushed away by the Coulomb field of the nucleus and, as it slides down the "volcano's slope" it acquires at its foot exactly the kinetic energy E_n it had stored before the alpha-decay. Then we observe its track in a Wilson chamber or a flash in Crookes's sphinthariscope. These flashes and tracks are not as impressive and picturesque as, say, a fuming Vesuvius is, but we shall see later that alpha particles are just silent harbingers of nuclear catastrophes, and the "eruption of the nucleus" is a much more dreadful picture than a view of molten lava.

There is hardly a quantum phenomenon whose "pedestrian" descriptions have been attempted more often than the tunnel effect. To make the picture more graphic laymen were asked to visualize a car that vanishes from a locked garage, a man who walks through the wall of a prison, and so on and so forth. Such attempted explanations, however, do not clarify the specific features of the phenomenon and, moreover, are fundamentally false. Any attempt to describe quantum phenomena without employing the key concepts of quantum physics is bound to fail, however sincere the motives and serious the intentions.

The tunnel effect is, above all, a consequence of the wave-particle duality of quantum objects. We can, of course, find some analogies for it in classical physics, but we will have to look for them not in the descriptions of the motion of particles (or cars) but rather in the descriptions of the propagation and diffusion of waves. Any radiation is known to be able to bend around an obstacle, i.e., to get within the object's geometrical shadow. This ability is especially apparent if the wavelength of the radiation is comparable with the size of the obstacle. For example, centimetre waves, used to transmit television images, cannot go around a mountain, hence the need to construct repeater stations. Such problems are unknown with radiowaves whose wavelengths can be as long as hundreds of metres, since they easily bend around all the obstacles on the terrain.

Take another example — total internal reflection. If a beam of light is passed, say, through a block of glass and it falls on the glass-air interface at an angle that is larger than a certain critical angle φ_{cr}, then the beam will be totally reflected from the boundary. If then we press tightly to this block from below another identical block, then the beam will propagate in a straight line, as if there were no interface. But what will happen if we move the glass blocks slightly apart?

To begin with, what is "slightly"? For example, the diameter of a hair — is it large or small? From the point of view of geometrical optics, the question makes no sense, it is formulated incorrectly: "slightly" in comparison with what? But it makes sense in wave optics, since here we have a natural scale — the wavelength. For instance, for visible light ($\lambda \approx 5000$ angstroms = 5×10^{-5} centimetre) the thickness of a hair (10^{-2} centimetre) is large, and for thermal radiation ($\lambda \approx 0.1$ centimetre) small. When the gap width d is comparable with the wavelength λ of the radiation, some of this radiation will still get through the air gap from the upper glass block to the lower one, and the more so the smaller the gap d. This phenomenon is well established and it can be observed with ease. It is this optical phenomenon that is the closest counterpart of the tunnel effect in quantum physics.

The particle aspects of alpha particles (momentum, mass, charge, etc.) are especially obvious outside the nucleus, e.g., as they move through a Wilson chamber. The picture within the nucleus is dominated by the wave properties of alpha particles, namely frequency and wavelength. Clearly, the wavelength of alpha particles in the nucleus cannot be greater than the nucleus's size ($\lambda \leq r_0 \approx 10^{-12}$ centimetre) and their velocity is approximately one hundredth of the velocity of light; therefore the frequency of their vibrations within the nucleus, $\nu = v/\lambda$, can be as high as $\nu \approx 4 \times 10^{20}$ cycles per second. Colliding with the walls of the potential well the waves of alpha particles, as a rule, undergo "total internal reflection", but sometimes — with a tiny probability — they get through the barrier, just as light gets through the air gap separating the two glass blocks. The higher the energy of an alpha particle in the nucleus, the less is the width of the potential barrier it will have to get through and the higher the probability of finding it outside the nucleus.

The probability for an alpha particle to break through the potential barrier is

$$w = |\Psi(r_1)|^2 \approx \exp\left(-\frac{2}{\hbar}\int_{r_0}^{r_1}\sqrt{2m(V(x)-E)}\,dx\right).$$

For those ungrounded in advanced mathematics this expression will undoubtedly seem to be all too cumbersome. But it follows immediately from the Schrödinger equation. And if we take into account that it explains virtually all the details of alpha decay, we should even recognize that it is quite simple. The above probability is extremely small. For the radium nucleus it is as low as $w \approx 3.3 \times 10^{-32}$, but it is still finite, which is the fundamental difference between quantum objects (alpha particles) and classical ones (magma). Each second an alpha particle strikes the wall of the barrier $\nu \approx 4 \times 10^{20}$ times and each time with probability $w \approx 3.3 \times 10^{-32}$ it can leave the nucleus, i.e., each radium nucleus may disintegrate each second with the probability

$$\Lambda = \nu \times w = 1.4 \times 10^{-11} \text{ per second.}$$

One gramme of radium contains $N_0/A = 6 \times 10^{23}/226 = 2.7 \times 10^{21}$ nuclei, and of these $2.7 \times 10^{21} \times 1.4 \times 10^{-11} = 3.7 \times 10^{10}$ nuclei decay each second. It is this number of decays per second that physicists agreed to adopt as unit of radioactivity, they called it the *curie* to commemorate the contribution of the Curie family to the science of radioactivity. Accordingly, the average lifetime of the radium nucleus is $\tau = 1/\Lambda = 7.4 \times 10^{10}$ seconds $\approx$ 2300 years, and the half-life of radium is $T_{1/2} = 0.7\tau = 1600$ years.

Now, at last, we are in a position to answer all the questions concerning the nature, cause, and laws of radioactivity that we put at the beginning of this chapter.

Why are alpha particles ejected from the nucleus? Because radioactive nuclei are intrinsically unstable; they, like people, at their birth are already doomed to die.

Why are alpha particles monoenergetic? An alpha particle in a nucleus has a fixed quantum energy, and so when it leaves the nucleus it moves

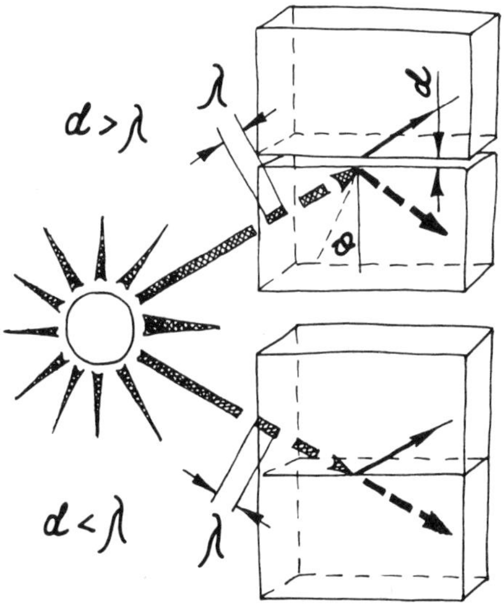

with a predetermined kinetic energy.

What affects the half-lives of nuclei? The half-life is largely determined by the energy of the alpha particles: the higher the energy, the narrower the barrier it has to get through, the higher the probability for an alpha particle to tunnel through it and the shorter the lifetime of the radioactive nucleus. This dependence is very strong; increasing the energy of alpha particles by only a factor of 1.5 changes their half-life billions of times.

Nucleus	E_n (MeV)	$T_{1/2}$
Uranium-238	4.2	4.5×10^9 years
Radium-226	4.8	1.6×10^3 years
Radon-222	5.5	3.8 days
Polonium-218	6.0	3 minutes
Polonium-214	7.7	1.6×10^{-4} seconds

The relation between the half-life of nuclei and the energy of the emitted alpha particles, known as the Geiger-Nuttall law, was discovered as early as 1909, but it had to wait twenty years to be adequately explained.

What dictates the time and place of a decay event? The laws of chance. A nucleus is a micro-object obeying the laws of quantum mechanics; therefore its description leans on the notion of probability. We can reliably predict the average lifetime of a nucleus and the average number of nuclei out of a large ensemble that will decay in a second. But we cannot predict the moment at which each individual nucleus will decay. The average lifetime of the nucleus of radium-226 is τ = 2300 years, but this does not mean that a radium nucleus that has just formed through the decay of thorium-230 will live precisely that long. It can disintegrate within a second or a million years. Unlike people whose mortality increases with age, radioactive nuclei do not age, and the probability of their decay is independent of how long they have lived at a given moment.

This feature of radioactive phenomena was noted by the Austrian physicist Egon Schweidler (1873-1948) as early as 1905. Essentially, this was the first evidence for the quantum nature of subatomic processes, although the profound implications of Schweidler's observations did not become clear until a quarter of a century later.

Looking back one cannot help thinking that alpha decay is a much simpler thing than the eruption of a volcano, and it is only the misconception that quantum mechanics is inherently difficult and unclear that prevents people from admitting it on the spot. In fact, no one has yet succeeded in predicting when a volcano will wake up and how many boulders it will throw out. On the other hand, the course of alpha decay can be predicted quite reliably.

The explanation of radioactivity, which followed so simply and naturally from the basic facts of quantum mechanics, impressed contemporaries a lot. The snow-ball of conflicting hypotheses and hopeless questions that had accumulated during thirty years around the phenomenon of radioactivity fell apart all of a sudden. Rutherford and Marie Curie lived to see it, and they witnessed the new knowledge illuminating the path that they had had to cover in the dark and explaining the significance of their enormous effort undertaken many years ago with the ardour and vigour of youth.

Modern nuclear physics has its origin in the work of Gamov, Gurney, and Condon. They made the world believe that quantum mechanics is not an esoteric science dealing with the structure of atoms and molecules (at first it was referred to exactly in this way — atomic mechanics), but rather a science dealing with all the phenomena of atomic and subatomic physics. (As luck would have it Edward Condon was born in the year that Rutherford and Soddy got an insight into the nature of radioactivity; and he was born at Alamogordo, at the edge of that desert where forty-three years later the fireball of the first atomic detonation would rise.)

Effective Cross-sections of Reactions

We now have a reasonable understanding of how alpha particles fly out of a radioactive nucleus. If and when they collide with the nuclei of other elements they can cause a nuclear reaction; in other words, they can get *inside* the nucleus. Quantum mechanics enables the probability to be worked out even for such processes. For example, it can explain why only one alpha particle out of 300,000 causes the famous Rutherford reaction

$$\alpha + {}^{14}\mathrm{N} \to {}^{17}\mathrm{O} + p\,.$$

To pick up the parlance of the trade, we will try for the time being to shake down the mesmerism of the word "quantum" and look at a simpler process. Imagine a magnified model of a crystal, like the one in the figure (nearly every school has one). Suppose then that each cubic centimetre of this "crystal" contains n_0 "nuclei" with radius r_0, the length of the "crystal" is l centimetres, the area of its end face is S square centimetres. Now we shoot at this end from a shotgun, the velocity of the shots being v centimetres per second. The cross-sectional area of one "nucleus" is $\sigma_0 = \pi r_0^2$ 0, and the total cross-sectional area of all the "nuclei" within the volume of the "crystal" is $\sigma_0 n_0 Sl$, i.e., it is the product of σ_0 by the total number of nuclei $n_0 Sl$ within the crystal volume Sl. The probability that each shot will hit any of the nuclei is

$$\frac{\pi r_0^2 \cdot n_0 Sl}{S} = \sigma_0 n_0 l\,,$$

which is the ratio of the total area $\sigma_0 n_0 Sl$ of the geometric cross-section of all "nuclei" within the volume of the "crystal" to the area S of its end face. The probability of hitting a nucleus within a unit time, w, can then be readily obtained by dividing the resulting value by the time $t = l/v$ taken by the shots to fly with velocity v through the crystal of length l, i.e.,

$$w = \sigma_0 v n_0\,.$$

Thus, if in each second through an area of 1 square centimetre one shot flies with a velocity v centimetres per second, the probability of its hitting one of the "nuclei" will be $w = \sigma_0 v n_0$.

This very important formula also holds in quantum mechanics, where by σ_0 is meant not the geometric cross-section of a nucleus $\sigma_0 = \pi r_0^2$ but some other, "effective", cross-section, which may be either smaller or larger than the geometric one — depending on the type of reaction we study. For example, if we are only interested in those collisions in which a nucleus disintegrates, it is clear that such collisions will always be fewer than simple collisions.

This decrease can be taken into account by some coefficient w_f, so that the formula becomes

$$w = w_f \sigma_0 v n_0 = \sigma v n_0 .$$

The quantity $\sigma = w_f \sigma_0$ is called the *effective cross-section* of a reaction. We might well visualize it as a certain "working" part of the geometric cross-section of nuclei. It is useful, however, not to forget its true physical meaning: the effective cross-section, or simply cross-section, is a measure of the probability of the nuclear reaction it characterizes.

In nuclear physics cross-sections are generally measured in special units, *barns*:

$$1 \text{ barn} = 10^{-24} \text{ square centimetre.}$$

The term may strike you as an incongruity, and there is a story behind it. During the war the papers concerned with uranium fission in the United States were classified. Therefore, even classified reports used not $^{235}_{92}U$ or $^{239}_{94}Pu$, but element-25 or element-49 — from the last figures of the atomic number and the mass number of the elements. Likewise, the values of cross-sections of nuclear processes were quoted in the secret units, barns. The originators of the term explained: "Because in nuclear physics the cross-section of 10-24 square centimetre is as large a value as a barn in everyday life." But the term stuck. The value of the cross-section unit is chosen, of course, not as accidentally as its name. Nuclear radii vary from $r_0 = 0.13 \times 10^{-12}$ centimetre (for hydrogen) to $r_0 = 0.8 \times 10^{-12}$ centimetre (for uranium), and hence their geometrical cross-sections $\sigma_0 = \pi r_0^2$ range from 0.05 barn

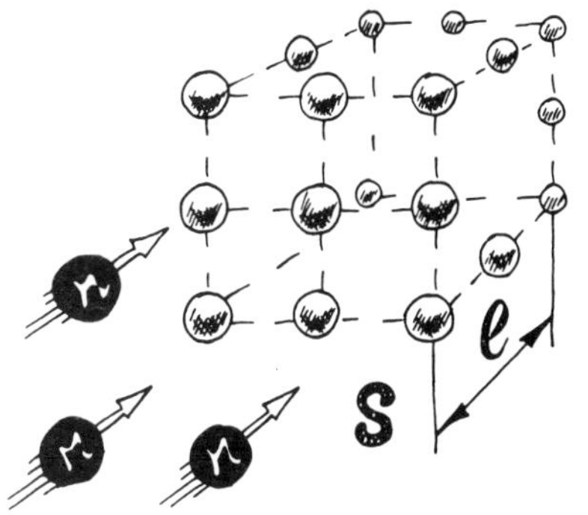

to 2.1 barns, i.e., they are comparable with the cross-section unit chosen.

Up to now we have tacitly assumed that effective cross-sections of reactions are independent of the energy of the oncoming particles. One might suspect that this assumption is fairly crude, and experimental evidence tells us that this is really so. In actual fact, effective cross-sections vary with the collision energy in a most involved sort of way, and for various reactions they may differ ten-, a thousand- or a million-fold. One merit of quantum mechanics is that it provides techniques of computing these cross-sections and thereby determining the relative probability of various nuclear reactions. It also follows from the relationships of quantum mechanics that the effective cross-section of elastic or inelastic scattering of nuclei is not equal to their geometric cross-section. This is an important point and we shall return to it.

Neutron Cross-sections

The fate of atomic energy hinged on the effective cross-sections of the interaction of neutrons with nuclei, or, for short, *neutron cross-sections*. This is no exaggeration: in fact, an error in their measurements influenced the fate of the world. In 1939 Germany decided to go ahead with an atomic bomb project. To accomplish this, as we shall soon learn, it was necessary to know the cross-section of the absorption of neutrons by carbon nuclei. The measurements were entrusted to the Nobel Prize winner Walter Bothe, whose experiments at one time contributed substantially to the discovery of the neutron. And he made an

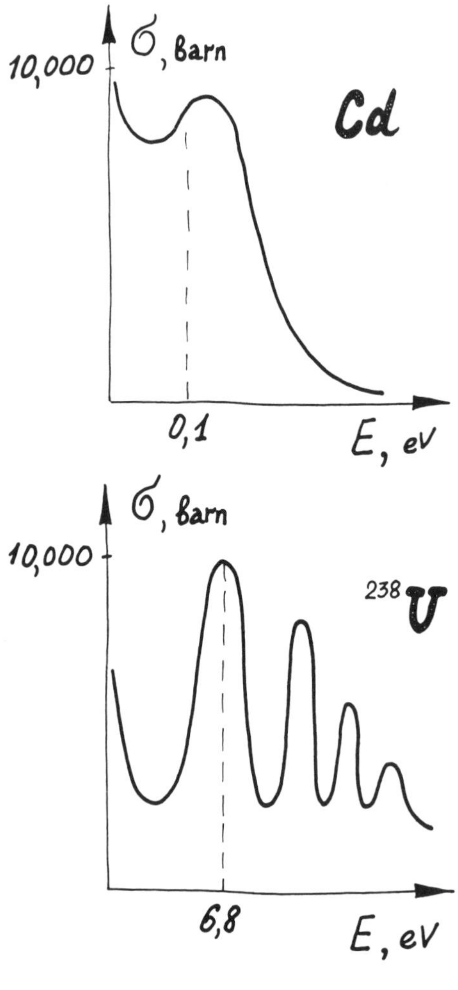

error. By a factor of 10 (hardly deliberately as many wished to believe at a later date). As a result, it was decided to construct an atomic pile using heavy water, which had to be imported from Norway, where the heavy-water plant was soon destroyed by Norwegian patriots.... The fate of the German uranium project was thus doomed.

Unlike the alpha particle, the neutron is devoid of electric charge and is always attracted by short-range nuclear forces. For the neutron, therefore, a nucleus is not a volcano but a basket, which can either retard it a little or capture it. If we stay with this analogy for a while, we can easily grasp that a slow neutron is more easily captured by the "nuclear basket" than a fast one. This is the case in practice: for neutrons with an energy of 1 MeV or more the cross-sections of nuclear reactions approximately coincide with the geometrical cross-sections of nuclei, but for lower collision energies the effective cross-sections behave in a bizarre manner.

The figure shows the cross-sections for the absorption of neutrons by nuclei of cadmium and uranium. For lower energies ($E < 100$ eV) they are very large — 10,000 barns or even more. Such peaks are known as *resonances* in cross-sections. Further, it is seen in the plots that at very low energies of the neutrons (such neutrons are called thermal, because the average energy of atomic motion at room temperature $t° = 20°C$ is about 0.04 eV) the cross-sections grow fast and monotonically as the energy is decreased. To understand better this behaviour of neutron cross-sections we should again remember the quantum nature of nuclear reactions, specifically the wave properties of the neutron.

In 1936, four years after the neutron was discovered, Walter Elsasser (even in "pre-quantum" days he pointed to de Broglie waves as a reason for the anomalies in Davisson and Germer's experiments) predicted that the neutron, like the electron, must possess wave properties. In that same year Peter Preiswerk (1907-1972) and Hans Halban (1908-1964) who worked at the Radium Institute in Paris confirmed his assumption experimentally. The neutron mass is $m_n = 1.66 \times 10^{-24}$ gramme, the velocity of a 1-eV neutron is $v = 1.4 \times 10^6$ centimetres per second, and the appropriate wavelength can readily be worked out from de Broglie's formula:

$$\lambda = \frac{h}{mv} = \frac{6.67 \times 10^{-27}}{1.66 \times 10^{-24} \times 1.4 \times 10^6} = 2.8 \times 10^{-9} \text{ cm.}$$

If, as usual, we take the "radius" of a quantum particle to be $r_0 = \lambda / 2\pi$, then at 1 eV the "radius of the neutron" turns out to be $r_0 \approx 4 \times 10^{-10}$ centimetre, i.e., it is 500 times larger than the nuclear radius of uranium.

We have so far supposed, implicitly and explicitly, that the neutron is smaller than the nucleus. We likened it, for example, to a shot that strikes a billiard ball. This all seemed to us to be so natural that we accepted this picture without any reservations and, as is clear now, without any grounds. If a "missile" is larger than a target, then it is only natural to interchange them and consider that the nucleus hits the neutron whose "geometrical cross-section" at 1 eV is $\sigma_0 = \pi r_0^2 = 1.5 \times 10^5$ barns, i.e., 150,000 barns. Real cross-sections are as a rule smaller than σ_0, but at times, namely at *resonance energies*, the effective cross-sections for reactions of neutrons and nuclei reach their upper limit σ_0, and the plots of the cross-sections show characteristic maxima — *resonances* at those energies.

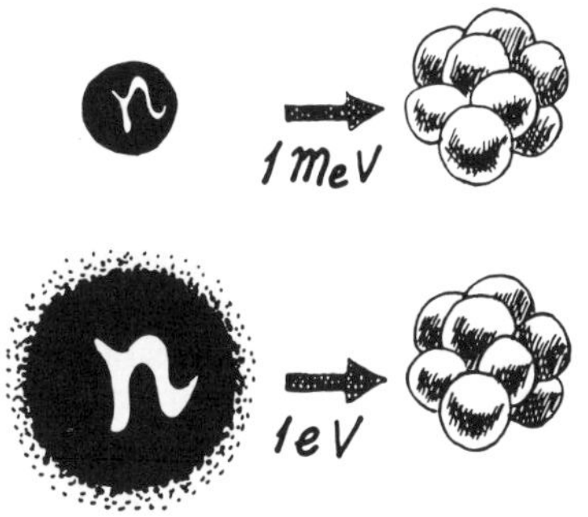

The why and how of these resonances can be understood fairly easily. Suppose that a neutron collides with a $^{238}_{92}U$ nucleus and is captured by it. In the process the neutron binding energy is liberated, which is about 6.8 MeV, and a new uranium isotope $^{239}_{92}U$ is formed. The nucleus, like the atom, is a complex quantum system whose energy takes on different, but always fixed, *quantum* values. Therefore, this system can go over from one state to another only by having absorbed a fixed portion (quantum) of energy.

Recall the famous experiment of Franck and Hertz: they irradiated mercury atoms with electrons but the atoms did not absorb the energy as long as it was insufficient to excite them. But when the electron energy attained a value of 4.9 eV the probability for the atoms to be excited shot up or, to use modern terminology, the cross-section of the excitation of mercury atoms with electron impact showed a resonance at 4.9 eV.

Likewise, when uranium is exposed to neutrons the reaction

$$n + {}^{238}_{92}U \rightarrow {}^{239}_{92}U^*,$$

which produces the nucleus of a new uranium isotope in an excited state, occurs with a large probability (has a large cross-section) only at certain *resonance energies*. A cross-section may have quite a few resonances like these, e.g., the absorption of neutrons by uranium-238 nuclei has a cross-section with eight resonances within the interval of neutron energies from 5 eV to 200 eV.

The uranium isotope immediately goes over to the ground state by emitting a gamma-quantum, i.e., in actual fact the reaction follows the scheme

$$n + {}^{238}_{92}U \rightarrow {}^{239}_{92}U^* \rightarrow {}^{239}_{92}U + \gamma.$$

This is the most common kind of reaction, the so-called (n,γ) reaction, one that at a later date, when a nuclear reactor was built, would cause much trouble.

For $E < 1$ eV there are no resonances, but as the neutron energy is decreased cross-sections themselves keep on growing, as one might expect from the expression

$$\sigma_0 = \pi r_0^2 = \pi \left(\frac{h}{mv} \right)^2.$$

However, rigourous quantum-mechanical calculations indicate that real cross-sections do not normally grow so fast, but follow the famous "law of $1/v$", i. e.,

$$\sigma = \frac{C}{v},$$

where v is the velocity of the neutron, and C is some constant. We now have a clue as to the effect observed by Fermi on that sunny day in October 1934, when he placed a neutron source in a goldfish pool. Colliding with hydrogen atoms, neutrons from the radon-beryllium source were slowed down and therefore absorbed by nuclei of elements with a higher probability.

It's all elementary, isn't it? Especially if we remember that here we are dealing with processes inside atomic nuclei, i.e., inside micro-objects that

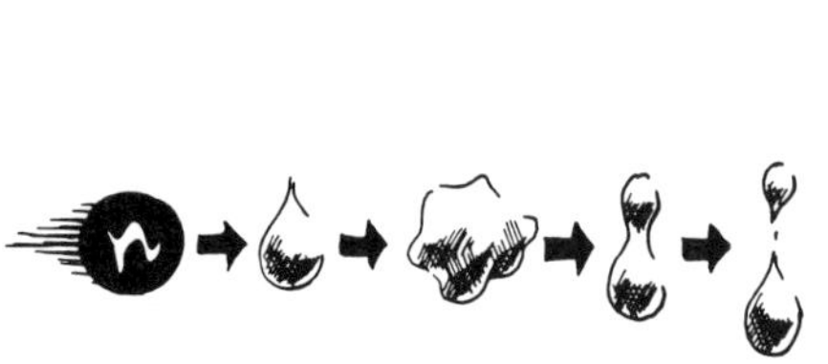

are about 10^{-12} centimetres across, whose architecture and behaviour have essentially been created by the power of our imagination. This is, of course, not an arbitrary picture — it rests on a large body of experimental evidence, but this does not detract from our amazement at the power of the human imagination.

Nuclear Fission

In 1936, two years before the fission of uranium was uncovered, Niels Bohr devised his "droplet model of the nucleus". This mental endeavour to gain an insight into the intranuclear processes took the author much scientific audacity. By that time the world had already seen that the enigmatic nucleus was an extremely strong entity; and then Niels Bohr came along and proposed that the nucleus be imagined as a liquid droplet, consisting of protons and neutrons. Central to his conception was the notion of the *compound nucleus*, which results when a neutron, a proton, or an alpha particle are captured by a conventional nucleus. Bohr went on to contend that the new particles lose their individuality, they, as it were, dissolve in the nucleus, and the energy they bring in is redistributed over all the nucleons of the nucleus. The nucleus gets excited and, to get rid of the excess energy, it may either emit a gamma-quantum, or throw out an electron and an antineutrino, or an alpha particle. Elaborating upon his ideas, Bohr was able to achieve a fusion of scattered facts from the physics of nuclear reactions and even to derive a formula for the binding energy of nuclei, which showed fairly good agreement with experiment.

When he found out about the discovery of uranium fission, Bohr immediately understood that it is accounted for in the most natural manner in terms of his droplet model of the nucleus. Sometimes an excited nucleus is deformed, so much so that it is easier for it to split in two to get rid of the excitation energy than to emit several particles. Together with his student John Archibald Wheeler (b. 1911), Bohr developed the theory of nuclear fission. (The problem was solved independently by the Soviet physicist Yakov Il'ich Frenkel (1894-1952).)

We know now a great deal about the process of fission of uranium. There are accurate measurements of the cross-section of the fission of ${}^{235}_{92}U$ by thermal neutrons; it was found to be rather large — 582 barns. The uranium nucleus is known to disintegrate in about fifty different ways; their probabilities vary widely but are generally not higher than 8 per cent. For example, barium, which was first found by Hahn and Strassmann in uranium fission fragments, results from the fission processes shown here (their probabilities are about 6 per cent):

$$\left\{\begin{array}{l} {}^{141}_{54}Xe \xrightarrow[1.7\ s]{\beta} {}^{141}_{55}Cs \xrightarrow[25\ s]{\beta} {}^{141}_{56}Ba \xrightarrow[18\ min]{\beta} {}^{141}_{57}La \xrightarrow[3.7\ h]{\beta} {}^{141}_{58}Ce \xrightarrow[33\ days]{\beta} {}^{141}_{59}Pr \\ {}^{140}_{54}Xe \xrightarrow[16\ s]{\beta} {}^{140}_{55}Cs \xrightarrow[66\ s]{\beta} {}^{140}_{56}Ba \xrightarrow[12.8\ days]{\beta} {}^{140}_{57}La \xrightarrow[40.2\ h]{\beta} {}^{140}_{58}Ce \end{array}\right.$$

$$n + {}^{235}_{92}U \to {}^{236}_{92}U^* \to (2 \text{ or } 3 \text{ neutrons})$$

$${}^{93}_{38}Sr \xrightarrow[7.9\ min]{\beta} {}^{93}_{39}Y \xrightarrow[10.3\ h]{\beta} {}^{93}_{40}Zr \xrightarrow[1.1 \times 10^6\ years]{\beta} {}^{93}_{41}Nb^* \xrightarrow[12\ years]{\gamma} {}^{93}_{41}Nb\,.$$

One fission option also yields ${}^{141}_{57}La$ with a half-life of 3.7 hours; it was first observed by Irène Curie and Pavle Savic. Oddly enough, the uranium nucleus almost never splits into equal halves. (As a rule, the mass ratio of fission fragments is 3:2.)

Fission fragments include more than one hundred various radioactive isotopes with a wide spectrum of half-lives. These isotopes are separated, sorted out, and used as "isotopic tracers", or "labelled atoms", in wide variety of fields of scientific research and technology.

In addition to fragments, fissile uranium nuclei also yield two or three neutrons — depending on the fission mode. Averaged over all fission modes one fission event produces ν_f = 2.42 neutrons with an energy of about 1.3 MeV. And if chance would have it that this number were a bare 10 per cent less, a natural-uranium reactor would be impossible.

Quantumalia

Labelled Atoms

In their summary of work in 1903 Rutherford and Soddy viewed radioactivity not only as a research object but also as a research tool. They wrote: "...radioactivity may be used to observe chemical transformations that occur in matter." They compared this scientific technique to spectral analysis.

Eight years later Rutherford proposed to George Charles de Hevesy (1885-1966), a young chemist from Hungary, who was his postgraduate student at the time, to separate radium *D* from lead. We now know that the task is insoluble by chemical means, since radium *D* is simply the radioactive lead isotope ^{220}Pb. But after many a futile attempt it occurred to Hevesy and the Austrian radiochemist Frisch Adolf Panet (1887-1958) to invert the problem, namely, to use radium *D* to study the chemical reactions of lead by taking advantage of the fact that lead and radium *D* are inseparable.

This technique, which was later called the "method of radioactive indicators", "tracers", or "labelled atoms" (1943 Nobel Prize for Chemistry), appeared to be rather effective as a probe into chemical reactions and the structure of matter. It is now in common use in physics and biology, medicine and metallurgy, archaeology and forensic science.

Artificial radioactivity yielded dozens of new radioactive isotopes, and uranium fission fragments from today's nuclear reactors contain more than one hundred different radioisotopes. Using them man has learned fascinating facts even about his own body: the total volume of blood in the body, the workings of the internal organs, and many other things. For example, if you drink some saline water to which a little of the radioactive sodium isotope ^{24}Na has been added, in two minutes you can detect it, using a Geiger counter, in your fingers; in an hour it will spread throughout the body, and in another three hours it will start to be discharged from your system. It appears that one cycle of blood circulation lasts only twenty-three seconds, that the atoms of the human body keep being replaced incessantly by new ones, which get into the system with food, and the total replacement cycle lasts about one year, i.e., each year a person is renewed nearly completely while retaining at the same time his individuality and wholeness.

The energy produced by the breaking down of the atom is a very poor kind of thing. Anyone who expects a source of power from the transformation of these atoms is talking moonshine.

Ernest Rutherford

Now, almost a century after the discovery of radioactivity, it seems strange that it had been overlooked before. It is everywhere around us and even within us. In the human body each second there occur more than 20,000 decays of potassium-40, about 300 decays of carbon-14, and four decays of radium-226, an element which was discovered with great difficulty by Marie and Pierre Curie. Indeed, there are many facts like this in the history of science. For example, radiowaves are all around us, but nobody suspected this before the work of Henrich Hertz, and it took Louis Pasteur to discover that each cubic centimetre of the air we inhale contains millions of various bacteria.

Radiocarbon Dating

In the wide expanses of the Universe each second there occur space catastrophes: new stars are born, old stars explode and throw out into interstellar wastes nuclei of hydrogen, helium, and other elements. The galactic magnetic fields accelerate these nuclei to incredible energies — hundreds and thousands of times higher than can be achieved today in our accelerators. These *cosmic rays,* as they were dubbed by Robert Millikan, enter the Earth's atmosphere and cause all manner of nuclear reactions when colliding with atmospheric atoms. They break nuclei into fragments, and knock out secondary protons and neutrons from them. These secondary particles cause new reactions; one of them in 1946 attracted the attention of the American physicist Willard Frank Libby (1908-1980).

He noticed that the secondary neutrons colliding with nuclei of nitrogen atoms, which account for 79 per cent of the terrestrial atmosphere, can knock out a proton from them and turn them into carbon nuclei:

I. Joliot-Curie

$$n + {}^{14}\text{N} \to {}^{14}\text{C} + p\,.$$

Unlike the conventional carbon isotope ^{12}C, the rare isotope ^{14}C is radioactive; it decays as follows:

$$^{14}\text{C} \to {}^{14}\text{N} + e + \bar{\nu}$$

and its half-life is $T_{1/2}$ = 5730 years. Each second about 4000 cosmic protons pass through an area of 1 square metre, which after a cascade of nuclear transformations yield about two nuclei of ^{14}C in each cubic centimetre of the upper atmosphere. These carbon isotopes soon combine with oxygen and as carbon dioxide they are taken up by plants and then by animals and man. As a result of the carbon turnover an equilibrium concentration of ^{14}C is achieved in the organic matter of all plants and animals. The radioisotope can be detected using a conventional Geiger-Müller counter. It turns out that 1 gramme of organic carbon — be it wood, grass, animal bones, or bacteria — contains about 70 billion atoms of ^{14}C, and each minute fifteen of them decay.

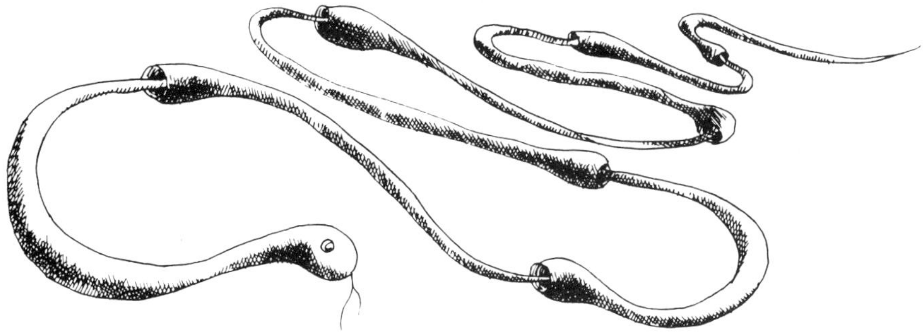

F. Joliot-Curie

But this picture is only valid for *living* matter. When a plant, an animal, or a man dies, this stops metabolism, which is the essence of life, and so in the remains the equilibrium concentration of ^{14}C atoms no longer persists. From the moment a living thing dies ^{14}C atoms in it decrease in number following an established law: in 5730 years their number will be halved, in another 5730 years their number will be halved again, and so on. And this implies that 1 gramme of carbon in a tree that has just been felled will in 5730 years emit not fifteen but only eight impulses per minute, in another 5730 years four impulses, and so on.

Now you will have surmised how you can date any archaeological artifact — be it the ashes of a primeval camp-fire, a mummy from an Egyptian pyramid, or the remains of the ship *Argo.* You just count the number of decays in one minute in 1 gramme of carbon from the sample (for convenience it is normally burned and the resultant carbon dioxide is tested for radioactivity).

The principle behind this dating method is fairly ancient: sand and water clocks use this idea, for instance. It is really fascinating to have in radiocarbon dating a quaint fusion of the most advanced achievements of nuclear physics and hoary antiquity, the cosmic rays from the world's depths and the minute work of living cells.

This discovery earned Libby the Nobel Prize for Chemistry in 1960, although the significance of this method goes far beyond the confines of chemistry or physics.

Atotarkho, Irokese

Chapter Sixteen

Chain Reaction • Nuclear Reactor

The Titan Prometheus — grandson of Uranus, son of Themis, and brother of Atlas — was especially helpful to Zeus in his battle with Cronus: he persuaded the great goddess Ge to side with Zeus and advised Zeus to throw the defeated Titans into the gloomy Tartarus. After his victory Zeus decided to destroy the human race and in its stead to create a new and better one. To begin with, he deprived people of fire as a punishment for Prometheus's tricking him in favour of people when sharing a sacrificial ox. In return Prometheus stole fire from Hephaestos's smithy and returned it to the people; he also taught them arts and crafts, land cultivation and animal husbandry, reading and writing.

The infuriated thunderer had Prometheus chained to a cliff over the sea in distant Scythia, and every day sent an eagle to tear at his immortal liver, which regenerated each night. Also, he ordered Hephaestos to manufacture out of clay and water a woman, Pandora — the first woman of the new generation — and to marry her to Epimetheus, Prometheus's brother. Aphrodite made Pandora look beautiful, Hermes instilled slyness into her, and Zeus gave her a closed jar containing the human woes. Tempted by curiosity Pandora took the lid off the jar and the human ills, hard work, and diseases flew out to wander among mankind. Hope alone remained within; it has served as happiness for people ever since.

This ancient legend has not faded even after so many years of the promiscuous use of its fragments in so many metaphors concerned with the discovery of atomic energy. The most quoted personage is the ill-starred Pandora with her jar and the atom bomb; another one is Prometheus, who produced atomic fire by paying the price of suffering and internal conflict between duty and morality. And the least remembered part of the legend is its end: the hero Heracles comes along and kills the eagle and shatters the chains. Zeus pardons the martyr, and the wise centaur Chiron dies instead, renouncing his immortality in favour of Prometheus.

The struggle of man for atomic fire even for us contemporaries is already a legend. The dramatic story of the mastering of atomic energy is inseparable from the tragic turn of fate of the entire human race during the most inhumane of wars. All the episodes of this history are important and significant, but to assess adequately their meaning we must first gain some understanding of the physical processes underlying the historic events.

Chain Reaction

Late in the evening of 21 December 1938 Otto Hahn and Fritz Strassmann finished writing the paper in which they had to confess that irradiation of uranium with slow neutrons then gave rise to barium, lanthanum, and cerium. That evening Hahn could hardly foretell all the implications of their discovery, although he had a feeling that they were important. He telephoned his friend Paul Rosbaud, the publisher of the weekly *Naturwissenschaften,* and asked him to publish their report as soon as possible. Rosbaud had the article inserted into the current issue and it was published within two weeks, on 6 January 1939. The article proved to be that last stone which starts an avalanche: during the year 1939 alone there appeared more than one hundred papers on the problems concerned with uranium fission. (Six years later this paper of Hahn would gain for him the Nobel Prize for Physics for 1944; he would learn about this in a fit of deep depression caused by the reports about Hiroshima and Nagasaki, while he was in custody in an English castle.)

At the time Paul Rosbaud was leafing through Hahn and Strassmann's paper, Enrico Fermi was boarding a ship to leave fascist Italy for ever. On 2 January 1939 he went ashore in New York harbour: "The Italian navigator has just landed in the New World"; with these words four years later Arthur Compton would announce to the directors of the US atomic bomb project (Manhattan) the successful start of the first nuclear reactor by Fermi's group. On 7 January, the day after the publication of Hahn and Strassmann's paper, Niels Bohr was leaving for the United States. On 16 January he arrived in New York, unaware that on that very day Otto Frisch had submitted two papers to *Nature.* In one of them, entitled "Fission of Uranium with Neutrons: New Type of Nuclear Reaction" (printed on 18 February) Frisch and Meitner gave an explanation of Hahn and Strassmann's discovery and introduced for the first time the term "nuclear fission"; in the other Frisch reported that he had recently observed fragments of uranium fission using a simple ionization chamber. (For years among friends Otto Frisch would be nicknamed "Fission".)

Niels Bohr and Enrico Fermi met on 26 January in Washington at a conference on theoretical physics, where Bohr reported on the discoveries of the last weeks. The reaction of the physicists was vehement and unanimous; it resembled very much the reaction of Bohr himself to the story of Frisch: "How could we possibly have overlooked this for so long?" he said, tapping himself on the forehead. Within the next two days or so the discovery of nuclear fission was confirmed in at least five laboratories in the United States; on 26 January this was borne out by Joliot-Curie in Paris (his note in the proceedings of the Academy was already published by 30 January), and by the time of the publication of the paper by Frisch and Meitner nuclear fission had already been observed by dozens of scientific researchers in Copenhagen and New York, Washington and Paris, Leningrad and Warsaw.

Speaking at the conference after Bohr, Enrico Fermi pointed out that in addition to two nuclear fragments a fissioning uranium nucleus must yield several neutrons, which in turn may cause further fission events, i.e., in uranium it is possible to obtain a *chain reaction of fission* accompanied by release of enormous energy. Fermi's conclusion was quite natural (although at the time it was not that obvious; for instance, Frisch and Meitner overlooked that sequel to their hypothesis). But it was in disagreement with the observed facts: nobody had ever seen a piece of uranium explode when bombarded with neutrons.

Pondering this contradiction, Bohr remembered that four years ago Arthur Dempster had found, using his advanced mass-spectrometer, the rare uranium isotope $^{235}_{92}$U, and that uranium appeared to consist of 99.28 per cent of $^{238}_{92}$U and only 0.72 per cent of $^{235}_{92}$U. Bohr supposed that slow neutrons split uranium-235 and the fast neutrons produced in this fission are immediately absorbed by uranium-238. And so the neutron flare-up goes out, like a match thrown on a pile of green firewood. Experiment confirmed this hypothesis of Bohr only a year later, in April 1940, but it was accepted as fact at once and was taken into consideration in subsequent research work.

Three new questions presented themselves:

How many neutrons and of what energy are produced by each fission event of uranium-235?

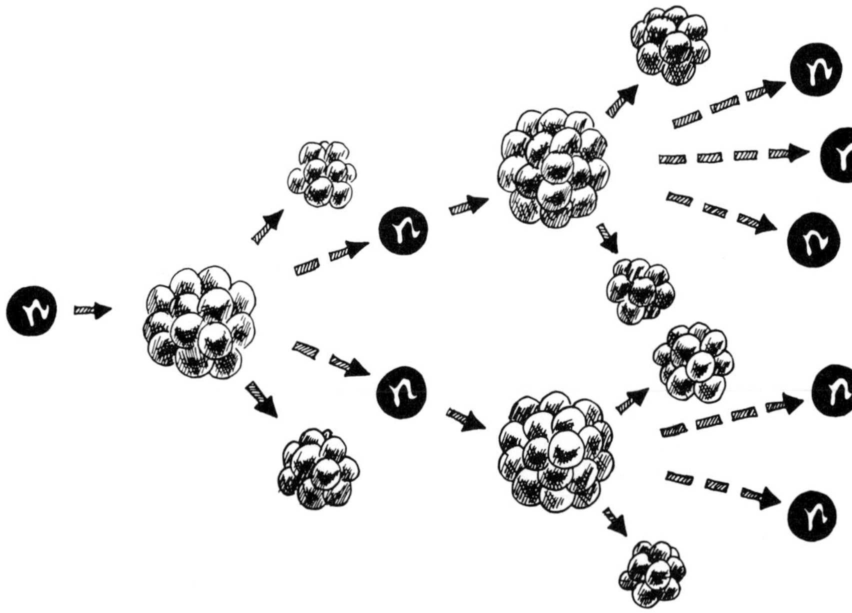

What happens to nuclei of uranium-238 after they have captured a neutron?

What conditions are required for a sustained chain reaction in uranium to occur?

An answer to the first question came in mid-March from four research groups: Frédéric Joliot-Curie, von Halban and Kowarski of France, Flerov and Rusinov of Russia, Fermi, Anderson and Hanstein, and also Szilard and Zinn of the United States. It appeared that each fissioning nucleus of uranium-235 emits two or three secondary neutrons with an average energy of 1.3 MeV. (The exact number of fission neutrons $\nu = 2.42$, measured later, remained a state secret until 1950.)

When investigators were seeking to answer the second question they recalled some work by Meitner, Hahn, and Strassmann. When in 1937 they repeated Fermi's experiments they noticed that uranium nuclei especially effectively absorb neutrons with an energy of 25 eV (so it was thought at the time; in reality the energy is 6.8 eV). This *resonance absorption* of neutrons is always accompanied by beta activity with a half-life of 23 minutes whereas when slow, or "thermal" neutrons with an energy of 0.04 eV are absorbed we observe a mixture of various half-lives. The reason for this now became clear: uranium-235 is best fissioned by slow neutrons (i.e., its fission cross-section is large), and the resultant fission products have various half-lives. In contrast, for electron volt neutrons the cross-section of the resonance absorption by uranium-238 is much higher than for the fission of uranium-235; therefore the prevailing process is the transmutation of uranium-238 into neptunium-239 by beta decay.

Consequently, in order that a chain reaction be possible it is imperative that a *neutron moderator* be employed, so that it would, first, reduce the energy of neutrons 10 million times — from an energy of 1 MeV, with which they are released as a uranium-235 nucleus disintegrates, to an energy of 0.1 eV — and, second, be so efficient that the neutrons would be moderated before they collided with a uranium-238 nucleus. Lastly, the moderator itself must not absorb neutrons, i.e., its neutron capture cross-section must be negligible.

The most efficient moderator is hydrogen (as was shown adequately by Fermi's experiments in 1934); unfortunately, it also appeared to be a good absorber. The reaction

$$p + n \rightarrow d + \gamma,$$

i.e., the collision of a neutron with a proton resulting in the formation of deuterium, a heavy hydrogen isotope, and the emission of a gamma-quantum, has a cross-section of 0.33 barns. This is far less than the cross-section for resonance capture in uranium-238, which is 10,000 barns, but still too much.

In the summer of 1939 the Soviet theoretical physicists Yakov Borisovich Zeldovich (1914-1987) and Yuli Borisovich Khariton (b. 1904) for the first time calculated the kinetics of the chain reaction of fission in an aqueous solution of uranium. Their conclusions were pessimistic: such a *homogeneous reactor* will only work provided the concentration of uranium-235 is brought up to 2.5 per cent from the 0.72 per cent to be found in any sample of natural uranium.

So emerged the first major problem of harnessing atomic energy — the *separation of uranium isotopes.* At first the problem was considered so intractable that for two years no more thought was given to it. It seemed highly unlikely that it would ever be possible to separate the chemically identical atoms of uranium isotopes whose nuclear masses differ by only 1.5 per cent. But war changes views on the possible and impossible. So already by 1944 uranium isotope separation plants were in operation; they were huge four-storey blocks half a kilometre wide and a kilometre long and they consumed electric power that amounted to the output of a large power station. Engineering details of the process are still classified but its underlying idea — gaseous diffusion — is well known.

If a mixture of two gases is passed through a porous membrane with holes slightly greater in diameter than the atoms themselves, beyond the membrane there will be a slightly higher concentration of the lighter gas than before the membrane; by repeatedly performing the process one can in principle achieve total separation of the gases. By some whim of nature, among about 200 uranium compounds there is only one that is gaseous — uranium hexafluoride, UF_6 — a toxic gas that at 56°C condenses into acidular crystals. If this gas is then passed through a porous membrane, one will find 0.14 per cent more uranium-235 after the membrane than before it, and by using several thousands of such membranes one can eventually separate uranium-235 from uranium-238. All this only became possible five years later but in the summer of 1939 scientists had another task on their hands — they were seeking ways of accomplishing a sustained nuclear reaction in natural uranium.

Another moderator was needed instead of

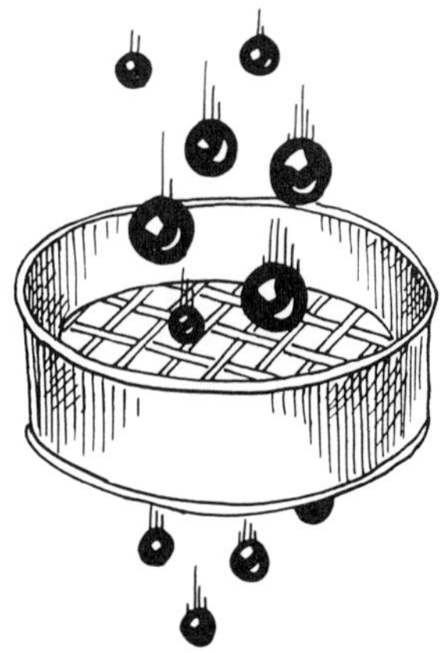

water. There was not too much choice: either carbon (neutron capture cross-section 0.0034 barn) or heavy water D_2O, i.e., water with hydrogen atoms replaced by deuterium atoms (neutron capture cross-section 0.0012 barn). It is easily seen that heavy water is preferable. But it is not readily available: 1 litre of common water contains as little as 0.15 gramme of heavy water. What is more, in 1939 heavy water was only obtainable from one small plant in Norway. Of the two options Joliot-Curie and Heisenberg independently chose heavy water, whereas Fermi, Szilard, and Kurchatov elected to try their luck with graphite.

The condition for a chain reaction to occur is generally given using the simple and famous "four-factor formula":

$$k_\infty = \eta\varepsilon\varphi\theta$$

or "kay-infinity" is the product of "eta", "epsilon", "phi", and "theta". The multiplication factor, k_∞, is the number of secondary neutrons produced in a reactor of infinite dimensions for each primary fission neutron. It is clear that a *divergent chain reaction* is only possible when $k_\infty > 1$, i.e., each successive "generation" of neutrons will be larger than the previous one.

The factor η is the number of secondary thermal neutrons in natural uranium, which is different from the average number of fission neutrons in pure uranium-235. This is because when uranium-235 nuclei capture slow neutrons they fission only in 84 per cent of cases; in the remaining 16 per cent of cases they emit a gamma-quan-

tum and turn into uranium-236 nuclei. In addition, thermal neutrons are also captured by uranium-238 nuclei, albeit to a lesser extent than those at resonance. But natural uranium contains 149 times as much uranium-238 as uranium-235; therefore for this mixture η = 1.34 is far smaller than the initial multiplication factor ν = 2.42, although it is still larger than unity.

The factor ε accounts for the fact that fast fission neutrons, while over 1.5 MeV in energy, are also capable of splitting uranium-238 nuclei, i.e., they make a sizable contribution to the generation of secondary neutrons. For uranium-graphite reactors ε = 1.03.

A source of trouble was the factor φ — the probability for neutrons to escape resonance capture as they are moderated in uranium-238. If all the fission neutrons were slowed down to thermal energies without being lost in uranium-238, the factor φ would be unity. Like any ideal, the value φ = 1 is unachievable, although it can be approached. A way of improving φ was thought up almost at once, in the summer of 1939; in France, the United States, and Germany the idea originated of the *heterogeneous reactor*. Its principle is simple: instead of a homogeneous mixture of uranium and moderator it uses blocks of uranium spaced at certain intervals, rather like atoms in a crystalline lattice, the gaps being filled up with the moderator. In that case, fission neutrons, upon leaving uranium blocks with an energy of about 2 MeV, will travel for most of their path through the moderator and by the time they reach another uranium block they will have passed the dangerous resonance energies. What is more, energetic fission neutrons pass through the bulk of the uranium block too quickly to moderate to the resonance energies, thereby further improving the efficiency of the reactor. (In the work of Isai Isidorovich Gurevich (b. 1912) and Isaak Yakovlevich Pomeranchuk (1913-1966), which became an important element of the Soviet uranium programme, this phenomenon was called the "blocking effect".) By adequately selecting the spacing between uranium blocks it became possible to achieve φ = 0.93, a significant improvement over

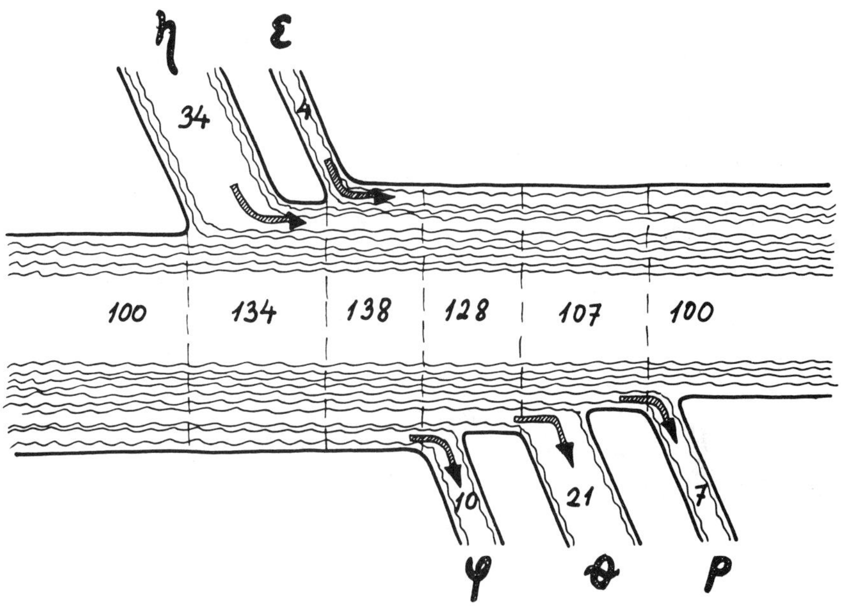

$\varphi = 0.65$ for the homogeneous reactor.

The factor θ in the four-factor formula is the probability for a neutron to escape being captured in the moderator and in various impurities. For pure graphite the greatest achievement is $\theta = 0.84$. It is extremely important that graphite be pure: impurities even in trace amounts, for example 3-4 atoms of boron per million atoms of carbon, make graphite unsuitable for moderation. (The neutron capture cross-section for boron nuclei is enormous, 755 barns; therefore, only when the boron concentration is of the order of 10^{-6} may we ignore the capture of neutrons in it in comparison with the capture in carbon, for which $\sigma_{cap} = 0.0034$ barn.)

Thus, for the natural uranium-graphite heterogeneous reactor we have

$$k_{\infty} = \eta\varepsilon\varphi\theta = 1.34 \times 1.03 \times 0.93 \times 0.84 = 1.07,$$

i.e., in an infinite reactor a nuclear chain reaction is possible.

In a real reactor of finite size some of the neutrons are lost, leaving the reactor volume through its boundary; therefore the real multiplication factor k is smaller than k_{∞} and is

$$k = k_{\infty} \times P,$$

where P depends on the size and shape of the reactor, but at all times it is smaller than unity. Clearly, there are some *critical dimensions* of the reactor for which $k = k_{\infty} \times P = 1$. It now remained to find these dimensions.

In December 1940 von Halban and Kowarski in England, where they had found refuge with a stockpile of uranium and heavy water after France had been occupied by the Nazis, established that for a self-sustaining nuclear reaction to proceed one should take 5 tonnes of heavy water and about as much uranium arranged in it in a certain manner. A similar conclusion was drawn by Heisenberg in Germany. But there was not so much heavy water in the world at that time, and one could hardly expect that it would be forthcoming in the very near future. And so Fermi set about

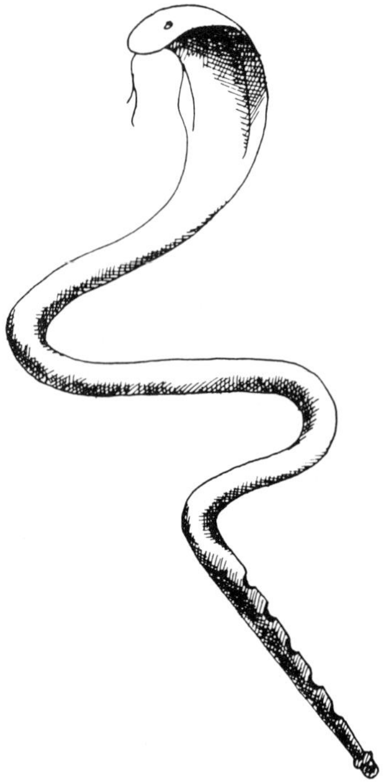

determining the critical dimensions of the uranium-graphite reactor and achieved his goal within a time span of two years.

To summarize, a nuclear reactor is feasible. But can it be controlled? Will it not explode once the value of the multiplication factor k exceeds unity? It turned out that here too nature cooperated with man.

In the niagara of publications following the discovery of uranium fission there was one that at first did not attract much attention. In March 1939 Roberts, Meyer, and Wong of Columbia University found that about 1 per cent of neutrons in uranium fission are not ejected simultaneously with the fragments but a short while later, in 0.2 seconds, 0.9 seconds and even 56 seconds. The physics of this soon became clear: the *delayed neutrons* come from the fragments, whereas *prompt neutrons* come from uranium nuclei.

The fragments — more than one hundred isotopes of about forty elements from the middle of the periodic table — are overburdened with neutrons and tend to shake them out. Most isotopes get rid of neutrons by beta decay, i.e., by turning them into protons, just as was found in Fermi's experiments on the absorption of neutrons by nuclei. But a small fraction of the resultant isotopes, after some hesitation, emit an extraneous neutron complete, without splitting into a proton and an electron. And it is precisely these delayed neutrons, which make up only 0.64 per cent of the neutron generation, that make it possible to control the operation of a nuclear reactor. In fact, if for some reason the number of neutrons in the reactor shoots up abruptly, then owing to the delayed neutrons they will start multiplying in an avalanche-like manner not immediately but only after several moments. This time will be sufficient to quench the "atomic fire" manually, without any automatic devices, simply by inserting into the reactor bulk some rods of boron or cadmium (as their neutron capture cross-sections are enormous).

Nuclear Reactor

The names "atomic energy", "atomic reactor", and "atomic bomb" are a tribute to historical tradition. In actual fact, we here always deal with *nuclear energy, nuclear reactor,* and *nuclear bomb*. And although nothing can be done now about this usage, we should remember this difference.

On Wednesday, 2 December 1942, at 15:25 on a tennis court under the stands of a stadium in Chicago, Enrico Fermi for the first time in the history of mankind accomplished a controlled nuclear reaction in a "nuclear pile". The world's first nuclear reactor was a flattened ellipsoid 8 metres in diameter and 6 metres in height, constructed out of 385 tonnes of graphite blocks with 2-kilogramme uranium blocks arranged between the graphite blocks at 21-centimetre intervals, the total weight of the uranium being 46 tonnes. The reactor resembled a crystal with a cubic lattice. The power of the reactor was 40 watts, the equivalent of a lit match, and after twenty-eight minutes of operation the nuclear reaction in it was discontinued using cadmium strips. There were no cheers; only Wigner uncorked a bottle of

Chianti, Fermi's favourite wine. The forty-three participants and witnesses of that event understood its significance: from then on there was no way back to the pre-atomic era.

On Wednesday, 25 December 1946, at 7 o'clock in the evening in Moscow, Kurchatov and his group started the first Soviet nuclear reactor.

On 15 December 1948, near Paris, Frédéric and Irène Joliot-Curie started the first French nuclear reactor.

On 27 June 1954 at Obninsk, near Moscow the first atomic power station in the world was commissioned. Its power was 5 megawatts.

Not many years have passed since those days, but now more than 300 reactors in thirty countries of the world produce 200 gigawatts of electric power — more than 10 per cent of the entire energy production on Earth, and more than all the hydropower stations of the world. In France nuclear power accounts for more than 60 per cent of electricity output and by the end of this century this share will reach 85 per cent (the world's target figure is 30 per cent).

The history of harnessing atomic energy is unique in many respects: unique are the significance of the problem, the circumstances of its solution, and the implications, some of which have not yet penetrated many minds. Science knows of many cases in earlier days when two investigators, working independently of each other, discovered one and the same thing. There is nothing special in this, if we believe in the objective nature of the laws of nature. But it was the first time that hundreds and thousands of workers, separated by oceans, warfare, and the walls of secrecy, had come step by step to the same conclusions, formulated and solved the same scientific, technological and engineering problems and in about the same sequence. It was only in 1955, after fifteen years of virtually total isolation, that scientists from seventy-nine countries, among them the United States and the Soviet Union, France and Great Britain, Canada and Japan, gathered at Geneva at the first International Conference on Peaceful Uses of Atomic Energy and found, much to their surprise, that their independent measurements and predictions coincided with great precision. What is more, even symbols in formulas obtained by different people at different times often coincided. It was as if the Book of Nature had opened for all of them at the same time and all had only to copy its precepts.

Looking back, one can hardly help being surprised at how narrow was the path and how rickety was the bridge that led from the age of steam and electricity to the age of the atom and the nucleus. To begin with, any of the four factors in the expression for the neutron multiplication factor might have happened to be 5-10 per cent smaller, with the result that a natural uranium reactor would have been impossible. Then if it were not for delayed neutrons the control of a reactor would have become quite a problem.

We now know all the details of the physical processes occurring within the nuclear reactor. In principle even one neutron is sufficient for a chain reaction to be triggered. And within the bulk of uranium, neutrons are available all the time: each second in 1 kilogramme of uranium seven nuclei fission spontaneously, and the resultant neutrons can be a "match" to light the "uranium bonfire". Neutrons thus produced within the reactor only live for less than one 1000th of a second before they die, giving birth to a new generation of neutrons. During that time they collide 114 times with carbon nuclei, having covered 54 centimetres, are moderated to thermal velocities, and cause fission of more uranium nuclei. The number of neutrons in the reactor builds up in an avalanche-like manner and in several seconds it becomes predetermined by the position of the controlling (absorbing) rods. Each cubic centimetre of a high-power reactor contains about half a billion neutrons, which are always in transit from one uranium nucleus to another. On the whole, within the reactor core, some stationary distribution of neutrons is established, the so-called *neutron field,* having a fairly complex configuration; it bears some resemblance to the distribution of the electric fields in electrolytic baths. It can be controlled; sometimes it undergoes oscillations and at all times it is closely observed by physicists and operators of the reactors.

By and large, although the physics of the "atomic pile" is complicated, its principal layout turns out to be exceedingly elementary. "The uranium reactor embodies the most brilliant and remarkable achievement of reason in the entire history of humanity," wrote Frederick Soddy late in his life, fifty years after he had begun his experiments with uranium and thorium.

Quantumalia

Spontaneous Fission of Uranium

E. Fermi

In February 1939 the papers by Hahn and Strassmann, and Meitner and Frisch reached Leningrad. As elsewhere, Kurchatov and his group started experimentation immediately. Before long they also were observing the fragments of uranium nuclei, and the following April Georgy Flerov and Lev Rusinov measured the number of secondary neutrons to be $\nu = 3 \pm 1$, and in May 1940 Flerov and Konstantin Petrzhak stumbled on something new and unexpected — the spontaneous fission of uranium nuclei. It appeared that even without neutrons, i.e., without an external impact, uranium nuclei explode spontaneously. This occurs quite rarely: the appropriate half-life is 10^{16} years, i.e., a million times longer than the age of the Universe, and two million times longer than the half-life of the alpha decay of uranium. Of the 3×10^{21} nuclei contained in 1 gramme of uranium only twenty-three nuclei decay each hour. This is next to nothing, but not infrequently it is such weak signals from nature that are indications of the most important and delicate details of its organization.

One can see the heavens through a needle's eye.

Japanese proverb

The Natural Nuclear Reactor at Oklo

On 7 June 1972 a routine mass-spectrometric analysis of a uranium sample at a French enrichment plant showed that the raw material contained 0.717 per cent uranium-235 instead of the 0.720 per cent, as is normal for all terrestrial rocks, samples of lunar soil, and meteorites. The source of the anomaly was a pit at Oklo, Gabon. At the Oklo deposit the concentration of uranium dioxide (UO_2) was on average no higher than 0.5 per cent (quite a common figure), but at times there occurred lenses about 1 metre thick and from 10 to 20 metres long where there was from 20 to 40 per cent UO_2. It was in these lenses that the content of uranium-235 appeared to be much smaller than usual, being in certain locations as low as 0.44 per cent.

Y. Zeldovich

For some time scientists were in a state of shock: until that time they had not known of a case where the isotopic composition of some element varied from sample to sample. Careful examination of the geology of the deposit revealed that it lies in the mouth of an ancient river, within the stratum of sedimentary rocks dated to 1.8 billion years ago, in the early Protozoic era. At that time the day was shorter than now by a factor of 1.5, Europe was then the floor of an ocean; instead of Asia there were several continents; and life was just making its appearance — in the sea there were already blue-green bacteria but the planet would have to wait about 1 billion years for them to master photosynthesis.

But the most significant thing about the Oklo phenomenon was that in the natural mixture of uranium isotopes there had been in those days about 3 per cent of uranium-235, just about what is needed

I. Pomeranchuk

for present-day water-moderated water-cooled reactors (recall that the half-life of uranium-238 is 4.5 billion years, whereas that of uranium-235 is about 0.7 billion years). Therefore, every time some water got into the lens of uranium ore (whose dimensions are also comparable with the volume of the core of today's reactors), a nuclear fission reaction started there that lasted until the liberated heat evaporated the water and the natural nuclear reactor ceased operating without its moderator. After the reactor had cooled down and filled with water, the reaction resumed. The Oklo reactor has been estimated to have worked in that mode for more than half a billion years, although its average power was under 25 kilowatts.

The Oklo phenomenon is a remarkable example of a natural occurrence that, while in sight of everybody, may for centuries remain hidden until the theoretical concepts have been devised that are needed for its adequate explanation. It is hardly necessary to prove that without such concepts as "atom", "nucleus", "neutron", "isotopes", "radioactivity", "half-life", "fission", "reaction cross-section", etc., the profound import of the difference between the two numbers 0.717 and 0.720 (only in the third decimal place) would simply be beyond everybody, let alone the fact that without a modern mass-spectrograph it would be impossible to detect such a minor difference in the content of uranium-235.

A man's faults all conform to his type of mind. Observe his faults and you may know his virtues.

Confucius

Also, the very existence of a natural nuclear reactor in the distant past is convincing evidence of the possibility of safe operation of the nuclear reactors of today. Lastly, the great headache of nuclear power, the problem of disposal of radioactive wastes, now appears to be not that hopeless: it turned out that at Oklo all the "nuclear ashes" remained where they had been produced about 2 billion years ago when the "nuclear fire" went out.

Meritsegher, Greek

Chapter Seventeen

Atomic Energy • Plutonium
The Atomic Bomb
The Atomic Problem

The idea of progress as a symbol of faith in the unrestricted moral and intellectual improvement of man took shape in the fifth century AD, in the treatise of Saint Augustine of Hippo. This idea that is now so common differs dramatically from the views of antiquity on the succession of ages — from the golden age to the iron age — or from yet more ancient teachings about the alternating rise and fall of humanity. In the seventeenth century the idea of progress was given a philosophical and scientific justification, and in the next century, complemented with the belief in progressive social development, it gained universal recognition.

The driving force of progress became science. *Scientia est potentia*, "knowledge is power", this motto of Francis Bacon has been repeated for four centuries now, although today the words are pronounced without the erstwhile pride: in our days there is an ominous literal aspect to them. The very science that has nourished and established the idea of progress throughout three centuries now delineates the confines of progress with reasonable precision. It testifies dispassionately that on Earth oil and gas will run out within 50-100 years, and in another 300-500 years the reserves of coal; it also predicts that with the present rates of ecological contamination our planet even in the next century will become unfit for life. It also maintains that in the world 55 per cent of the arable land and 15 per cent of the fresh water have now been utilized and that the planet is capable of supporting only two, or at most three, times its current population.

For the first time man finds himself confronted with such issues of global, or even cosmic, scale, and there is no saying how he is going to handle them. One thing is beyond doubt: his first priority is going to be energy, since at all times — from the first bonfire to the nuclear power station — about one third of human effort has gone into producing energy. Even now it is clear that without nuclear power the problem cannot be solved. And if science were in need of redemption, this breakthrough alone would provide it.

Atomic Energy

A modern atomic power station is a rather complex technical structure the height of a nine-storey building. It is essentially a nuclear reactor, where the energy of nuclear fission is released, a steam generator, and an electric generator, where this thermal energy is converted into electrical energy. The heart of the station is the nuclear reactor. Reactors come in dozens of types, such as uranium-graphite, water-moderated water-cooled, heavy-water, thermal-neutron, intermediate-neutron, and fast-neutron reactors, and many others. All of them are fuelled by the same material, uranium (natural or enriched in ^{235}U), and their design differences stem from differences in the neutron moderators and coolants.

The uranium-graphite reactors of today have essentially the same underlying idea as the first reactors of Fermi and Kurchatov. For example a reactor with an output of 1000 megawatts (or 1 gigawatt) is a graphite cylinder of mass 600 tonnes, of height 7 metres and of diameter 12 metres, through which are bored about 2000 vertical channels 15 centimetres in diameter. About one hundred channels accommodate control rods made of steel with some boron, and the remaining channels house about 200 tonnes of uranium. The uranium fuel in the channels is in the form of long rods, fuel elements that are assembled from tablets of uranium oxide enriched in uranium-235 to 1.8 per cent. (The purpose of the enrichment is clear: the more dry firewood is added to a campfire of wet wood the more stable the flames.) Further, there are tubes through the bulk of the reactor along which water is pumped at a temperature of 300°C and pressure of 150 atmospheres. The water supplies heat to the steam generator, which in turn powers huge steam turbines and electric generators. The operating temperature of the graphite is about 700°C, and of the fuel elements about 2000°C.

Water-moderated water-cooled reactors are yet simpler structures. They are essentially large tanks with water in which fuel elements and control rods are immersed. In this reactor water doubles as both the moderator and coolant. It is striking how much energy is stored in such a reactor: in a water reservoir the size of a conventional railway engine tank the energy produced each second is a hundred times higher than the energy release of an average volcano or is half the output of the Bratsk hydropower station in Siberia.

On the technical side, this idea is, of course, not so simple to realize. Here are just some of the problems to be solved: control of environmental contamination by radioactive fission fragments (with a sophisticated system, of filters), protection of personnel from radioactive radiations (with concrete shielding 3 metres thick), enrichment of uranium, and manufacture of fuel elements. Nevertheless, even now the costs of electricity produced at nuclear stations are lower than at fossil fuel stations, and the difference shows a tendency to grow — fossil fuels on Earth are becoming more and more scarce.

Another, no less important, advantage of nuclear power is its minimal impact on the biosphere. A 1-gigawatt nuclear power station "burns" only 1 kilogramme of uranium a day. Even if we consider that the expendable uranium accounts for only 2-3 per cent of the total uranium, this is still far less than the train-load of oil or coal a day needed to fuel a conventional thermal power station of equal output. Clearly, the mining operations and transport costs are reduced accordingly.

Much has been written about the safety of atomic power stations, especially after Chernobyl and Three-Mile Island. Be that as it may, the risk of dying of irradiation near an atomic station is lower than the risk of being killed by lightning or a large meteorite. What is more, in terms of radiation, a thermal station is a much worse place to be near: each tonne of coal contains about 80 grammes of uranium, and so the radioactivity of the smoke plume is hundreds of times higher than that of releases from an atomic power station. And the sulphur dioxide in the plume soon kills all the vegetation and animals in the locality.

And still people have instinctive fears of nuclear power. They protest at rallies and referenda, governments resign over this issue. The reason for this is not only the universal lack of knowledge of the nature of atomic energy — it is commonly identified with the atomic bomb. The emotional perception of atomic energy is one of the many psychological tricks of our mind that often prompt us to make decisions that may do harm to our own causes. For example, many city-dwellers long for peace and quiet, but few will want to take residence in a deserted castle, even if they do not believe in ghosts and vampires. But whatever the freaks of our psychology, the logic of life wins out.

So on the ashes of Hiroshima new houses have been erected and new children have been born. And even the tragedy of Chernobyl cannot change the logic of the development of nuclear power for long: humanity has otherwise no long-term chance to survive on Earth. Atomic energy cannot be "undiscovered" now. Just as we cannot do without cars, ships, and planes, which kill thousands of people in accidents each year.

The power industry of the future will not be able to do without atomic energy — and this is a fact with which nearly all agree. But how long will "uranium fuel" last? Each gramme of the terrestrial crust contains on average 3.5×10^{-6} gramme of uranium, i.e., 3×10^{-8} gramme of uranium-235. In terms of fission energy this corresponds to 600 calories, i.e., only one tenth of the chemical energy stored within 1 gramme of coal. Incidentally, one gramme of coal contains more uranium, namely 0.8×10^{-4} gramme, and when coal is burned it for the most part stays in the ash, which makes up 20 per cent of the original coal. We can readily estimate that 1 gramme of ashes has stored in it about 60 kilocalories of the energy of fission of uranium-235, i.e., about ten times the energy of combustion of the coal itself. Consequently, the earth under our feet is one large deposit of nuclear fuel — we will only have to learn to extract it.

Uranium deposits in the world are believed to be commercial if they contain more than 10^{-3} gramme of uranium per gramme of rock, or 1 gramme per kilogramme of rock. Such deposits normally store about 5 million tonnes of uranium or 50,000 tonnes of uranium-235, the current annual world production being about 300 tonnes. Today's annual electricity consumption corresponds to about 500 tonnes of uranium-235, and with the current rates of growth of the nuclear power industry the world's uranium-235 stores will not last long — not more than a hundred years. It follows that to solve the energy problems of the future we must find a way of using uranium-238. The way was found quite early: Fermi came up with the idea of the "breeder-reactor", which generates more nuclear fuel than it burns.

Plutonium

An atomic reactor can be likened to a bonfire, in which there is 1 per cent of dry firewood (uranium-235), the balance being wet logs (uranium-238). Well, this bonfire produces heat all the same, but after it has gone out there remains a heap of smouldering fire-brands. Perhaps there are ways to make use of them. One way is that used by charcoal burners: they make a pile of wet firewood, blanket it with some sort of covering, and after the firewood has been partially burned with limited access of air what is left is excellent charcoal. A similar process can be realized in the atomic reactor by turning "incombustible" uranium-238 into "combustible" plutonium-239.

Pure plutonium is a steel-grey heavy metal of density 19.82 grammes per cubic centimetre and melting point 640°C. Its chemistry is said to be now known better than that of iron. Plutonium is virtually nonexistent in nature; its content in uranium ores is one 400,000th that of radium; on the other hand, hundreds of tonnes of plutonium are available in the arsenals of the nuclear powers.

We now know fifteen plutonium isotopes, from plutonium-232 to plutonium-246; all of them are radioactive with half-lives ranging from 20 minutes to 76 million years. The most famous of them is plutonium-239. Its half-life is 24,360 years, i.e., on the scale of a human lifetime it may be regarded as stable. Like radium, it emits alpha particles with an energy of 5.1 MeV to become uranium-235:

$$ {}^{239}_{94}\mathrm{Pu} \xrightarrow[\text{24,360 years}]{\alpha} {}^{235}_{92}\mathrm{U} \, . $$

Just like uranium-235, plutonium-239 has the rare ability to fission when exposed to slow neutrons. Its fission cross-section is 742 barns and on average it yields 2.92 neutrons per fission, even more than uranium-235 (σ_{fis} = 582 barns and ν = 2.42, respectively); therefore, plutonium-239 is the best nuclear fuel and nuclear explosive. This was understood quite early, within only two years of the discovery of fission.

In that fateful summer of 1939, on the East Coast of the United States Fermi was looking for a way to reduce absorption in uranium-238 so as to accomplish a chain reaction; on the West Coast, in California, Edwin McMillan decided to take a closer look at what happens to uranium-238 after it has incorporated a neutron. He had at his disposal a brand-new cyclotron that could accelerate deuterons to 16 MeV. By aiming them at a beryllium target he caused the reaction

$$d + {}^{9}\text{Be} \rightarrow {}^{10}\text{B} + n\,,$$

yielding a powerful flux of neutrons. To obtain such a flux using the standard radon-beryllium source one would need several kilogrammes of radium, i.e., more than was available in the entire world. By bombarding a thin uranium target with neutrons, McMillan, like many before him, observed a multitude of energetic fission fragments ejected from the uranium target. The target itself also became radioactive and emitted electrons with a half-life of 23 minutes, i.e., exactly as had been observed by Hahn, Meitner, and Strassmann back in 1937.

By that time hardly anybody had any serious doubts that the decaying isotope here was uranium-239, which formed when a uranium-238 nucleus captured a neutron following the scheme

$$n + {}^{238}_{92}\text{U} \rightarrow {}^{239}_{92}\text{U} \xrightarrow[23\ \text{min}]{\beta} {}^{239}_{93}\text{Np}\,.$$

The *transuranic element* with atomic number 93 formed as a result of the beta decay of uranium-238 would be called neptunium in 1946. But to prove its existence some quantity of it would have to be isolated in pure form.

In May 1940 Philip Abelson, a physical chemist from Washington University, visited McMillan at Berkeley to carry on his uranium studies working on this unique cyclotron. It took them a week to separate the new element from uranium. It appeared to be an emitter of electrons as well, but its half-life was 2.3 days. This strongly suggested that neptunium-239 transformed into a new element, number 94, which would then be called plutonium:

$${}^{239}_{93}\text{Np} \xrightarrow[2.3\ \text{days}]{\beta} {}^{239}_{94}\text{Pu}\,.$$

At the time it was just a hypothesis, awaiting proof, but many accepted it at once.

In March 1941 four American researchers, Joseph Kennedy, Glenn Seaborg, Emilio Segre, and Arthur Wahl, proved that neptunium-239 really does decay into plutonium-239, having a half-life of 24,360 years, which in turn emits alpha particles and turns into our old acquaintance uranium-235. Two months later they made sure that plutonium-239 is split by slow neutrons, like uranium-235, as predicted by the theory of fission of Bohr-Wheeler-Frenkel. A year later, on 18 August 1942, Burris Cunningham and Louis Werner of Berkely for the first time isolated 0.1 milligramme of plutonium. (For their discovery of plutonium Edwin McMillan and Glenn Seaborg would be awarded the Nobel Prize for Physics in 1951.)

Only now, seven years after Fermi's experiments on the irradiation of uranium with neutrons, did their meaning become clear. He had simultaneously observed more than one hundred fission fragments of uranium-235 and, in addition, the entire chain of transmutations of uranium-238:

$$n + {}^{238}_{92}\text{U} \rightarrow {}^{238}_{92}\text{U}^{*} \xrightarrow[10^{-17}\text{s}]{} {}^{239}_{92}\text{U} + \gamma$$

$${}^{239}_{92}\text{U} \xrightarrow[23\ \text{min}]{} {}^{239}_{93}\text{Np} + e + \tilde{\nu}_e$$

$${}^{239}_{93}\text{Np} \xrightarrow[2.3\ \text{days}]{} {}^{239}_{94}\text{Pu} + e + \tilde{\nu}_e$$

$${}^{239}_{94}\text{Pu} \xrightarrow[24{,}360\ \text{years}]{} {}^{235}_{92}\text{U} + \alpha\,.$$

By and large, he was right when he spoke about his having observed transuranic elements, although at the time he did not realize all the complexity of the observed phenomenon.

Returning to the happenings of those not too distant, although already historic, days, one cannot help thinking that Fermi's decision to cease the investigation into the reactions of neutrons with uranium (for which he would never forgive himself) in the long run turned out to be an unexpected blessing for mankind. It is bloodchilling to think how the history of humanity would have turned out if uranium fission had been discovered not in 1938 but in 1934 — shortly after the Nazis came to power. There is no doubt that the Nazi war-mongers would have made use of the entire scientific potential of Germany in order to develop and use nuclear weapons.

The discovery of plutonium changed the very approach to the uranium problem. It became clear, above all, that the absorption of neutrons in uranium-238 is a useful process since it yields much "nuclear fuel", which is no less effective than uranium-235. Besides, plutonium can be separated from uranium, from which it derives, by chemical means, and this is much simpler than the separation of uranium isotopes. But that beautiful scheme could only be realized provided that a nuclear chain reaction is feasible in natural uranium. (The wet firewood can only be dried as long as the fire keeps going.) The commissioning of the first nuclear reactor in December 1942 removed that last doubt. Now nuclear energy was within reach, with only engineering difficulties to be tackled. But of these there were also quite a few and to overcome them was not at all simple. Suffice it to remember that to get 1 gramme of plutonium requires the processing of approximately 1 kilogramme of irradiated uranium, which involves about thirty chemical reactions and more than one hundred operations. But as early as August 1944 at Hanford, Washington, huge "uranium piles" were started and in the spring of 1945 their daily output was nearly 1 kilogramme of plutonium.

Reviewing the history of atomic energy brings out the sharp contrast between the simplicity of the final result (uranium rods in a water tank) and the sophistication of the physical ideas that give an insight into the workings of the reactor. The mastering of nuclear energy required the use of all the major results of twentieth-century science: relativity and quantum mechanics, atomic and nuclear physics, the theory of radioactivity and accelerator technology. Perhaps never before had the everyday life of people been so glaringly dependent on advances in a most abstract field of learning.

The Atomic Bomb

The words "atomic bomb" first appeared in 1913 in the science fiction novel by H.G. Wells entitled *The Liberated World.* (Oddly enough, in this novel Wells predicted the discovery of artificial radioactivity in 1933 and the start of the first atomic power station in 1953, and in both cases he only erred by a year.) Over the years people got used to these words:

The world was exploding
In Curie's experiments
As an atomic, bursting bomb,
Into electronic streams...

which the Russian poet Andrey Bely wrote in 1921.

In 1933 the Hungarian scientist Leo Szilard fled from the Nazis to England, where for the first time he read the novel by Wells. His impressions of the novel, the recent discovery of the neutron, the presentiment of an approaching war and his hatred of Nazism — all of that complex tangle of emotions and new knowledge led Szilard in the spring of 1934 (soon after artificial radioactivity had been discovered) to patent the first atomic bomb. It would explode through the chain reaction of neutron multiplication in beryllium:

$$n + {}^{9}\mathrm{Be} \rightarrow {}^{8}\mathrm{Be} + 2n\,.$$

This seems to have been the first attempt to hypothesize a sustained nuclear chain reaction. However, it was doomed to failure, since in this reaction energy is not liberated but, to the contrary, absorbed. Therefore, when Szilard came with his idea to Rutherford the latter simply refused to discuss it.

Nevertheless, the idea that it is in principle possible to manufacture an atomic bomb and, much worse, that it could get into the criminal hands of Nazis, haunted Szilard and, when in January 1939 he learned from his compatriot Eugene Wigner about Hahn and Strassmann's discovery, he plunged into frantic activities: he borrowed money to get some radium for experimentation, wrote to the Joliot-Curies and persuaded Victor Weisskopf to send them a telegram urging them to cease further scientific publications on uranium, and also called upon his colleagues to exercise voluntary self-censorship. Lastly, after Joliot-Curie had nevertheless published his results, which suggested that a chain reaction in uranium is in principle realizable, in August 1939 Szilard persuaded Albert Einstein to write the famous letter to Franklin D. Roosevelt, President of the United States, stating that world civilization was in grave danger. (In May 1945, after the defeat of Nazism, Szilard would write letters in another vein; he would appeal for a ban on the uses of atomic weapons — in truth, Szilard was destined to become the Cassandra of the atomic era.)

Throughout 1939 the atomic bomb was much in the news, even in the evening papers — probably, because physicists themselves were dubious as to its feasibility, especially after Niels Bohr had explained that the project would require separating uranium isotopes, a possibility that never occurred to scientists even in their wildest dreams. (Bohr himself was prepared to advance "fifteen weighty arguments proving that this is impossible", and Otto Hahn reiterated in hope, "Undoubtedly, this would be contrary to God's will".) No wonder then that the first talks between Fermi and officers from the US Department of the Navy in March 1939 concerning the atomic bomb ended in polite mutual mistrust.

The beginning of the Second World War in September 1939, the invasion of Belgium by German divisions in the spring of 1940, the fall of Paris in June 1940, the attack of Germany on the Soviet Union in June 1941, Japan's bombing of Pearl Harbor in December 1941, and other developments all made statesmen listen at last to the warnings of physicists. Also, it became known that all the leading German physicists had been brought together in the Uranium Society, that 1200 tonnes of uranium concentrate from the Belgian Congo (half the world's stock) had been confiscated by Germany from the defeated Belgium and that the world's only heavy water plant in Norway was under the guard of SS troops.

Especially aware of the danger were *émigré* scientists: Leo Szilard and Eugene Wigner of Hungary; Albert Einstein, Victor Weisskopf, Hans Bethe, Franz Simon, and Rudolf Peierls of Germany; Enrico Fermi of Italy; Otto Frisch of Austria; Francis Perrin, Hans von Halban, and Lew Kowarski of France; Joseph Rotblat of Poland — they all became the initiators of the US military atomic programme. And up to the summer of 1940 the obvious difficulties of the separation of uranium isotopes still held out little hope that the atomic bomb could be fabricated in the foreseeable future.

Things changed dramatically after Philip Abelson and Edwin McMillan reported on 15 June 1940 that the bombardment of uranium-238 seemed to have given rise to a fissile isotope of a new transuranic element, later called plutonium-239. Since plutonium can be separated from uranium by chemical methods, this automatically eliminated the problem of isotope separation. To Sir James Chadwick the implication of this work appeared so obvious that he sent to the United States a protest against publishing these results in the open literature. His fears were not without grounds: in July 1940 Carl von Weizsäcker in Germany had already figured out that uranium-235 could be replaced in an atomic bomb by plutonium-239, and soon the same conclusion was arrived at by Fritz Houtermans. Somewhat earlier, on 27 May 1940, in the United States the importance of ^{239}Pu was stressed in a secret report by Louis Turner. (It was at that time that all the publications on uranium in the United States were classified.)

That an atomic bomb was feasible was first realized in England where Frisch, Peierls, Perrin, and Chadwick in 1939-1940 made the first estimates of its critical mass. These results were instrumental in persuading the US Government to initiate on 6 December 1941 an atomic bomb effort. At 5:30 on 6 June 1945 the first atomic bomb was detonated. By the end of that year there were already about 200 of them.

An atomic bomb is simply a piece of uranium-235, or plutonium-239, and the major problem is the acquisition of these fissionable isotopes. The minimum mass of an atomic bomb is determined by the critical dimensions of the piece of uranium or plutonium, i.e., dimensions that make a chain reaction possible, despite the fact that some of the neutrons escape through the surface. As opposed to an atomic reactor, the bomb does not include uranium-238, which absorbs neutrons, and so there is no need for a moderator, with the result that it becomes a tiny thing about 1 litre. The critical mass of a ball-shaped piece of uranium-235 is 47.8 kilogrammes and of plutonium-239 only 9.65 kilogrammes. The balls may be reduced in mass

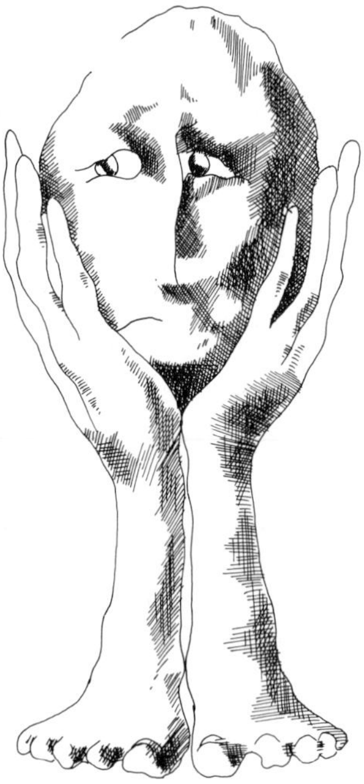

if compressed using conventional explosives.

For an atomic bomb to go off it is sufficient to bring together its parts, each of which is subcritical. The yields of the bombs dropped at Hiroshima (about 20 kilogrammes of uranium-235) and Nagasaki (about 5 kilogrammes of plutonium-239) were equivalent to 13,000 and 21,000 tonnes of trinitrotoluene (TNT) respectively. An explosion lasts a split second and so not all of the material becomes involved in the chain reaction (at Hiroshima only 0.7 kilogramme of uranium; at Nagasaki 1.2 kilogrammes of plutonium) and most of it just evaporates. Using the formula $E = mc^2$ we can estimate the nuclear masses lost by the bombs: 0.7 gramme for the Hiroshima bomb and 1.2 grammes for the Nagasaki bomb. The temperature within the fireballs was higher than within the Sun, and the mushroom-shaped clouds of radioactive dust ascended to 15 kilometres.

Some of the modern nuclear bombs, called *thermonuclear* or *hydrogen* bombs, use the fusion energy evolved in the synthesis of deuterium nuclei and tritium nuclei by the scheme

$$d + t \rightarrow {}^4\mathrm{He} + n + 17.6\ \mathrm{MeV}.$$

The idea of this bomb was discussed by Fermi and Teller in 1942, and by 1952 it had already been exploded.

The hydrogen bomb uses a plutonium bomb as a primer. When the primer fires, the temperature attains 100 million degrees — seven times as high as within the Sun. At such temperatures two hydrogen atoms can already overcome the Coulomb repulsion barrier and merge to yield a helium nucleus. The tremendous energy liberated in the process is three times that yielded by an equal mass of uranium-235.

Real thermonuclear weapons use lithium deuteride ^{6}LiD instead of a mixture of deuterium and tritium. Tritium is obtained in the neutron flux produced by the explosion of the nuclear primer:

$$n + {}^6\mathrm{Li} \rightarrow {}^4\mathrm{He} + t\,.$$

The thermonuclear bomb has no critical mass, and the largest one detonated so far was 5000 times more powerful than the Hiroshima bomb.

All in all, the arsenals of the nuclear powers now include more than 40,000 hydrogen bombs, each 10-15 times more powerful than the first nuclear bomb. Five countries have possessed nuclear arms for several years and, according to estimates, eight more countries are close to, or have recently succeeded in, producing them. In a word, to manufacture an atomic bomb now is no problem; what is harder is to perceive how we are to continue on Earth if for each inhabitant of the planet, including old people and children, we now have 5 tonnes of TNT equivalent of nuclear explosives.

The Atomic Problem

With the discovery of radioactivity and an insight into its nature and energetics came the fears that, motivated by the thirst for knowledge, scientists unwittingly might become like the ill-starred Pandora.

In 1903 Rutherford noted: "It may well be that some idiot in a laboratory will inadvertently explode the entire world."

In that same year Pierre Curie said when he received his Nobel Prize: "It can even be thought that radium could become very dangerous in criminal hands, and here the question can be raised whether mankind benefits from knowing the secrets of Nature, whether it is ready to profit from it or whether this knowledge will not be harmful for it. The example of the discoveries of Nobel is characteristic, as powerful explosives have enabled man to do wonderful work. They are also a terrible means of destruction in the hands of great criminals who are leading the peoples towards war. I am one of those who believe with Nobel that mankind will derive more good than harm from the new discoveries."

In 1936 Francis Aston wrote: "Accessible sources of intraatomic energy are undoubtedly available everywhere around us, and the day will come when man will liberate and put under control its nearly infinite force. We will be unable to prevent this from happening and can only hope that man will not use it with the exceptional purpose to explode his neighbour."

The feeling of that inherent antinomy between the logic of cognition and the moral imperative was with scientists even at the moments of their highest triumph. "We, all of us, are sons of a bitch now," said Kenneth Bainbridge to Robert Oppenheimer, staring at the ominous nuclear mushroom in the Alamogordo desert. (He was thus able to witness the validity of Einstein's formula $E = mc^2$, which he had himself confirmed quantitatively in 1933.) "We've made a devil's contrivance," Oppenheimer would say ten years later.

Well, witnessing an atomic glow, men of science perhaps experienced pride in the power of human reason, but they also sensed how helpless they were to prevent any criminal uses of the forces they had discovered. The awareness of that impotence was the cause of many a personal tragedy. In 1949 Frederick Soddy wrote: "It is an awful thought to imagine into what unprepared hands science has given, so prematurely, the forces that slightly over four years ago seemed inaccessible." And the mild, gentle and kind Otto Hahn after Hiroshima and Nagasaki was close to suicide.

Not all scientific workers took the new situation so tragically. "Perhaps life will now become less happy, but it will not discontinue. We have not yet a force that could destroy our planet," wrote Fermi after a hydrogen bomb test. Einstein echoed his thought: "The discovery of uranium fission threatens civilization no more than the invention of the match. Further development of mankind depends on its moral principles, and not on the level of technical advances." And Robert Oppenheimer in the closing years of his life compared the fears of the atomic bomb to the horror of ancient man confronted by lightning and the stormy sea. Modern man too cannot control those elements but he invented the lightning rod and constructed ships that can survive in any storm. In much the same manner, although it is hard to believe that the chaos of human passions will be brought under strict control, we can hope that with time man will find a protection against these disastrous implications. Paradoxically, since the coming of the atomic bomb Europe has lived for over forty years without a major war — European history has witnessed this only once before. Since 1945 there have been more than one hundred armed conflicts in the world but not a single one has involved the territory of a nuclear power.

Man first encountered fire as forest fires. As distinct from animals, he learned to overcome his horror of fire and to control it. Likewise, the terror of the atomic bomb should not overshadow our path into the future and paralyse our will to live and our faith in our ability to prevent self-destruction. The Temple of Artemis even now is helpless against a modern Herostratos, but this does not mean to suggest that it should not have been erected. And to conclude: "there is hardly a thing that bothers a wise man as little as death."

Quantumalia

A Chronology of the Atomic Era

Now that many facts kept secret during the war have become widely known it is instructive to trace when and what was done to fabricate atomic weapons by various countries separated by fronts and oceans.

R. Oppenheimer

The tradition of all the dead generations weighs like a nightmare on the brain of the living.

Karl Marx

1939

6 January — Hahn and Strassmann publish their paper concerned with uranium fission.

25 January — Fermi announces the possibility of a chain reaction in uranium at the meeting of the American Physical Society in Washington.

17 March — Enrico Fermi meets US naval officers.

30 April — Germany's Ministry of Science convenes the first meeting on uranium problems.

May — Henry Tizard attempts to secure for Great Britain the sole right to purchase uranium from the Belgian company Union Minière, the then only producer of uranium concentrate.

2 August — Roosevelt receives Einstein's letter.

1 September — Germany attacks Poland, World War II begins.

26 September — On the initiative of the German Ministry of Armaments the Uranium Society is instituted, which included Bothe, Heisenberg, Heiger, Weizsäcker and others.

21 October — The Advisory Committee on uranium, established on orders from Roosevelt, meets for the first time.

16 December — The problem of plutonium is discussed for the first time by Fermi, Segre, Lawrence, and Pegram.

1940

7 March — Einstein writes a second letter to Roosevelt.

16 March — Joliot-Curie receives 185 litres of heavy water from Norway.

8 April — Germany attacks Denmark. Uranium stockpile in Belgium is confiscated. Heavy water plants in Norway come under Nazi control.

10 April — A meeting under the chairmanship of G.P. Thomson on the atom bomb issue is held in London. For the first time fast-neutron fission is discussed.

May — P. Abelson and E. MacMillan at Berkeley establish that uranium-238, when bombarded with neutrons, gives plutonium-239.

14 June — Paris falls.

July — V.I. Vernadsky and V.G. Khlopin send a letter to the Deputy Chairman of the Council of People's Commissars N.I. Bulganin. Vernadsky talks with V.M. Molotov. A.E. Fersman heads a commission to prospect for uranium deposits.

30 July — Presidium of the USSR Academy of Sciences establishes the Uranium Commission under Khlopin.

September — Another portion of the world's stock of uranium concentrate (about 1200 tonnes) comes from Katanga to New York.

15 October — Presidium of the USSR Academy of Sciences approves the first programme of uranium studies.

December — J. Chadwick correctly estimates the critical size of the atomic bomb.

1941

The greatest difficulties lie where we are not looking for them.

Johann Goethe

May — Plutonium-239 is proved in Berkeley to be a good substitute for uranium-235 as nuclear fuel and explosive.

22 June — Germany invades the USSR.

30 June — Fermi reports to the Advisory Committee on Uranium on future prospects of atomic energy.

11 June — Lawrence reports to the US Government on the feasibility of a plutonium bomb.

15 July — G.P. Thomson reports to the Government of Great Britain on the feasibility of the manufacture of a uranium-235 bomb by the end of the war.

24 September — Secret report to War Cabinet of Great Britain on the uses of uranium to produce explosives.

3 October — G.P. Thomson and M. Oliphant acquaint their US colleagues with the results of British atomic bomb studies.

11 October — Roosevelt proposes to Churchill a joint effort in making an atomic bomb.

18 October — Within the British atomic project the company Tube Alloys is organized.

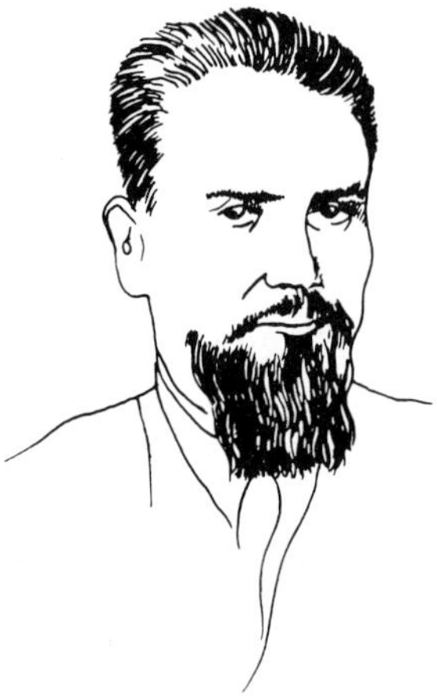

I. Kurchatov

6 December — US Government orders that work be started on the design and manufacture of an atomic bomb.

6 December — The Red Army begins a counter-offensive near Moscow.

7 December — Japan bombs Pearl Harbor; the US enters the war.

1942

January — A project is initiated to construct in England a prototype plant to separate uranium isotopes using the gas-diffusion method (due to Simon, Peierls, and Dirac).

26 February — A conference is held in Berlin on questions concerned with the uranium bomb; among the participants are Heisenberg, Hahn, and Bothe.

March — USSR's State Defense Committee considers a proposal of the intelligence service to establish a centre to coordinate work on nuclear weapons.

7 April — G.N. Flerov writes to Joseph Stalin stressing the need to resume work on uranium.

4 June — Minister for Armaments and War Production, generals and admirals of the Third Reich discuss matters relating to the atomic bomb.

20 June — Roosevelt meets Churchill; it is decided to join forces in the atomic bomb effort.

23 June — A. Speer, Minister for Armaments and War Production, briefs Hitler on the future outlook of the development of atomic weapons. The work in this domain is virtually stopped.

There is a coherent plan in the universe, though I don't know what it's a plan for.

Fred Hoyle

8 July — A meeting at the headquarters of the Japanese Navy to discuss the problem of atomic weapons with the physicists I. Nishina and R. Sagane.

13 August — A project is established under the code name of Manhattan Engineer District to accomplish the US military atomic programme.

18 August — At Berkeley, USA., the first sample of plutonium is obtained, mass 0.1 milligramme.

2 December — Fermi's group achieves the first self-sustained nuclear reaction, offering promise of commercial production of plutonium.

Who will eat the kernel of the nut must break the shell.

Anonymous

1943

2 February — Nazi troops are routed at Stalingrad.

11 February — I. Kurchatov is appointed the head of the Soviet uranium programme.

27 February — Norwegian patriots explode the heavy water plant, thereby leaving the Nazi uranium project without a moderator.

March — At Los Alamos Oppenheimer becomes the director of Project Y, a group that is to design the actual weapon.

5 May — Nishina reports to the Navy of Japan on the technical feasibility of creating an atomic bomb. A research programme is approved.

All sunshine makes a desert.

Arabian proverb

August **1944** — Commercial production of plutonium is started at Hanford, Washington.

1945

16 July — The first plutonium bomb is tested in the desert area near Alamogordo, New Mexico.

6 August — The city of Hiroshima in Japan is destroyed by a uranium-235 atomic bomb.

9 August — The city of Nagasaki is destroyed by a plutonium-239 bomb.

25 December **1946** — Kurchatov's group in Moscow accomplishes sustained nuclear reaction.

10 June **1948** — The first commercial nuclear reactor to produce plutonium is commissioned in the USSR.

29 August **1949** — The first Soviet atomic bomb (plutonium-239) is tested.

30 October **1952** — The first British atomic bomb is detonated on the west coast of Australia.

1 November **1952** — A US thermonuclear device is tested at Eniwetok Atoll.

12 August **1953** — The first Soviet thermonuclear bomb (lithium deuteride) is exploded in the USSR.

13 February **1960**	—	The first French atomic bomb is tested in the south-western Sahara.
16 October **1964**	—	The first Chinese atomic bomb is tested.
17 May **1974**	—	India detonates a "peaceful nuclear device".

All in all, in the last forty years in the world more than 2000 atomic bombs have been exploded and hundreds of tonnes of plutonium have been accumulated in the world.

Soddy on Atomic Energy

The same fire purifies gold and consumes straw.

Italian proverb

Frederick Soddy (1877-1956) was an outstanding scientist, a man of all-round culture and education, and a deep and provocative thinker. These facets of his personality have clearly manifested themselves in his popular science books, which at the beginning of our century enjoyed merited recognition and renown. (Frédéric Joliot-Curie, so he confessed, was influenced in his choice of career by Soddy's book *The Interpretation of Radium.*) In his books he stressed repeatedly the extent to which human life came to be subjected to irreversible changes by the new science of radioactivity. Few people at the time spoke with such definiteness and concern about the future of atomic energy, about ecology and the energy crisis — about all that only now, about a century later, has become so obvious to many. Already in his lecture of 23 February 1904 Soddy was explaining to the audience that "artificial transmutation of elements will forever liberate mankind from the problem of energy".

The following excerpts from the many books by Soddy are interesting in many respects — as documentary evidence of his epoch, as an imprint of his personality, and as a rare opportunity to take a look with the eye of a contemporary at the first steps of the atomic era, with the advantage of hindsight.

Let your soul stand cool and composed before a million universes.

Walt Whitman

In 1907, in his famous book *The Interpretation of Radium* Soddy wrote: "In a few years the elementary principles of radioactivity would be taught in all schools as belonging to the very beginnings of physical science."

"Last century civilization may be said to have attained its majority and to have entered upon the control of an inheritance of energy stored up by the sun in fuel during the long ages of the past, and now it is dissipating that inheritance as quickly as it can. With the light-heartedness and irresponsibility of youth, it is taking no thought of the future, but confidently assumes that the supply of natural energy, upon which at every turn it is now entirely dependent, will continue indefinitely...

It is, indeed, a strange situation we are confronted with. The first step in the long, upward journey out of barbarism to civilization which man has accomplished appears to have been the art of kindling fire... One can imagine before this occurred that he became acquainted with fire and its properties from naturally occurring conflagrations.

With reference to the newly recognised internal stores of energy in matter we stand to-day where primitive man first stood with regard to the energy liberated by fire. We are aware of its existence solely from the naturally occurring manifestations in radioactivity...

It does not require much effort of the imagination to see in energy the life of the physical universe, and the key to the primary fountains of the physical life of the universe to-day is known to be transmutation (of the elements)...

F. Soddy

Of course life depends also on a continual supply of matter as well as on a continual supply of energy... The same matter, the same chemical elements, serve the purposes of life over and over again, but the supply of fresh energy must be continuous... The same energy is available but once. The struggle for existence is at bottom a continuous struggle for fresh energy...

The aboriginal savage, ignorant of agriculture and of the means of kindling fire, perished from cold and hunger unless he subsisted as a beast of prey and succeeded in plundering and devouring other animals. Although the potentialities of warmth and food existed all round him, and must have been known to him from natural processes, he knew not yet how to use them for his own purposes. It is much the same today. With all our civilisation, we still subsist struggling among ourselves for a sufficiency of the limited supply of physical energy available, while all around are vast potentialities of the means of sustenance...

By the achievements of experimental science Man's inheritance had been increased, his aspirations had been uplifted, and his destiny had been ennobled to an extent beyond our present power to foretell. The real wealth of the world was its energy, and by these discoveries it, for the first time, transpired that the hard struggle for existence on the bare leavings of natural energy in which the race had evolved was no longer the only possible or enduring lot of Man. It was a legitimate aspiration to believe that one day he would attain the power to regulate for his own purposes the primary fountains of energy which Nature now so jealously conserves for the future. The fulfilment of this aspiration is, no doubt, far off, but the possibility altered somewhat the relation of Man to his environment, and added a dignity of its own to the actualities of existence.

Four years later in another of his books, *Matter and Energy*, he goes on to say:

A needle is sharp only at the end.

Chinese proverb

"The horror of the waste of food is inborn, the horror of the waste of fuel has still to be acquired... The upward progress of the race has ... been classified into succeeding eras, each designated by the name of a material. Thus are distinguished the Stone Age, the Bronze Age, the Iron Age, and the Steel Age... Fundamental as materials are in shaping the broad lines of progress, it is necessary to go but very little deeper to come upon something equally fundamental... Materials are employed merely as weapons, tools, or instruments for the utilisation of power or energy... Not until the law of conservation of energy was established... could energy be regarded as one of the fundamental existences. Its recognition, as a separate entity, distinguishes the present age from all its predecessors. This is the Age of Energy...

"The age we live in, the age of coal, draws its vivifying stream

This is the best of all possible worlds.

Gottfried Leibnitz

from a dwindling puddle left between the comings and goings of the cosmical tide...

"Civilization, as it is at present, even on the purely physical side, is not a continuous self-supporting movement... It becomes possible only after an agelong accumulation of energy, by the supplement of income out of capital. Its appetite increases by what it feeds on. It reaps what it has not sown, and exhausts, so far without replenishment. Its raw material is energy, and its product is knowledge...

"The triumphs of science over Nature till now resemble somewhat schoolboy successes. This period is passing away... Thoughts of economy and conservation will inevitably replace those of development and progress, and the hopes of the race will centre in the future of science. So far it has been a fair weather friend. It has been generally misunderstood as creating the wealth that has followed the application of knowledge. Modern science, however, and its synonym, modern civilisation, create nothing, except knowledge... It looks, therefore, as if our successors would witness an interesting race, between the progress of science on the one hand and the depletion of natural resources on the other...

"Sooner or later, but certainly not indefinitely later, nothing known will remain to supplement the natural rate of supply of energy, save the primary stores of atomic energy..."

In 1920, after he had lived through the death of Moseley and the barbarisms of World War I, he would add: "Let us hope that the achievements in the field of sources of physical forces harnessed by man will not share the fate of those which in the past have turned the blessing deserved by science into damnation. Just as the very discovery of radioactivity came unexpectedly and suddenly, so any moment some lucky member of a small group of researchers in that field of science may find a clue and use it. Thus, the primary source of natural energy would be uncovered for human knowledge and would be used at its wellspring — for good or ill — judging from how well the long and bitter lessons of the heavy past and present have been absorbed."

Among the discoverers and pioneers of radioactivity, Soddy was almost the only one to live to see the nightmare of Hiroshima and the start of the first atomic power station. In his declining years he wrote another book, *The Story of Atomic Energy,* in which he summarized the highlights of a half-century of a truly fantastic history of the harnessing by man of atomic fire.

Pan, Greek

Chapter Eighteen

Solar Light • Crucibles of Elements
The Fate of the Sun
The Sun, Life, and Chlorophyll

Gazing at the starry skies has at all times evoked some obscure discomfort in the human soul. Advances in the latest knowledge have changed little here: the mysteries of the skies have not vanished, they have just become more distant.

About the heavens we now know quite a lot. Our Sun is a run-of-the-mill star among 100 billion other stars that inhabit our Galaxy, and the latter is simply one of many, many billions of similar galaxies scattered all over the visible part of the Universe. Our Earth revolves at 30 kilometres per second around the Sun, which in turn moves at 200 kilometres per second around the centre of the Galaxy, which also flies along at 600 kilometres per second to God knows where. Space distances are unfathomably gigantic and they overpower our sensibilities by their unimaginability: the light from the nearest star Proxima Centauri travels to us for 4.3 years; from the Galactic centre, 30,000 years; from the Andromeda nebula — the nearest large galaxy — 2 million years; from the visible boundaries of the Metagalaxy, more than 10 billion years. We know now the size, mass, temperature, and composition of stars, why they glow, how long they live and why they explode — we know quite a lot really.

But perhaps it is for this reason that at midnight, when he silently opens up the window into the starry abyss, man suddenly becomes gripped by the keen sensation of being forlorn on a tiny island in the ocean of the Universe. He all of a sudden comes to comprehend the precariousness of the phenomenon of life, which by some miracle has stuck to the thin frozen crust of a planet that is burning hot inside and is rushing around the Sun twenty times faster than a cannon ball. In such desperate moments man is only helped by the ancient heat of the hearth, the eyes of his children, and the hand of his friend.

In olden days salvation from the fear of the heavens was sought in religion. In our enlightened age the ends of the logical inferences and the obvious consequences of the exact sciences are generally not drowned in the bottomless well of faith. We derive strength from the awareness of our belonging to the human race and from faith in its yet unclear predestination, from admiration at the power of the human reason and the recognition of the laws that man has observed.

Solar Light

Strange as it may seem, the question of whence the Sun derives its energy seems to have been of no interest to the philosophers of antiquity. And not only to them but even to such men of modern science as Laplace and Herschel. It is common knowledge these days that beyond the Earth's atmosphere solar rays bring an energy of 0.135 joules per square centimetre per second, i.e., 0.135 watts per square centimetre. The Sun's distance is $R = 150$ million kilometres, or 1.5×10^{13} centimetres, i.e., the total power of the solar radiation is $4\pi R^2 \times 0.135 = 4\pi(1.5 \times 10^{13})^2 \times 0.135 = 3.8 \times 10^{26}$ watts. This is a tremendous amound of energy. To obtain it under terrestrial conditions we would have to burn 1.3×10^{16} tonnes of coal per second — a thousand times the known coal reserve on Earth. Therefore, if the Sun shone by coal burning, then, its mass being 2×10^{33} grammes, it would only last 5000 years. It is sheer nonsense, but it was not until 1845 that it attracted the attention of the discoverer of the law of energy conservation Julius Robert Mayer (1814-1878) of Germany. Mayer himself considered in 1848 that the Sun derives its energy from collisions with meteorites. There were other explanations: Hermann Helmholtz in 1854 suggested that the Sun derives its luminous energy from slow contraction, and Sir James Jeans thought that the luminary is powered by the merging of protons and electrons.

The discovery of radioactivity has changed the line of argument of scientists, and although the spectroscope has testified impassively that there is no radium on the Sun, the idea of some "subatomic" source of solar energy became quite common at the beginning of the century. The spectroscope has also shown that the Sun consists mainly of hydrogen and helium, and so when Aston's precise measurements of atomic masses became available the English astrophysicist Sir Arthur Stanley Eddington (1882-1944) pronounced that solar radiation is the energy of the synthesis of four hydrogen nuclei to form a helium nucleus.

In 1920 this hypothesis had many opponents, among them Rutherford. At the Cavendish laboratory he had just accomplished his first nuclear reaction, and he, none better, knew how difficult it was. "Stars are insufficiently hot for this," objected Rutherford. "Find a hotter place," retorted Eddington (a hint at hell). He went on to say, "What is hard for the Cavendish laboratory cannot be too hard for the Sun." Before the coming of quantum mechanics, however, emotions alone were of little help in settling the dispute.

In 1929, after Gamow had explained the quantum nature of alpha decay, the graduates of the University of Göttingen Rudolf Atkinson and Fritz Houtermans pointed out that at temperatures over 20 million degrees protons may, due to the tunnel effect, penetrate the Coulomb repulsion barrier of light nuclei and combine with the nucleus, forming a new nucleus and releasing in the process a rather large binding energy, which is quite likely to be responsible for the long-term glow of the Sun. But even this conjecture was a bit premature. Another ten years would have to elapse until Hans Bethe constructed his consistent theory of nuclear burning in stars.

But during those years some important advances were made, without which his theory would have been impossible:

1931 — Wolfgang Pauli put forward the hypothesis of the neutrino, a neutral massless particle ν.

1932 — The US chemist Harold Clayton Urey discovered the heavy hydrogen isotope deuterium d (1934 Nobel Prize for Chemistry).

— James Chadwick discovered the neutron n (1935 Nobel Prize for Physics).

— Carl David Anderson of the US discovered the positron e^+ (1936 Nobel Prize for Physics).

— Dmitry Dmitrievich Ivanenko of the Soviet Union suggested the hypothesis of the proton-neutron structure of the nucleus.

1933 — Enrico Fermi developed his theory of the beta decay of nuclei and introduced a new kind of interaction, the *weak interaction*.

1934 — The Joliot-Curies discovered artificial radioactivity and the beta-decay of protons in nuclei into a neutron, a positron, and a neutrino (1935 Nobel Prize for Chemistry).

Furthermore, in ten years quantum mechanics had become an indispensable and common tool in atomic and subatomic physics, in crystal physics and in theoretical chemistry, and the concepts of "cross-section of a nuclear reaction" and "resonance" became commonplace in science.

Drawing on these advances Gamow and Teller could in 1938 repeat Atkinson and Houtermans's estimates and take them seriously. In April 1938 Gamow convened in Washington a small conference of astrophysicists and nuclear physicists, in which Weizsäcker and Bethe took part. Quite soon there appeared a series of famous papers on stellar energy sources, which was concluded in 1939 by a comprehensive work by Bethe (1967 Nobel Prize for Physics). Bethe's theory was tested and refined well into the 1950s, and now it can be put in a nutshell quite simply. ("There is nothing simpler than a star," was Eddington's favourite quip.)

Inside the Sun the pressure is as high as hundreds of billions of atmospheres, the density 100 grammes per cubic centimetre, and the temperature 13-14 million degrees. A chain of reactions occur there, which are now known as the proton-proton or hydrogen cycle of stellar nuclear reactions:

$$p + p \rightarrow d + e^{+} + \nu + 1.442\,\text{MeV}\,(1.3\times 10^{10}\,\text{years})$$

$$d + p \rightarrow {}^{3}\text{He} + 5.494\,\text{MeV}\,(6\,\text{seconds})$$

$${}^{3}\text{He} + {}^{3}\text{He} \rightarrow {}^{4}\text{He} + 2p + 12.86\,\text{MeV}\,(10^{6}\,\text{years})$$

Net result:

$$4p \rightarrow {}^{4}\text{He} + 2e^{+} + 2\nu + 26.73\,\text{MeV}.$$

The longest stage is the first one: it takes a proton 13 billion years to find a pair and to form a deuterium nucleus — a weakly bound state of a proton and a neutron. No wonder — to achieve this the proton must first turn into a neutron, but the rate of the nuclear reaction $p \rightarrow n + e^{+} + \nu$ is rather low, since it is dictated by the weak interaction.

The resultant deuteron within only six seconds enters into a reaction with a proton to form the nucleus of the light helium isotope ^{3}He, which then wanders for about 1 million years until it meets an identical nucleus and merges with it to form an alpha particle (nucleus of ^{4}He) and to emit again two protons. But it takes several more million years for the energy liberated within the Sun to reach its surface and to be radiated into space. In another eight minutes solar light will reach the Earth.

In actual fact the "burning" of hydrogen nuclei is a more complicated process, since in addition to the main cycle there are some side cycles, which reduce the energy release in the proton-proton cycle to 26.171 MeV. Moreover, along with the main cycle (its contribution is 86 per cent) hydrogen is synthesized into helium according to the so-called carbon-hydrogen cycle, where the energy release is 24.97 MeV. And so, on average, as four protons fuse to yield a helium nucleus the energy yield is $26.17 \times 0.86 + 24.97 \times 0.14 = 26.00$ MeV, i.e., 6.5 MeV per hydrogen nucleus burned. It follows that its mass $m_{\text{H}} = 1.007825$ amu now becomes smaller by $\Delta m = 6.5/931.5 = 0.006978$ amu, i.e., by 0.69 per cent, or seven times as much as when a uranium nucleus splits.

One gramme of hydrogen contains $N_{\text{A}} = 6.02 \times 10^{23}$ nuclei (Avogadro's number) and as these nuclei burn they yield $6.02 \times 10^{23} \times 6.5$ MeV $= 3.91 \times 10^{24}$ MeV $= 6.27 \times 10^{18}$ ergs $= 6.27 \times 10^{11}$ joules. Therefore, to provide a radiation output of 3.8×10^{26} watts, the Sun must burn within its bowels 0.607×10^{15} grammes per second or 607 million tonnes of hydrogen. The mass lost in the process will be $607 \times 0.0069 = 4.2$ million tonnes per second and the energy released will be dispersed as light in space. (Some 5 per cent of the energy is carried away by neutrinos and is not included in this balance; therefore, in actual fact a bit more hydrogen will be lost, namely 630 million tonnes or 0.63×10^{15} grammes per second.)

With this "burning" rate the Sun's mass (2×10^{33} grammes), which is 75 per cent hydrogen, would last 0.8×10^{11}, or 80 billion, years. But in real stable burning a mass loss of no more than 10 per cent is allowed, so that the Sun will shine as it does now for 5 to 7 billion years. Although it emits so much energy, the Sun is exceedingly economical. So its specific power is only 1.9×10^{-7} watt per gramme — one 10000th that of the human body (2×10^{-3} watt per gramme) and one 50,000,000,000th that of a burning match (about 10,000 watts per gramme).

Crucibles of Elements

In the depths of the Sun each second 630 million tonnes of hydrogen turn into helium. But how have all the other elements come about, the elements of which earth, plants, and we ourselves are made up? Quantum physics and its derivative, nuclear astrophysics, can now answer this question as well.

When all the hydrogen in the Sun's core has burned up, the flux and pressure of radiation will decrease immediately. But it was the solar radiation that stopped the Sun from collapsing under gravity and so now the core will decrease in volume. In the process of this gravitational collapse the density in the centre of the Sun will be as high as 10^5 grammes per cubic centimetre, and the temperature over 100 million degrees. Exactly at this moment helium will start to "burn". The corresponding nuclear reaction — the triple-alpha process (3α process) — is remarkable in many respects and is worth considering in more detail.

First and foremost, the simple reaction of the fusion of two helium nuclei to yield a beryllium nucleus

$$^{4}\text{He} + {}^{4}\text{He} \rightarrow {}^{8}\text{Be}$$

is impossible, since there is no such beryllium isotope in nature. Happily, the cross-section of this reaction has a resonance at 0.1 MeV, which can be thought of as a highly unstable nucleus $^{8}\text{Be}^*$. This "nucleus" lives for only 10^{-16} second, but on a nuclear time scale this is not so small. The two colliding alpha particles manage, before disintegration, to undergo about one million oscillations as $^{8}\text{Be}^*$. Within this short time span a third alpha particle can come close to them and form with them a nucleus of ^{12}C.

This possibility, however, would remain unrealized, if it were not for a second piece of luck contributing to the success of the triple-alpha process. The point is that the three alpha particles are 7.28 MeV larger in mass than a ^{12}C nucleus, and so the direct formation of carbon nuclei out of three alpha particles is highly unlikely. But the ^{12}C nucleus has an excited state $^{12}\text{C}^*$ with an excitation energy of 7.66 MeV, i.e., the mass of the $^{12}\text{C}^*$ nucleus, unlike ^{12}C, is no smaller than that of the three alpha particles. On the contrary, it is even larger by $7.66 - 7.28 = 0.38$ MeV. And this suggests that at sufficiently high energies of collision of the alpha particles the resonance reaction

$$^{8}\text{Be}^* + {}^{4}\text{He} \rightarrow {}^{12}\text{C}^*$$

is possible.

The excited nucleus $^{12}\text{C}^*$ does not live long — only 10^{-12} second, and it returns to its ground state with the emission of gamma-quanta or an electron-positron pair. But this time appears to be long enough for the three alpha particles to merge irreversibly.

Even at temperatures over 10^8 K the mean kinetic energy of an alpha particle (0.02 MeV) in a helium star is far lower than 0.38 MeV, the resonance energy for the reaction $^{8}\text{Be}^* + {}^{4}\text{He} \rightarrow {}^{12}\text{C}^*$. Within such a star, however, there always exist some very energetic particles (one particle per billion) for which this condition is met, which is enough for the triple-alpha process

$$^{4}\text{He} + {}^{4}\text{He} \rightarrow {}^{8}\text{Be}^*$$

$$^{8}\text{Be}^* + {}^{4}\text{He} \rightarrow {}^{12}\text{C}^* \rightarrow {}^{12}\text{C} + \gamma$$

to occur with a rate thousands of times higher than that of hydrogen burning.

The triple-alpha process was predicted in 1952 by the US theoretician Edwin Ernest Salpeter (b. 1924) and it was not until much later that it was supported by the entire body of observational evidence. Now it is well established, but that has not detracted from its fascination. If helium and carbon nuclei differed in mass from their actual ones by only 0.1 per cent, then the rare coincidence of two resonances in the triple-alpha process would be disturbed and the conditions for nucleosynthesis in stars would be different.

Carbon is the basis of all living things and one of the most necessary elements on Earth. But only now has it become clear on what tiny details of the structure of nuclei and what vicissitudes of their combination is life itself ultimately dependent, including us, an intelligent life form, capable of understanding and evaluating the meaning of these details.

After carbon has formed, the helium nucleus forms other elements, such as oxygen, neon, and magnesium:

$$^{12}C + ^{4}He \to ^{16}O + \gamma,$$

$$^{16}O + ^{4}He \to ^{20}Ne + \gamma,$$

$$^{20}Ne + ^{4}He \to ^{24}Mg + \gamma.$$

By the magnesium stage all the helium in the star has run out and for further nuclear reactions to become possible a new compression of the star and further growth of its temperature are required. This is, however, not possible for all stars. It is only possible for those stars whose mass is larger than the so-called Chandrasekhar limit $M = 1.2M_{sol}$., i.e., for stars with a mass that is at least 20 per cent larger than the solar mass M_{sol}. (The existence of this limit was established in the 1930s by the Indian scientist Subrahmanyan Chandrasekhar (b. 1910).)

The evolution of stars with $M < 1.2M_{sol}$ terminates at the stage of the formation of magnesium and then they turn into white dwarfs — stars with a mass of about $0.6M_{sol}$, the size of our Earth and having a density of about 1 tonne per cubic centimetre. In white dwarfs electrons are separated from nuclei, so that the star resembles one huge crystal, whose properties can only be described using the equations of quantum mechanics, notably the famous Pauli exclusion principle, which forbids two electrons to have the same quantum numbers. (The theory of white dwarfs was constructed as early as 1926 by the British mathematician Ralph Howard Fowler (1889-1944).)

In heavier stars at temperatures from 5×10^8 to 10^9 K silicon is synthesized in the reactions

$$^{24}Mg + ^{4}He \to ^{28}Si + \gamma,$$

$$^{16}O + ^{16}O \to ^{28}Si + \alpha.$$

After another stage of gravitational collapse the temperature reaches 2 billion degrees and the average energy of the emitted gamma-quanta reaches 0.2 MeV, and now they are capable of splitting silicon nuclei into alpha particles:

$$^{28}Si + \gamma \to 7\,^{4}He.$$

These particles are then successively pressed into the silicon nuclei, so forming heavier elements — up to iron. This depletes the sources of nuclear energy within stars, since heavier elements are formed that do not produce but absorb energy. The evolution of stellar matter enters a new phase.

Now nuclear reactions occur on the surface of the iron core of the star, where there remains some unburnt nuclei of ^{4}He, ^{12}C, ^{20}Ne, and also a small amount of hydrogen. Some of these reactions produce free neutrons, which are absorbed by iron nuclei and, precisely as in Fermi's experiments, the beta decay of neutrons gives rise to new nuclei with the next number, i.e., cobalt nuclei

$$^{58}Fe + n \to ^{59}Fe^* \to ^{59}Co + e + \tilde{\nu}..$$

Likewise, cobalt transmutes into nickel, nickel into copper, and so on to the bismuth isotope ^{209}Bi.

This exhausts the possibilities of the *s process* (*s* for slow neutron capture) for the formation of chemical elements, and all the elements heavier than bismuth are formed through the so-called *r process* (*r* for rapid process), i.e., through stellar explosions.

Explosions are possible if a star's mass is large enough for gravitational forces to be able to compress and heat its iron core upwards of 4 billion degrees. Under these conditions each nucleus of iron ^{56}Fe decays into thirteen alpha particles and four neutrons and absorbs in the process an energy of 124 MeV. The stellar core cools down and begins to be compressed catastrophically under gravitational forces, which are not counteracted by the pressure of radiation. Thus occurs an *implosion,* i.e., an explosion directed inwards, or a collapse. At first alpha particles fall apart into protons and neutrons, and then electrons are pressed into protons, forming neutrons and emitting neutrinos:

$$p + e \to n + \nu.$$

The complex interplay of processes within the bowels of a star and its shell (yet imperfectly understood) causes the entire star to explode, shedding its shell. (The remains of the latter we then observe as cosmic rays.) At that moment we observe in the sky the birth of an exceedingly bright supernova star, or simply a *supernova.* This, not very convenient, name was coined in 1934 by the eminent German astronomers Walter Baade (1893-1960) and Fritz Zwicky (1898-1974), who advanced the idea that supernova explosions result in the formation in the centre of the star of a small *neutron star* with a mass equal to about a solar mass, and a radius as small as 10-13 kilometres, i.e., its density may be billions of tonnes per cubic centimetre.

The possibility of the existence of neutron stars was first indicated by Lev Davidovich Landau (1908-1968) in 1932, immediately after the discovery of the neutron, but for a long time the hypothesis was dismissed as another daft idea of theoreticians. Thirty-five years passed and then Jocelyn Bell, a postdoctoral fellow in Sir Martin Ryle's group at Cambridge, detected in the sky a periodic source of radio emission with a period of 1.3 seconds. When the first fears of encounters with extraterrestrial civilizations were allayed, all agreed that this *pulsar* is just a rapidly rotating neutron star. To date well over 300 pulsars are known. (For this work Ryle (b. 1918) and another British radio astronomer, Antony Hewish (b. 1924), were awarded the Nobel Prize for Physics in 1974.)

Stars of 1.5-3 solar masses can end their life in a different way: after a star has run out of its nuclear fuel, it does not explode but starts condensing irresistibly. The star undergoes *contracting gravitational collapse* and turns into a *black hole*. The size of the black hole is determined by its gravitational radius, which for the Sun, say, is not more than 3 kilometres. The gravitational pull of the black hole is so powerful that nothing — not even light — can overcome it to betray its existence. Therefore, it is only possible to observe the formation of a black hole through its action on other celestial bodies. And still, even these stellar "tombs" linger on, radiating energy into space. The spectrum of their radiation is essentially that of a black body. This conclusion was made in 1974 by the English scientist Stephen William Hawking (b. 1942), a man of great talents and difficult fate.

The neutron fluxes that emerge in supernova explosions are enormous, so much so that one nucleus picks up dozens of neutrons before at least one of them beta decays. It was in this way that all the radioactive elements originated, including uranium and thorium; when they started forming (around 10 billion years ago) there was 1.5 times more uranium-235 than uranium-238.

Chemical elements keep on synthesizing in stars even now. A dramatic proof of this was found in 1952, when in the spectrum of one star technetium lines were found: the presence of the lines was indicative of the fact that the element is forming there non-stop, since all the technetium isotopes live less than 3 million years and so during the lifetime of the solar system (about 5 billion years) it has decayed completely. (It was synthesized only in 1937 by one of Fermi's "boys", Emilio Segre.)

The idea that stars are crucibles in which elements are transmuted was put forward at the end of the nineteenth century by Sir Joseph Norman Lockyer (1834-1920), who discovered helium on the Sun and coined the name of the element. His ideas, notably his book *Inorganic Evolution*, had a decisive influence on the turn of mind of the young Rutherford and the direction of his investigations later in life. But at the time it was just a daring conjecture, like Prout's hypothesis that all elements originated from hydrogen. Eventually, Prout and Lockyer turned out to be right, although real life is far richer than their speculations.

Real clues to the story of the creation and evolution of the elements could only be found using the ideas and techniques of quantum physics, and only quite recently, at the end of the 1950s and the beginning of the 1960s, owing to the contributions of William Fowler, Fred Hoyle, Margaret and Geoffrey Burbridge, and many others. (William Fowler and Chandrasekhar were awarded the Nobel Prize for Physics in 1983 for their work on astrophysics.) Not all details of this grandiose picture have yet been established with equal credence, but the underlying ideas and general outlines are beyond doubt. In any case we can now with sufficient accuracy work out abundances of chemical elements in space and convince ourselves that prediction is in agreement with observation. (The famous prediction of the French philosopher Auguste Conte (1798-1857) that it is impossible to find out the composition of stars, which he made a bare three years before the advent of spectral analysis, now looks rather empty in this context.) But we can now do much more — we can sketch the general picture of the birth and death of stars.

The Fate of the Sun

A modern book of Cosmogenesis, written by astrophysicists, begins thus: "In the beginning there was the Big Bang..."

When the Universe was 0.01 of a second old it was an entity with a density 4 billion (4×10^9) times that of water and a temperature of 100 billion (10^{11}) degrees. This entity expanded at nearly the speed of light and consisted largely of photons and neutrinos, with some electrons and positrons, which kept emerging out of light and re-annihilating into radiation — light and matter were one. At that moment protons and neutrons were but in trace amounts, one nucleon per billion of lighter particles.

In one tenth of a second the temperature decreased three-fold, in a second ten-fold, and in another fourteen seconds it dropped down to 3 billion (3×10^9) degrees, and then the irreversible annihilation of electrons and positrons into photons began.

By the end of the third minute the temperature of primordial matter was already 1 billion degrees, the density dropped to that of water, and the formation of helium nuclei from protons and neutrons started.

In another hour the temperature had fallen to 300 million degrees and a mixture of hydrogen and helium had formed, which we still observe everywhere throughout the Universe.

About a million years passed before the temperature dropped to 3000 degrees and electrons combined with the nuclei of hydrogen and helium — from then on photons could not split them, and radiation separated from matter.

And now the Universe was largely a homogeneous mixture of three-quarters hydrogen and one-quarter helium. Its density, compared with that of space today, was quite large — thousands of atoms in a cubic centimetre (today only one or two atoms per cubic centimetre), and admixtures of other atoms (mainly deuterium and lithium) were less than 0.0001 per cent. It was at that time in the evolution of the Universe that, obeying the law of universal gravitation, the primordial mixture of hydrogen and helium began to concentrate into clusters that later on gave birth to galaxies and stars.

Even at the time of Newton it was known that a homogeneous mass of a large amount of matter is unstable, but it was not until two centuries later, in 1902, that Newton's ideas about the gravitational instability of distributed masses of matter were elaborated in the work of James Jeans (1877-1946) (the Jeans of the Rayleigh-Jeans distribution in the theory of thermal radiation).

The hydrogen-helium clusters were condensing fairly quickly: in a million years the density and temperature within such clusters reached values at which the nuclear burning of hydrogen sets in. And then the contraction stops owing to the counter-pressure of radiation and the resultant star is stabilized until the reserves of nuclear energy are exhausted. The lifetime of stars depends on their mass — the lighter a star the longer it lives. Stars with $M < M_{sol}$ live tens of billions of years, and with $M > 5M_{sol}$ a thousand times less.

In the early Universe massive stars formed more often; they quickly consumed their resources of nuclear fuel and exploded as supernovae. A supernova explosion is quite an event, even on the scale of space. It is essentially a detonation of a thermonuclear bomb the size of our Sun. The energy released in that explosion is hundreds of billions times higher than that emitted by the Sun, and the supernova shines for a time like a whole galaxy. The mass of the shell the supernova sheds in the explosion is comparable with the solar mass, and its matter is rich in heavy elements, which have formed under the intense flux of neutrons of the thermonuclear explosion.

It is these explosions that have changed the primeval chemical composition of matter in the Universe: now, in addition to hydrogen and helium it developed from 1 to 3 per cent admixtures of heavy elements. This is like salt in the space soup, but it is out of this salt that our Earth is made up. The relative abundances of elements on Earth are to within a considerable accuracy like those on the Sun (helium and free hydrogen having volatilized early in the geological epoch of the Earth). The composition of the tissues of the human body is essentially like that of sea water and is in keeping with the abundances of elements on Earth, so that we are all, in the final analysis, the remnants of stars that exploded eons ago.

Supernova explosions in our Galaxy are now exceedingly rare occurrences, they happen with a frequency of one or two in a century. During the last 1000 years chronicles have only recorded four of them: in 1006, in 1054, in 1572 (the Tycho

Brahe star), and in 1604 (the Johan Kepler star). They flared up to become brighter than the planet Venus and were visible even in daytime. (The last two flare-ups shattered the medieval dogma concerning the invariability of the heavens and in no small degree motivated the invention of the telescope in 1608.) But during the first billion years it was such explosions that shaped the present picture of the Universe.

Our Sun is a third-generation star. It was formed out of cosmic dust that had already passed through space crucibles twice, about 5 billion years ago. In 5 to 7 billion years the Sun will run out of its stock of hydrogen and will reach the stage of helium burning — and then its days will be numbered. The outer shell of the Sun around the helium core will start expanding fast, it will reach the Earth's orbit and turn it into an incandescent waste planet. The Sun will reach the state of a red giant, just like Aldebaran or Betelgeuse. This phase will not last long: in a mere 10,000 years the solar shell will scatter and the Sun itself will turn into a white dwarf, just like Sirius B, the satellite of Sirius A, the brightest star in our skies (which by that time will most likely have exploded). But it is hardly likely that any of the terrestrial inhabitants will witness this apocalypse.

The Universe is still expanding and cooling even now. This conclusion follows from the equations of Einstein's general theory of relativity. But when in 1922 the Russian physicist Alexander Alexandrovich Friedmann (1888-1925) obtained solutions of these equations suggesting that the Universe was expanding, Einstein was opposed to that inference and only much later recognized that he had been wrong. Seven years later, in 1929, the American astronomer Edwin Hubble (1889-1953) found the "recession of galaxies" and even measured its rate. It appeared that distant galaxies recede the faster the farther away they lie from us — in full accord with the consequences of the Big Bang. At the boundaries of the visible Universe (10-15 billion light-years) galaxies recede with velocities close to the velocity of light.

A relic of the Big Bang epoch, which took place about 15 billion years ago, is the so-called *background radiation* discovered by Arno Penzias and Robert Wilson of the Bell Telephone Laboratories in 1965 (1978 Nobel Prize for Physics). This radiation in 0.01 second after the Big Bang had a temperature of 10^{11} K and a million years later had cooled down to 3000 K. The background radiation now uniformly fills the entire Universe (about 500 quanta per cubic centimetre), its average temperature is 2.82 K, and its spectrum is described by that famous formula of Planck's for black-body radiation with which the story of quantum mechanics begins and with which we started this book.

The circle has closed. The ancient Chinese said, "Man's roots ascend to his forefathers, but the roots of everything that exists ascend to heaven."

The Big Bang hypothesis was put forward by George Gamow in 1948 — twenty years after he had published his theory of alpha decay as a consequence of Friedmann's theory and Hubble's experiment. It is now generally referred to as "the theory of the hot Universe", or more often as the "standard model" of the Early Universe. These deliberately prosaic terms do not accord with the magnificence and grandeur of the subject at hand — just imagine, we are discussing the beginnings of the world and, whatever the desire to be scientifically precise and objective, it is hard to avoid some feeling of unreality. A single human life (less than one hundred years), the entire history of science (300-400 years), the history of civilization (10 millenia), the emergence of man (a million years), and even the genesis of the Solar System (about 5 billion years) are all just insignificant occurrences in the life of the Universe.

But why should we believe in this, nearly Biblical, picture of the Cosmogenesis? The answer is simple, although not all that convincing: for the same reason that we believe in the picture of the atom produced by our imagination. So far all the astronomical observations agree with the predictions of the "standard model".

"Truth is more astounding than any fantasy," and, to use the words of Faraday, "nothing is so beautiful as the truth". The truth about stars and the Universe appear to be more amazing than all the poetic figments of the imagination about them. And this truth, while elevating our spirit, suppresses our imagination. We know now that starlight, which for centuries has been a symbol of peace and serenity, harbours in itself memories about deaths and births of atoms, about magnetic storms, and cosmic explosions. And this knowledge is not without consequences for us.

The fervour of scientific search and the inevitable concomitant trifles not infrequently overshadow the grandeur of discoveries. And yet, when alone with the star-filled abyss, we cannot

help feeling some primeval dread, which is very like the feeling of man standing on the brink of a precipice.

"Astronomy is a happy science, it has no need for decorations," said François Arago. Yes, astronomy is one of the most ancient sciences, but the real laws of the heavens we learned only quite recently, since quantum physics was created. We can only hope to gain an insight into the architecture of the Universe by first probing into the structure of the atom and the nucleus.

We live in an amazing time: only half a century ago we found out our address at the edge of our Galaxy. We are witnessing the greatest revolution in astronomy since the days of Copernicus and Galileo, and astrophysics, a new science that is currently replacing astronomy, can change the world outlook of man even more than the discovery of atomic energy.

The Sun, Life, and Chlorophyll

This was the name given to his book by Kliment Arkadyevich Timiryazev (1843-1920), an eminent Russian scientist, who was once struck by the miracle of photosynthesis and devoted his life to studying it. "Hardly any other process that goes on on the surface of the Earth is worthy of universal attention to such a degree as that as yet unravelled process which occurs in a green leaf when sunlight shines on it", he wrote in one of his papers.

What feeds plants? What are leaves for? And why are they green? — These and similar "children's" questions have been asked since time immemorial. But answers to them came only in the last century, after an international constellation of men of science, notably the Flemish physician and chemist Jan Baptist van Helmont (1579-1644), the English botanist Stephen Hales (1677-1761), the English chemist Joseph Priestley (1733-1804), the Dutch physician Jan Ingenhousz (1730-1799), the Swiss naturalists Jean Senebier (1742-1809) and Nicolas de Saussure (1767-1845), had established that sunlight causes in green leaves the transformation of carbon dioxide and water into sugar, starch, and wood fibre with the release of oxygen.

Man and the entire animal kingdom on Earth are in all respects dependent on this process: we breathe atmospheric air, eat bread baked from cereals, drink milk produced on pastures. But just as we are generally unconscious of the air we breathe, so we rarely trouble ourselves with the thought of the cosmic role of plants, for they are the only living things on Earth capable of catching the energy of solar radiation and turning it into the chemical energy of organic compounds needed to support the life of animals and man.

In centuries past people seem to have been more fascinated by this. "I see my blood form in a spike of wheat... and wood in winter gives back the heat, fire and light it has stolen from the sun," wrote Senebier in 1791. His ideas were echoed in 1845 by Mayer: "Nature undertook to intercept in flight the light that comes to Earth and to turn this most fluid of forces into solid form."

In 1817 the pharmacists Pierre Joseph Peltier (1788-1842) and Joseph Bieneme Cavantu (1795-1877) of Paris isolated from leaves a substance, the "green blood of plants", and called it *chlorophyll*. For the first time green bubbles of this substance were first observed by Antonie van Leeuwenhoek (1632-1723), the inventor of the microscope, back in the late seventeenth century, but it was not until well into the nineteenth century that it became clear that chlorophyll is the main link in the complex sequence of transformations of water and carbon dioxide into starch.

In 1906 the Russian botanist Mikhail Semenovich Tsvet (1872-1919), the inventor of chromatography, found that there are at least two forms of chlorophyll. In 1913 the German biochemist Richard Martin Willstätter (1872-1942) determined their chemical compositions: blue-greenish chlorophyll *a* consists of 137 atoms ($C_{55}H_{72}N_4O_5Mg$), and a yellow-greenish one, or *b*, consists of 136 atoms ($C_{55}H_{70}N_4O_6Mg$).

But only in 1940 did Hans Fischer (1881-1945) succeed in establishing the structural formula of chlorophyll, i.e., the sequence in which its atoms are interconnected. It turned out that this structure was similar to that of haem, the core of haemoglobin in the blood of all animals. Only instead of an iron atom, responsible for the red colour of the blood, the chlorophyll molecule has at its centre a magnesium atom, which colours it green. (Thus the metaphor "the green blood of plants" suddenly appeared to be a strict scientific statement. No wonder then that it was Hans Fischer who in 1929 deciphered the structure of haem and was awarded the Nobel Prize for Chemistry in 1930.)

Another 20 years passed and in 1960 the American biochemist Robert Woodward (b. 1917) synthesized chlorophyll. (He also in 1962 synthesized tetracycline; 1965 Nobel Prize for Chemistry.) But even these advances have not revealed all the details, although the general outlines of these complex phenomena have been established quite reliably now — and the science of quanta has been instrumental in this undertaking.

The chemical aspect of photosynthesis is extremely simple: a water molecule (H_2O) combines with a carbon dioxide molecule (CO_2) giving off an oxygen molecule (O_2) and "forming a building block" CH_2O for many organic compounds (for example, glucose $C_6H_{12}O_6$ or $(CH_2O)_6$ consists of six blocks). That is:

$$CO_2 + H_2O \rightarrow CH_2O + O_2 - 5\,eV.$$

This atomic restructuring requires much energy: 3.32 eV to break the bonds between the hydrogen and oxygen in the water molecule, and 1.68 eV more to remove the oxygen atom from CO_2. This atom then combines with another oxygen atom from the molecule H_2O to yield the molecule O_2. The required energy is taken by the green leaf from the flux of sunlight quanta.

Each chemical bond is formed by a pair of

electrons. Therefore, when two hydrogen-oxygen bonds are broken four electrons have to be shifted. It was found that for this purpose at least eight quanta of red light are needed, i.e., two quanta per electron. Therefore, the detailed photosynthesis equation has the form

$$CO_2 + H_2O + 8h\nu \rightarrow CH_2O + O_2 .$$

A quantum of red light with a wavelength of 7000 angstroms has an energy of 1.8 eV, and the total energy of eight quanta is 14.4 eV. One third of this energy is stored as the energy of chemical bonds in the glucose molecule.

When we drink sweet tea and breathe, the oxygen molecules picked up by haemoglobin combine in the presence of enzymes with glucose molecules in the reverse reaction

$$CH_2O + O_2 \rightarrow H_2O + CO_2 ,$$

liberating the energy of sunlight stored by chlorophyll, which in the final analysis supports our life. (To use the words of Hermann Helmholtz, each of us "has a right, along with the Chinese Emperor himself, to call himself a son of the sun".)

The simplicity of the photosynthesis equation is misleading: this is not just another reaction but, rather, a complex biochemical process, which incorporates several stages and dozens of various reactions.

In green leaves chlorophyll molecules (which are 10^{-7} centimetre across) are packed into special structures, *chloroplasts,* which are essentially flakes 10^{-3} centimetre across and 10^{-4} centimetre thick. These structures are covered with a shell and are rather involved inside: suffice it to say that they include up to ten various forms of chlorophyll and more than 200 other compounds.

In chloroplasts chlorophyll molecules are combined into cells (with about 300 molecules in each) together with other pigments, whose purpose is to collect light and communicate its energy on to the reaction centre of the cell. The structure of this centre has not yet been clearly understood, but it is assumed to consist of a pair of chlorophyll *a* molecules, which in some manner are fixed to pigment molecules and which absorb red light with a wavelength of 700 nanometres = 7×10^{-5} centimetre. The energy of these quanta (1.8 eV) is sufficient to remove an electron from chlorophyll *a* and to move it through a chain of intermediate compounds to a place where the carbon of CO_2 is combined with the protons of the disintegrated molecule H_2O. In one second the reaction centre (it is called the P_{700} centre) can "process" up to fifty light quanta, i.e., provide for the synthesis of one molecule of glucose and the release of six molecules of oxygen.

There are actually two types of reaction centres: photosystem I and photosystem II. In photosystem I (reaction centre P_{700}), when an electron is removed from chlorophyll *a*, intermediate unstable compounds are synthesized that store the energy of the quanta absorbed by chlorophyll *a*. (Among these compounds stands out the fairly complex adenosine triphosphate (ATP), or $C_{10}H_{16}O_{13}N_5P_3$, which is a universal energy accumulator in all living things.)

Photosystem II includes a reaction centre, P_{680}, which absorbs red light with a wavelength of 680 nanometres and uses the light energy to remove electrons from some system *S*, which is assumed to be a protein complex that contains a manganese (Mn) atom. By successively giving away four electrons it becomes a catalyst in whose presence water molecules split into hydrogen and oxygen.

The two steps of photosynthesis — the formation of ATP and the splitting of H_2O — are fast (about 10^{-9} second) and they take place only under light. Then follows a fairly long stage (about 0.05 second) that does not need light. It includes about twenty reactions, the so-called Calvin cycle, in which protons, using the energy accumulated in ATP, through a chain of intermediate complexes add themselves to the carbon of carbon dioxide to form the structural unit CH_2O of any kind of wood. This stage was investigated in much detail fairly recently, in the years 1946-1951, using ^{14}C techniques in the laboratory of Melvin Calvin (b. 1911) (the 1961 Nobel Prize for Chemistry). But the two-stage nature of the photosynthesis process was only proved in 1958.

Photosynthesis studies go on. The problems dealt with grow in complexity, experimental techniques grow in sophistication, and the body of accumulated evidence grows in volume. But, as in past millenia, from sunrise to sunset plants continue their silent perpetual work — they capture sunlight and lay it up in store. But that was not always so: photosynthesis on Earth is a fairly recent development; it emerged in the process of evolution of plants nearly a billion years ago, when the terrestrial atmosphere contained less than 1 per cent oxygen — it was all nitrogen and carbon dioxide. This was a decisive turn in evolution that changed the face of the planet: the simplest blue-green algae started to transform carbon dioxide into oxygen, with the result that the planet acquired an ozone layer, or ozonosphere, which even now shields all living matter on Earth from murderous exposure to ultraviolet radiation, and with this protection life ventured from the oceans on to land to give rise to land creatures and man, who now pay back their debt to plants by supplying them with carbon dioxide.

In an hour 1 square metre of foliage takes up 6-8 grammes (3-4 litres) of CO_2 from the air and gives off the same volume of oxygen. Man consumes about 500 litres of oxygen per day (the output of three mature trees) and returns the same volume of carbon dioxide to plants. All the carbon dioxide in the atmosphere passes through plants within 300 years, and all the oxygen through animals within 2000 years.

Red light, which alone is responsible for photosynthesis, only makes up 2 per cent of the total solar radiation flux and then only 30 per cent of that tiny fraction, i.e., about 0.5 per cent of the total flux, is absorbed by plants. Not much, perhaps, but this narrow path is the only link between the realm of inanimate nature and the world of living creatures in which the energy of thermonuclear solar furnaces, after an infinite variety of transformations, reaches the nerve cells of the human brain, capable of grasping all this.

Quantumalia

The crisis of today is the joke of tomorrow.

H.G. Wells

Life under the Sun

Worship of the Sun is the most ancient of religions and, if it makes sense to speak thus about a faith, the most understandable among them. "The Sun... is an inexhaustible source of physical force — that continuously wound-up spring which sustains in motion the mechanism of all the activities on Earth," wrote Robert Mayer in 1845.

The Earth's radius is 6350 kilometres, and so if viewed from the Sun, i.e., from a distance of 150 million kilometres, it looks like a two-pence coin from 100 metres. Out of 4.2 million tonnes of photons radiated by the Sun every second, the Earth receives only 0.45×10^{-9} part of it, or 1.85 kilogrammes. In the cold waste of the cosmos it is these less than 2 kilogrammes of photons per second that keep the oasis of our planet warm and green. Owing to them rivers flow, winds blow, forests rustle, and the human race flourishes.

Confidence is a plant of slow growth.

New Testament

Knowledge without sense is two-fold folly.

Spanish proverb

A. Friedmann

Two kilogrammes of photons is not that small: from Einstein's formula $E = mc^2$ their energy is 1.7×10^{17} joules, 20,000 times the power of the world's power industry (about 10^{13} watts). About a half of that energy (0.8×10^{17} watts) reaches the terrestrial surface, which is 5×10^{14} square metres in area, i.e., the average power of the solar radiation at ground level is 160 watts per square metre.

The bulk of this energy (99.9 per cent) is absorbed by soil, and goes into the evaporation of water, causing winds, thunderstorms, and all that we loosely call weather. And only 0.1 per cent of the radiant energy of the Sun (around 10^{14} watts) is captured by plants through photosynthesis of organic substances from carbon dioxide and water. It is this energy that supports all the living things on Earth, from bacteria to animals and man, since the life of animals is essentially a process opposite to photosynthesis, i.e., the decomposition of organic substances into carbon dioxide and water.

The primary annual output of photosynthesis is 10^{14} watts, or 10^{11} tonnes of dry organic matter — that is all that man can count on in his long-term plans and predictions. These figures cannot be drastically increased, since photosynthesis needs fresh water, and now 60 per cent of the world's reserves of fresh water are involved in the turnover of organic matter.

About 10 per cent of the photosynthesis energy, or 10^{13} watts, falls on arable land, meadows, and pastures, and about one half of this energy is consumed by man for his needs.

The last figure might be worked out differently, by recalling that to lead a normal sort of life a human being must every day consume food containing about 3000 kilocalories, i.e., about 1.26×10^7 joules of energy. In twenty-four hours there are 8.6×10^4 seconds; therefore the average power of the life process in man is $1.26 \times 10^7 / 8.6 \times 10^4 \approx 140$ watts — less than the power of a burning match. To feed the world population of 4.5 billion requires $140 \times 4.5 \times 10^9 = 0.63 \times 10^{12}$ watts. Given the efficiency of land use of about 13 per cent, we obtain exactly 5×10^{12} watts.

This is a sizable share (about 5 per cent) of the total photosynthesis output. Man consumes about as much as this in the form of wood, thus bringing the total up to 10^{13} watts, i.e., 10 per cent of the primary production. And if we made allowance for the fact that meadows and arable lands are one-third as productive as the forests they have replaced, the share of the photosynthetic production consumed by man will increase up to 17 per cent. Humanity derives about the same amount from fossil fuels, by burning coal, oil, and gas. Accordingly, the human race, whose total biological mass is under 2×10^8 tonnes (5×10^{-14} part of the Earth's mass), consumes annually 2×10^{10} tonnes of organic matter — 100 times its mass. This is a great deal and implies that man — only one of the many millions of biological species — during the last century has turned into a decisive factor in the further evolution of life on Earth. Man is the master of nature — we have heard these words from very early days, but we seem to overlook the fact that power presupposes responsibility.

Essentially all the photosynthesis production (10^{11} tonnes of dry organic matter per year) is again decomposed by living organisms into carbon dioxide and water. Only a tiny fraction (10^{-4} part or 10^7

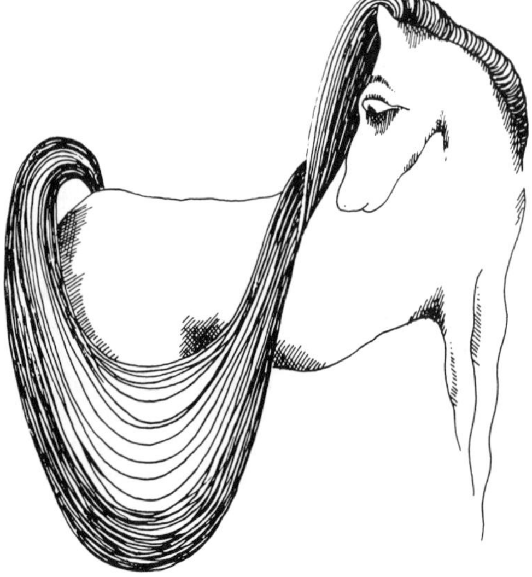

A journey of a thousand miles starts with a single step.

Chinese proverb

tonnes per year) remains unused and stored. It follows that during the previous 300 million years — the period the Earth has had lavish vegetation — 10^{15} tonnes of coal, oil, and gas have been stored in its bowels. The accessible resources are far more meagre: 10^{13} tonnes of coal, 3×10^{12} tonnes of oil and as much gas. About 5 per cent of these stores have already been burned since the American Edwin Drake drilled the first oil hole in 1859. Each year the world burns up to 3×10^9 tonnes of coal, 4×10^9 tonnes of oil and 3×10^9 tonnes of gas, i.e., 10^{10} tonnes of organic matter — about 0.1 per cent of the prospected reserves, 10 per cent of the annual production of plants and a thousand times faster than the rate at which plants make their stores.

Literary intellectuals at one pole — at the other scientists.... Between the two a gulf of mutual incomprehension.

Charles Snow

Each day the world's population increases by 200,000 people (two people per second) and during the last 300 years every thirty-five years the population has doubled. Simple calculations show that, should the demographic growth rate be sustained, in only 1500 years the total mass of people would exceed the mass of the Earth itself. The suggestion is obviously absurd and this implies that the years to come must see some qualitative changes in our way of life. All estimates of the Earth's resources indicate that the planet cannot support more than 10-12 billion people, three times the present population. But even then fossil fuels will last only another 300-500 years, and no longer. This is the most optimistic estimate and humanity, it seems, is not yet prepared to draw the correct conclusions from it. One thing is clear:

I sometimes think that God in creating man somewhat overestimated His ability.

Oscar Wilde

without nuclear energy — be it the energy of uranium fission or the energy of the synthesis of hydrogen isotopes — the human race has no long-term prospect of maintaining the current level of development. And our duty to generations to come is not to keep coal and oil reserves intact for them, but rather to bequeath them a higher culture, which will in time help them to find new paths and new energy sources. Otherwise, however economical they are with natural resources, sooner or later they will find themselves in a pre-industrial epoch, their numbers having declined ten- to twenty-fold.

In the science-fiction literature of recent years we somehow no longer encounter star ships, galaxy travellers, and cosmic Odysseys. The limited resources of our small Earth do not allow man to leave his blue planet. In the face of this simple truth space fantasies fade — there is hardly a more important task for mankind than to continue on Earth as a biological species.

The two big tricks of the twentieth century are: technology instead of grace, and information instead of virtue.

Ulysses Canrois

A Sun on Earth

"If it is intraatomic energy that supports the fire in the gigantic furnaces of stars, then we, it seems, find ourselves much closer to the realization of our hope to master that latent power for the prosperity of the human race, or its suicide," wrote Arthur Eddington in 1920. His predictions came true in a mere thirty-two years — having detonated a thermonuclear bomb man has got a Sun on Earth for the first time.

Unfortunately, only one half of Eddington's foreboding came true. Up until the present time we can only use the energy of nuclear fusion as a tool for suicide. It only remains for us to hope that in the course of time we will learn how to tame the thermonuclear explosion and to put it to work for the benefit of mankind.

On the issues of thermonuclear fusion, e.g., the magnetic confinement of a plasma in thermonuclear reactors, and the grandiose outlook for thermonuclear power we can now find many books and papers. Since the day when some hope emerged to "catch a sun in a magnetic bottle" forty years have elapsed. But the goal has not yet been reached, although thousands of scientific workers all over the world still hope some day to find a way to the fabulous valley of Eldorado with its inexhaustible resources of thermonuclear energy.

Quanta Around Us

The development of civilization makes work more effective, leisure more varied and sophisticated, and knowledge more extensive. In our system of knowledge quantum mechanics occupies a special place: no previous single physical theory has explained such a wide variety of phenomena and achieved such brilliant agreement with experiment.

When we watch television we never waste a thought on the fact that the TV tube relies for its operation on the tunnel effect, and that we see the TV image only because in our eye there is rhodopsin, a substance that like chlorophyll captures light quanta and transforms their energy into the energy of an impulse propagating along the optic nerve. When we customarily click our cameras, it hardly enters our heads that it is a sequence of photographic reactions that allows us to "freeze the moment". And in the frenzy of the morning haste, when we cast a glance at our electronic watches, hardly anyone remembers that their principle is based on quantum effects.

No aspect of a substance — colour, hardness, melting point, magnetic and electric properties — can be understood and visualized without the quantum theory. Quantum chemistry, without any additional hypotheses, has given us an insight into the nature of the chemical bonds in atoms and molecules, the quantum theory of condensed matter has explained superconductivity and the behaviour of semiconductors, without which there would be no modern television, air liners and space flights. Invariably, when we wish to obtain an understanding of a truly profound fact of nature — from nuclear fission to the workings of cells — we are bound to slip into the language of quantum pictographs. Who knows, perhaps this is exactly the native language of Nature, a knowledge of which is as indispensable to modern man as was a knowledge of the "language" of wild animals to his prehistoric ancestors.

4. Reflections

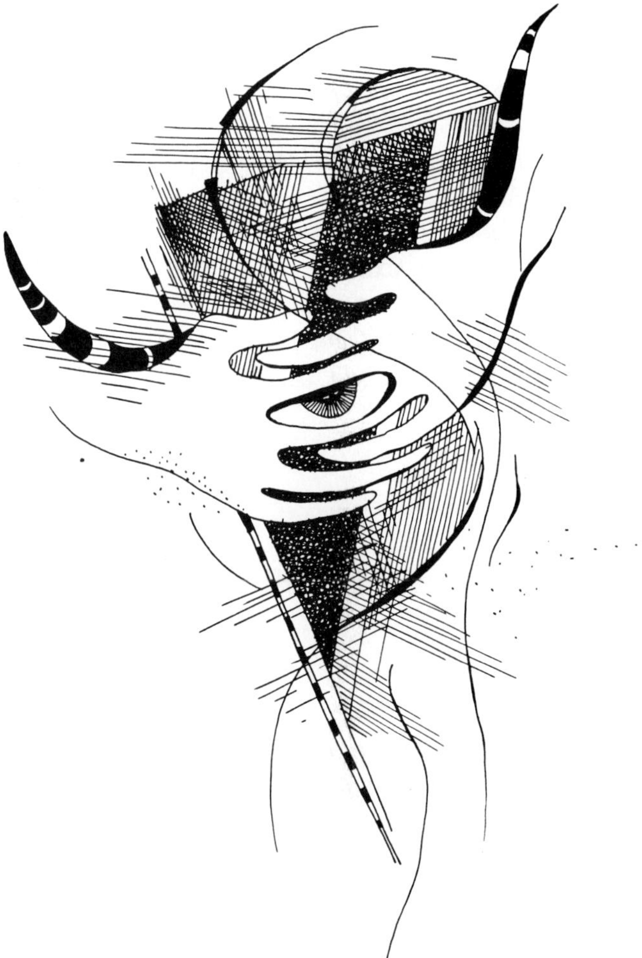

Between the idea
And the reality
Between the motion
And the act
Falls the Shadow...
Between the conception
And the creation
Between emotion
And the response
Falls the Shadow...
Between the desire
And the spasm
Between the potency
And the existence
Between the essence
And the descent
Falls the Shadow...

T.S. Eliot

Study without reflection is a waste of time,
reflection without study is dangerous.

Confucius

Basiliscus, Greek

Chapter Nineteen

Inception of the Scientific Method
Essence of the Scientific Method and its Development
Truth and Completeness of the Scientific Picture of the World
Science and Humanity • Boundaries of the Scientific Method
Science and Art • Future of Science

Science is always wrong. It never solves a problem without raising ten more problems.

Bernard Shaw

"Humanity is bewitched by the irresistible advance of science, and art alone is capable of bringing it back to reality," George Bernard Shaw once said with his characteristic brilliance.

The powers of modern science impress even an experienced mind: science has split the atomic nucleus, landed man on the Moon, uncovered the laws of heredity... But in the atmosphere of universal admiration it is not always possible to understand the gist of the scientific method, perceive the origins of its might, and, even more, to see its boundaries.

Recently, in the heat of an argument, a physicist declared: "In principle, a knowledge of Coulomb's law and the equations of quantum mechanics alone is enough to describe even such a complex system as man." Such statements have been made before in science. A lever was all Archimedes asked for to move the Earth. Laplace undertook to predict the future of the world if furnished with the initial coordinates and momenta of all the particles in the Universe. Although such faith in the completeness and omnipotence of science is alluring, it does not pay to forget the warning of Friar Roger Bacon (1214-1294), which is just as true today as it was 700 years ago. He wrote: "If a man lived in this mortal vale even a thousand centuries, he would still never reach perfection in knowledge; he doesn't know the nature of a fly, but certain presumptuous scholars believe that philosophy has been completely developed."

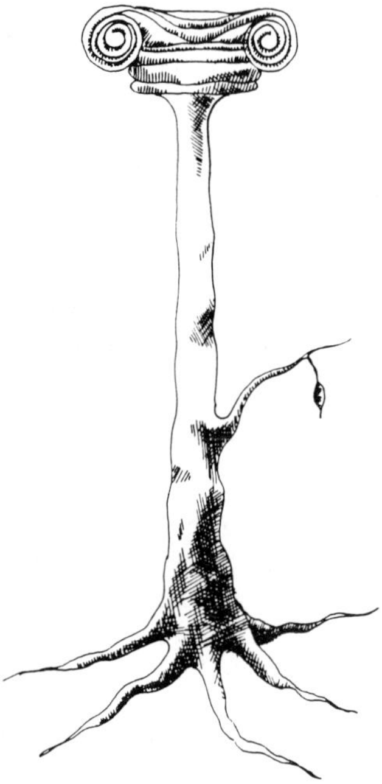

We witness how bedazzlement with the triumphs of science, ubiquitous even a quarter of a century ago, has been replaced with a sober analysis of scientific results. Quantum physics is a rare piece of human knowledge, one that is held in especial esteem in the twentieth century. We should remember, however, that this is just a small, although exceedingly important, part of general human culture, and it is only in this context that we can understand its true place and role in the evolution of modern civilization.

Inception of the Scientific Method

Only fragments of the knowledge of the ancients have reached us. But they are scattered, outlandish, and naive. The ancient Greeks are thought to have originated the science of today. They are our forerunners, both in time and in spirit. The Greeks invented *proof.* The idea did not emerge in either ancient Egypt, or Mesopotamia, or China — probably because in those lands there reigned tyranny and absolute subordination to the authorities. In those conditions the very thought of reasonable proof sounded seditious.

In Athens for the first time in the history of mankind a republic was established. We should not perhaps idealize it inasmuch as it thrived on the labour of slaves. One way or another, the conditions that evolved in ancient Greece made a free exchange of opinion possible, and this led to an unprecedented growth in science.

In the Middle Ages the inherently human need to seek a rational cognition of nature was suppressed by attempts to comprehend the place of man in the world within the context of various religious dogmas. For almost ten long centuries religion furnished exhaustive answers to all the questions of life, answers that were not to be criticised or discussed.

Euclid's works were translated into Latin and became known in Europe in the twelfth century. At that time, however, they were considered to be simply a set of witty rules to be learned by heart, so alien were they to the spirit of medieval Europe, which was accustomed to believe rather than to look for the roots of truth. But the scope of knowledge widened rapidly and it could no longer be harmonized with the trends of medieval thought.

The late Middle Ages are usually associated with the discovery of America in 1492. Some indicate an even more precise date — the 13th of December 1250 — the day when Frederick II Hohenstaufen, Emperor of the Holy Roman Empire, died in the castle of Florentino near Lucera. We should not take these dates too seriously, of course, but several such dates, when taken together, impart an unquestionable feeling of authenticity to the fundamental change that occurred in the minds of people at the turn of the fourteenth century. The period that followed is called the Renaissance. Without any apparent reason and driven only by some latent laws, Europe took only two centuries to revive all the achievements of ancient knowledge that had been neglected for over ten centuries and that would later on come to be known as scientific knowledge.

The Renaissance brought about a drastic change in the minds of people. Now instead of seeking their place in the world they tried to understand its workings without miracles and divine revelation. At first

The most beautiful thing we can experience is the mysterious. He to whom this emotion is a stranger, who can no longer wonder and stand rapt in awe, is as good as dead: his eyes are closed.

Albert Einstein

the mainstay of this upheaval was the aristocracy, but with the coming of printing it spread to include all sections of society. This period saw the transition from the faith of the Middle Ages to the knowledge of modern times. The church resisted these new trends in every possible way. It severely persecuted philosophers who maintained that there were things true from the viewpoint of philosophy but false from the viewpoint of faith "as if in contradiction to the truth in the Scriptures, there could be truth in the books of the heathens".

If we put aside the political passions of the time in which Galileo lived, it becomes clear that he was tried and condemned not only for his sympathy with the Copernican theory. The same ideas had been advocated a century earlier by Cardinal Nicholas of Cusa (1401-1464), who had gone unpunished. But the learned cardinal *asserted* these ideas referring to the pertinent authorities as was befitting to a true believer; Galileo went about *proving* the same idea, as is befitting to a scientist. He suggested that each person could check the truth only by relying on his experience and common sense. This was just what the church could not forgive him. But it soon became impossible to shore up the collapsing bulwark of faith, and so the liberated spirit began searching for new paths of development.

The principles of scientific learning and ways to realize them had been sought long before modern science came into being. As far back as the thirteenth century Roger Bacon wrote in his treatise *Opus tertium*: "There exists natural and imperfect experience that does not realize its power and is unaware of its procedures: it is used by craftsmen, not by scientists... Higher than all speculative knowledge and art is the skill of conducting experiments, and this science is the queen of sciences...

Philosophers should know that their science is helpless if they do not apply powerful mathematics to it... It is impossible to distinguish a sophism from a proof without checking a conclusion by means of an experiment and application."

In 1440 Cardinal Nicholas of Cusa wrote a book called *De docta ignoranta* ("On Learned Ignorance") in which he insisted that all knowledge of nature should be written in numbers and all experiments with nature should be conducted with scales.

The new views were slow in striking root. For example although Arabic numerals were put into general use as far back as the tenth century, even as late as the sixteenth century calculations were normally carried out using special counters rather than on paper. These counters were even less efficient than the abacus.

It is customary to begin the history of scientific method with Galileo and Newton. According to the same tradition Galileo is regarded as the father of experimental physics and Newton of theoretical physics. In their time, of course, physics was considered a united and indivisible science; indeed, there was not even physics as such for it was then called natural philosophy. But the division has a profound meaning. It enables us to bring out the two aspects of the scientific method, and it amounts to the division of science into experiment and mathematics as formulated by Roger Bacon.

Essence of the Scientific Method and its Development

Why explore the Universe? It is almost ironic that we should have to ask this question because it is almost as though we have to apologize for our highest attributes.

Norman Cousins

Man has the faculty of knowing things, i.e., the faculty of bringing to light links between them and establishing a chain of cause and effect. Different epochs saw different ways of realizing these human endowments — depending on what the epoch regarded as being of primary importance and what kind of answers it expected from its best representatives. Our age is an age of science. We have become so used to identifying the concepts "knowledge" with "science" that we cannot conceive of any other knowledge than scientific knowledge. What then are its essential and distinctive features?

The essence of the scientific method can be explained quite simply. The method enables knowledge about phenomena to be obtained that can be checked, stored and passed on to another person. It follows that science does not study phenomena in general, but only recurrent phenomena. It is mainly interested in seeking laws governing these phenomena. And at different times science accomplishes this task in different ways.

The ancient Greeks were keen observers of phenomena. Their observations helped them to gain insights into the harmony of nature relying on the power of their intellect and the perceptions accumulated in their memory. During the Renaissance it became evident that knowledge cannot be gained using the five senses alone, i.e., without instruments, which are just extensions and intensifiers of our sensory organs. Then the question arose: to what extent can we trust the readings of instruments and how can we store the information obtained using them?

The second problem was soon solved through the invention of printing and the systematic application of mathematics to natural sciences. It was harder by far to answer the first question. Generally speaking, even now it has no definitive answer, and the entire history of the scientific method is one of the continual extension and modification of this question.

It did not take long for scientists to understand that the readings of instruments can be trusted, as a rule; that is, that they represent something real in nature, something that exists independently of the instruments. (So astronomers were at last satisfied that sunspots are really spots on the sun and not defects in the telescope used to observe them.) During this golden age of experimental physics were accumulated all those facts that became the basis for the vigorous growth of technology at the turn of this century.

Those to whom everything is clear are unhappy people.

Louis Pasteur

But the volume of knowledge was snowballing and there came a time when people could no longer understand how they were to relate the numbers obtained using their instruments with real phenomena in nature. This period in the history of natural science is known as the crisis in physics.

The reasons behind the crisis were two-fold. On the one hand, the instruments were already widely detached from direct human sensations, and so intuition, deprived of the imagery which at all times formed its basis, no longer provided a vivid picture of things studied. Thereby the possibilities of graphic interpretation of experimental findings were exhausted. On the other hand, no scheme of logic was in existence that could be of help in achieving an orderly arrangement of scientific facts

and could yield such observable consequences that would satisfy common sense.

The crisis was overcome. Physicists continued to trust the readings of their instruments, but they devised new concepts and new treatments that enabled them to interpret these measurements in a new manner. (It was at this moment that theoretical physics came to the fore.) Quantum mechanics was instrumental in breaking down established concepts. It did not just endow us with a power to control the entirely new world of atomic phenomena. It did more than that. Quantum mechanics convinced us that the readings of instruments are not simply photographs of events in nature. The measurements only reflect and describe using numbers some aspects of them and only when combined with our views on nature do they acquire some meaning and significance. In the course of time our knowledge becomes more profound and enables us to predict ever finer details. This seems to be accepted now by nearly all physicists. But, like all people, they want to grasp something more: just how complete is the picture of the world sketched by physics? This question is a philosophical rather than a physical one. It arose many a time throughout the history of learning but was first clearly formulated in Plato's dialogues.

Plato likened scientists to prisoners chained in a cave with their backs to the opening. They see no light, only shadows moving on the opposite wall. He held that even under such conditions it was possible, by closely observing the motion of the shadows, to learn to foresee the behaviour of the bodies casting the shadows. But knowledge acquired in this manner is still infinitely distant from that gained by a freed prisoner when he comes out of the cave.

One of Plato's personages asks: "Don't you think that, recalling his earlier life, the wisdom and the inmates, he would deem his change a happy one, and would pity the others, the ones who remained in the cave?... Remembering, also, the homage and praise that had been rendered to each other and the rewards to those who keenly observed the goings-on and were the best at noticing what comes first, what comes after, and what comes simultaneously and used this knowledge to foresee what would come next — do you think he would desire the same and envy the people who were there considered respectable and influential?"

There is no objecting to Plato's argument. The world around us is really much more splendid that the one we can conceive of only on the basis of physical evidence. A person who was born blind can learn all the facts of optics, and yet he will never have the slightest idea about what light is, let alone the riot of colours in spring. When we enter the world of atomic phenomena we are very much like people who were born blind, since we are completely deprived of "quantum sight" and can only grope for our way in this unusual world.

What is complementary to the notion of truth? Answer: clarity.

Niels Bohr

Analogies like this can be readily multiplied, and each one of them teaches physicists to be more modest. Now we know that questions concerning the completeness of physical knowledge and the essence of phenomena are beyond the scope of physics and cannot be answered by physical means. Physics only studies the laws governing these phenomena. On this account, it follows exactly the "theory of shadows".

But how true is even this restricted knowledge of nature?

Truth and Completeness of the Scientific Picture of the World

Logic alone cannot give an answer here. We *believe* in science because it enables us correctly to predict things and is independent of the likes or dislikes of the investigator. We may question the structure of scientific images — they depend on the method of communicating our ideas. But we are certain, say, that all terrestrial and celestial bodies consist of the same elements taken in approximately the same proportions. Moreover, we are certain that the laws of nature are the same throughout the Universe, and so an atom of sodium will always radiate the same *D* line wherever it is, on Earth or Sirius. This is now agreed almost universally, and no one doubts the truth of *this* knowledge. Doubt only arises when on the basis of particular, although firmly established, facts we attempt to create a general and coherent picture of the world, one that agrees with the totality of evidence and the general nature of human reason. The most frequent among the questions that present themselves is how unique is the *form* of physical laws?

Increasing knowledge of science without a corresponding growth of religious wisdom only increases our fear of death.

Sarvepalli Radhakrishnan

There is no definite answer to this question. Those acquainted with the history of science know that there were times when two physical theories existed side by side, each considering itself to be true and each providing an equally good explanation of the phenomena known at that time. Fresh experiments favoured only one of the two theories or later in the development of scientific thought they merged together on the basis of more significant principles, as was the case with the corpuscular and wave theories of light, when quantum mechanics came on the scene.

Facts and concepts of science may seem to be accidental, if only because they have been established at random times by chance people and often in chance circumstances. But when taken together, they form a single regular system with so great a number of cross-links that you cannot touch a single link without affecting all the others. This system is continually changing under the pressure of new facts, becoming ever more accurate, but it never loses its integrity and distinctive completeness. The present array of scientific concepts is the product of a long evolution. Over the centuries old links have been replaced by new, improved ones, and even truly revolutionary discoveries have always emerged "on the shoulders" of previous knowledge. In other words, science is a living, growing organism, not something frozen and dead. Although all the concepts of science are creations of human reason, they are nevertheless arbitrary to the same extent that the very existence of intelligent life in nature is arbitrary too.

Like a child from the womb,
Like a ghost from the tomb,
I arise and unbuild it again.

Percy Shelley

In one of Ray Bradbury's science-fiction stories the hero takes a trip on a time machine into the dim past. During his short visit he accidentally squashes a small butterfly. When he returns to the present time he does not recognize the world he departed from. It turns out that his awkward and seemingly negligible intrusion into the course of biological evolution has completely altered its end results. This example is evidently no more than a spectacular extreme, one that is only possible in science fiction. Admittedly, all things are interrelated in nature — but not with such rigid causality, which borders on determinism. The links are more resourceful and flexible, just like the statistical causality of quantum mechanics.

There is nothing either good or bad but thinking makes it so.

William Shakespeare

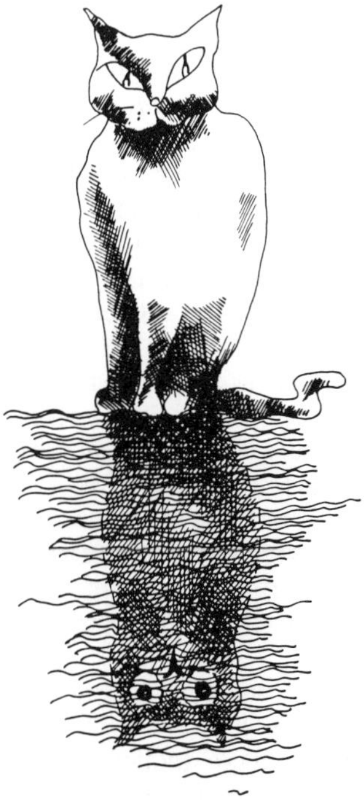

The evolution of the system of scientific concepts is as uncontrolled, but regular, a process as the evolution of the animal world. We can imagine it as deviating in particulars and even wonder at its strange whims, but we cannot imagine it as being entirely different. We do not know how the first organism and the first concept emerged, nor do we know what would have happened if they had been quite different. But we do know that each step in evolution was conditioned by all the previous ones. Therefore, we can readily imagine, say, a horse with a tiger's paws or an atom the shape of a doughnut, but to think up an entirely different animal world and a system of scientific concepts is beyond our powers. Both the process of biological evolution and the formation of scientific concepts obey their respective intrinsic laws, which we cannot change and of which we do not yet have complete knowledge.

We are born into a world of species that have already been formed and of concepts that have already been established. We can breed a new strain of horses or replace one concept with another one that will agree better with scientific truth. But the question of whether or not the whole system of human knowledge is true is beyond our understanding. Moreover, the question makes no sense. Science has been created by man for man, and the entire system of its concepts has been devised to conform to the nature of the human mind. The ultimate purpose of all concepts is to explain and predict things that affect our senses or their extensions — instruments.

Almost certainly somewhere at the far reaches of the Universe there exist intelligent beings; their sensory organs will differ from ours and their minds will be structured differently. Their system of concepts may also differ markedly from ours. But even if we were able to understand their system sufficiently well to compare it with ours, we would be unable to deduce from this comparison that their system would be false. On the contrary, it is always true if it provides its sensory organs with correct predictions. Our scientific knowledge of the world is no more than shadows that are produced by the light of our mind. Just as an object casts various shadows depending on the angle from which it is illuminated, so a system of scientific knowledge produced by the intelligent life of another planet may differ from ours. It may well be that some day in the future we will have an opportunity to compare these "conscious shadows" and, like Plato's prisoner after escaping from the cave, use them to restore truth in all its completeness and magnificence, just as a skilled craftsman makes a three-dimensional machine component from several plane projections on a blueprint. But meanwhile we must refine our science of today. However imperfect, it is the only method available so far for penetrating deep into observed phenomena.

The world is objective and so it exists independently of our consciousness. The world is indifferent to how we, a part of it, conceive of the internal mechanism of its external manifestations. This is of importance only to us ourselves. What is important is how far we can advance along this path. And down to what level can we keep refining our conceptions about the causes of things we observe. And so instead of the problem of physical reality, we should concern ourselves with the limitations of the scientific method. This question has become quite urgent since quantum mechanics assumed prominence.

Science and Humanity

The most dangerous thing in the world is to leap a chasm in two jumps.

Lloyd George

Science, as we understand it now, has existed for no more than 400 years, and the word "scientist", in its present meaning, appeared for the first time as recently as 1840, in the works of William Whewell. Within this extremely short period man has colonized the Earth, tamed the ocean, leaned to fly and to talk with friends on the opposite side of a planet, stepped on the Moon and glanced from there at the Earth. Science has totally changed the way of life of civilized peoples, their world outlook, their thinking, and even their moral attitudes. The main feature of the new philosophy of life is an awareness of continuous motion in the world, and, as a result, a striving to understand the surrounding world so as to respond appropriately to its changes. Modern man regards with scepticism the principles that purport to be established once and for all; he just does not believe that any knowledge can be final and so he is perpetually in quest of optimal solutions. His insatiable thirst for knowledge, whetted especially during the Renaissance, has not been satisfied yet.

The scientific method has transformed the world we live in. It has populated this world with machines, it has given sufficient food to people, and protected them from diseases. Triumphs of the scientific method have engendered and strengthened a new faith — a faith in science. The change wrought by the scientific method in the minds of people is only comparable with such great religious upheavals as Buddhism, Christianity, and Islam. Science has taken the place of religion both in form and in essence. It is now expected to give answers to all the questions of one's life; its verdicts are regarded as final; its models on which to base one's life are sought for among its votaries, and its followers grow in number faster than did the army of Buddhist monks in the East in ancient times. The countries of the West that adopted this new religion have left the formerly flourishing countries of the East far behind. This all became possible due to a simple discovery: many things in nature can be described using numbers with equations establishing the relationships between the numbers. Like any consistent method, the scientific method is not without its problems; it has its range of applicability and its constraints.

In ancient times it was not considered an unworthy occupation for a man to sit on the seashore and watch the sun describe its great circle across the heavens. Much has changed since then. Inductive sciences came to replace pure speculation and have undertaken to "verify harmony by algebra". Science now took its stand on the firm ground of experiment but lost its aspects of serene wisdom and leisurely contemplation. We can lament this but we can hardly change it.

In its earliest days science was an occupation of lone devotees, who undertook it at their own peril. For a long time the results of science were not considered obligatory for everybody. Even as late as the middle of the last century Faraday made an appeal to have science recognized as an element of general education. In our day science has acquired a mass character, and scientific work has become a most ordinary, and frequently humdrum, pursuit. Science has become a productive force, not just a means of cognition. At the same time it has given rise to human passions far less noble than those to which it owes its origins.

Be that as it may, in an age of science it is inconceivable to deny oneself the results of science merely from moral considerations. In our striving for cleanliness we should not induce sterility. So far science has been doing its work well. It builds machines, feeds mankind, produces energy, fights diseases. This, of course, does not release scientists from their moral responsibilities for their, sometimes deadly, discoveries.

It is well known that biological evolution always yields some mutations, some of which rapidly take hold and supplant less fitting traits. But some of the mutations remain latent and they only become apparent when a change in external conditions threatens to destroy the biological species. Evolution has not ceased even now; it has only changed in form. For thousands of years the body of man has remained almost the same, but his mind has changed beyond recognition and irreversibly so. Perhaps, science is exactly that source of new ideas that will save humankind some day from impending catastrophe.

Science is a convention between people, one that is fruitful enough to become universal. On the basis of this convention a kind of collective intellect has grown up. It may not be immortal, although it seems to be comparatively long-lived. How long it will last and where the boundaries of the scientific method lie, we do not yet know. But the fact that such boundaries exist is beyond all question.

Boundaries of the Scientific Method

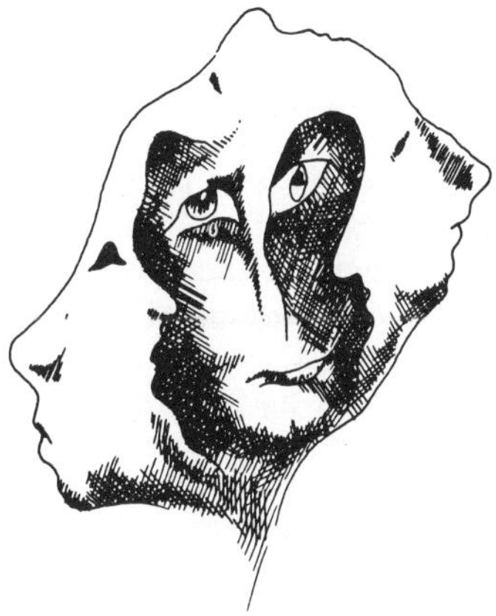

When people will be taught not that they should think, but rather how they should think, all the troubles will vanish.

Georg Lichtenberg

Man has always been preoccupied with the "eternal questions" of life and death, good and evil, God and eternity, the ultimate aim of life and the place of man in the Universe. Religion failed to give all the answers; all it managed was to allay for a time the longing for the answers, and to provide mankind with a brief consolation through forgetting the complications of life on earth.

Science is not equipped to answer questions about the meaning of life; it has more modest tasks. Dazzled by the successes of the exact sciences, we often forget this and overlook one simple possibility, namely that to future generations our rationalism and belief in science may be just as absurd and incomprehensible as the ceremonial rites of ancient Egyptian priests are to us. Only cognition itself, not its historical forms, is without limit.

Science is only capable of knowing those phenomena whose properties can be assessed by numbers. The performance of a hypnotist cannot be described by mathematical formulas, and yet its results are doubtless and reproducible. The accomplishments of Hindu yogis are a fact; they have been checked many times, but they cannot, however, become objects of exact science because they do not lend themselves to a quantitative description by numbers and formulas.

By no means can all things be broken down into elements and described in mathematical terms. We should not grieve over this, however. This simply means that our world is more varied and complex than the image provided by science. For science a genius and a murderer, for example, are indistinguishable, because they can be proved rigourously to be composed of identical atoms. And no science can explain the joy we

A strong conviction that something must be done is the parent of many bad measures.

Daniel Webster

experience when viewing the smile of a child. We should always remember this in order not to become steeped in "learned ignorance".

With such discourse as a background, quantum mechanics should seem to be an entirely simple science. Indeed, we know so much about the hydrogen atom that we can predict all its observable properties. It is considerably more difficult, but still possible, to calculate the behaviour of a hydrogen molecule. We are unable, however, to predict the properties of a protein molecule. There are not all that many proteins, but each person, unique as he is, is composed of them.

In short, science is beneficial and even necessary, but necessity should not be made into a virtue, and there is no need to subordinate everything to science simply because we cannot do without it for the time being.

Science and Art

The limitations of science are most evident in attempts to use scientific methods to unveil the secrets of art. Science "knows everything" about the grand piano: the number, quality, and length of its strings; the species of wood used; the composition of the glue, and the finest details of its design. Nevertheless, it is unable to explain what happens to this polished box when a virtuoso sits down to play. Perhaps this is even unnecessary. A person crying over a book does not usually concern himself with the means the author used to achieve this effect. He can, of course, then read a critical work, twice as thick, on the book that has so impressed him. All this, however, will resemble an autopsy, a thing necessary for specialists but extremely unpleasant for most people. Marcus Aurelius wrote that "to despise songs and dances, it is sufficient to decompose them into their component elements". But art is wise — through all the ages it has guarded the intangible truth of sensory perceptions from the persistent intrusions of probing science. Art has always been valued precisely for its capacity to "remind us of harmonies inaccessible to systematic analysis".

Nominally a great age of scientific inquiry, ours has actually become an age of superstition about the infallibility of science: of almost mystical faith in its non-mystical methods.

Louis Kronenberger

Anyone can understand the construction of a nuclear reactor even if he has never seen one. But it is absolutely impossible to explain to a person what rapture is if he has never been enchanted.

The might of science lies in its universality. Its laws are free from the arbitrariness of people; it only represents their collective experience, independent of age, nationality, or frame of mind. The secret of art is its inimitability. The power of its influence depends on the whole body of the previous experience of a person, on the wealth of his associations, on elusive changes in his mood, on a chance glance, word, or touch — on all that constitutes the individuality, the beauty of the transient, and the power of the inimitable.

The highest achievement for a scientist is to have his findings confirmed, i.e., repeated by another scientist. On the other hand, sameness kills art, and so a great tragic actor "dies" on the stage in a new way each night.

Cases are known of symphonies composed by persons without even the rudiments of formal musical education. These works may have been unusual but should be accepted as they are if at least a small section of the public liked them. In science such a situation is inconceivable. It has a criterion of truth and its language does not contain the words "like" and "dislike".

In science truths are proved and phenomena are explained. In art they are interpreted. Logical reasoning is alien to art, which substitutes the spontaneous cogency of images for rigourous proofs.

As a rule, science can explain why this formula is good and why that theory is bad. Art can only show the fascination of music and the brilliance of a sonnet, never explaining anything completely.

Science started when people learned to single out simple regularities from the chaos of random facts. But art begins only when a combination of things that are simple and natural suddenly produces a miracle.

Science is thorough and unhurried; it keeps on solving its problems for years on end, and many of them are often passed over from generation to generation. It can afford this luxury because of an unambiguous method that has been devised for recording and storing the facts established by science. In art the intuitively precise world of images is fluid. (Great actors are sometimes called "heroes of the fleeting moment".) One keen but split-second perception, however, may awake in the heart of a person a response that will stay with him for years and may even alter the whole course of his life.

Then would I hail the fleeting moment
O stay — you are so fair!

was Faust's passionate longing that could only be fulfilled by the magic of art. It is this magic that after a lapse of many years can bring back with frightening clarity the nuances of remote thoughts and moods that defy any words.

Notwithstanding the seeming fragility of the ambiguity of artistic images, art is more durable and ancient than science. The Gilgamesh epic and Homer's poems stir us even now because they tell us something that is vital in man and that has remained unchanged for thousands of years. As for science, it has hardly had time to consolidate the new possibilities of research. It is almost impossible today to read books on physics written in the last century, so obsolete have they become and so much has the whole style of scientific thought changed since then. The importance of scientific works is, therefore, determined by their productivity, not their longevity. They have already done their bit, if they helped to promote science in their time.

Do you believe then that sciences would ever have arisen and become great if there had not beforehand been magicians, alchemists, astrologers, and wizards who thirsted and hungered after secret and forbidden powers?

Friedrich Nietzshe

We could go on searching for and finding endless shades of distinction between art and science. The benefit of such an exercise is doubtful, for the two human endeavours only differ in their ways of gaining knowledge of the surrounding world and human nature. Ancient Greeks did not distinguish between the two notions and called them by a single word τεχνε (techne), meaning "skill", "art", "craft", and "refinement" (hence "technology"). And the first laws of physics established by Pythagoras were laws of harmony.

Poets have long been searching for a "poesy of thought". And not simply poets — scientists, for their part, speak about "poetry in science". Both clans, it seems, are now eager to break down the age-old barriers between them and to forget their ancient feuds. There is no sense in

Truth is not as benevolent as its semblance is malicious.

La Rochefoucauld

arguing about which hand, right or left, is the more important, even though they develop and function differently.

Any actor understands that he cannot reach the acme of his art without first mastering the sciences of diction, mime, and gesture. And only then (provided he is talented, of course!) can he create something unique and wondrous quite unconsciously.

In exactly the same manner, a scientist, even though he has mastered the trade of a physicist, will create no real physics if he only trusts formulas and logic. All profound truths of science are paradoxical at birth and cannot be attained by only leaning on logic and experiment.

To cut a long story short, real art is impossible without the most rigourous science. Likewise, deep scientific revelations only in part belong to science, the other part lying in the domain of art. But there are always boundaries to the scientific analysis of art, and there is always a limit to grasping science by an impulse of inspiration.

There is an apparent complementarity in the methods utilized by art and science to know the world. Science relies routinely on the *analysis* of facts and search for cause-effect relations; it strives to "...find an eternal law in the marvellous transmutations of chance", endeavours to "...find a fixed pole in the endless train of phenomena". Art, on the other hand, is largely an unconscious *synthesis,* which finds among the same "transmutations of chance" the unique ones and among the same "endless trains of phenomena" infallibly selects only those that enable one to sense the harmony of the whole.

The world of human perceptions is infinitely diverse, although chaotic and coloured with personal emotions. Man has a way of putting his impressions in order and comparing them with those of others. To this end, he has invented science and created the arts. Art and science have thus had common beginnings. They are united by the feeling of wonder they evoke — how did this formula, this poem, this theory or this music come into existence? (The ancients said, "The beginning of knowledge is wonder.")

The creative aspect of all arts and sciences is the same. It is determined by one's intuitive capacity to group facts and impressions of the surrounding world so as to satisfy our emotional need for harmony, a feeling one experiences when out of the chaos of external impressions one has worked up something simple and consummate, e.g., a statue out of a block of marble, a poem out of a collection of words, or a formula out of numbers. This emotional satisfaction is also the first criterion of the truth of the product, which of course is to be tested later on — by experiments in science and by time in art.

"A scientist studies nature not because it is useful; rather he studies it because it is a source of pleasure for him, because nature is beautiful. If nature were not beautiful, it would not be worthy of the effort that goes into knowing it, and life would be not worthy of the effort it takes to live it."

These words belong to Henri Poincaré. Aesthetic perception of the logical beauty of science is inherent in some form or other in each true scientist. But perhaps nobody spoke about this better than Poincaré. "He loved science not only for the sake of science. For him it was a source of the spiritual joys and aesthetic delights of an artist who has mastered the art of couching beauty in real forms," commented the Russian translator of his famous books *Science and Method, Science and Hypothesis, The Value of Science, Mathematics and Science: Last Essays,* which were instrumental in deciding the scientific careers of Louis de Broglie, Frédéric Joliot-Curie, and many, many others.

Future of Science

I know not whence I came nor whither I go nor who I am.

Erwin Schrödinger

He who is only wise lives a sad life.

Voltaire

That is the best part of beauty which a picture cannot express.

Francis Bacon

Thinking of the future of science we normally imagine a world of machines, push buttons, and transparent domes — in a word, a world of *things* controlled by a man in a spotless white outfit. The same mistake is made by most people after a superficial acquaintance with quantum physics. As a rule, they are amazed by the concrete, tangible facts, such as the atomic bomb, nuclear-powered ice-breaker, or atomic power station. Few understand that all these items are but simple consequences of quantum mechanics. What really should amaze us is the wonderfully simple and harmonious system of the scientific ideas of quantum mechanics which made possible the ice-breakers, the atomic power stations, and, unfortunately, the bomb.

No one can speak of the future of science without the risk of lapsing into naivety or overstatement. It is easy to demonstrate the limitations of the scientific method in a field where it does not work, but it is impossible to predict its potentialities. The scientific mode of thinking is, no doubt, but one of the faculties of the human mind which has not yet been exhausted, however. It is quite probable that in the future man will discover in himself new resources for probing into the deep recesses of the world around him and will better understand his place in this world. But this new, more refined knowledge will, almost certainly, incorporate all the main achievements of science.

It is anybody's guess what this new knowledge will be like, for man can always do more than he thinks he can. It may well be that in time man will develop a new capacity for synthetic intuitive knowledge. It was this capacity that distinguished the wise men of ancient times, which has now become almost extinct, being suppressed by the successes of scientific analysis. Perhaps in the future intuitive power will, instead of a means of scientific foresight, become a tool of scientific proof. There is nothing incredible about this — we trust the power of the eyes of a diamond sorter, although they only differ from the eyes of an average person in that they are trained for the job. Maybe the time will come when we will likewise learn to train our intuition and to cultivate it in different people. If we succeed in this, all questions about the ambiguity of scientific concepts and in general the whole cumbersome apparatus of logic will become unnecessary. It is impossible to foresee the implications of such a revolution in thought.

Hypotheses on the future of science are plentiful. At one extreme of the spectrum is unbridled optimism; at the other, the most gloomy pessimism. Some believe that humankind will survive even when our Sun becomes extinct. Others foretell the imminent downfall of our civilization through incompetent employment of the forces it has called forth.

But whatever the path taken by mankind it will at all times recall our stormy and vigorous age of science with amazement, much as we now recall the Renaissance or the time of the sages of antiquity.

EPILOGUE

Sorcery helped the heroes of an old fairly-tale to learn to understand the language of birds and beasts. This made them more powerful, but at the same time more vulnerable, for now they became responsible for much of what they could have overlooked earlier just because they did not know.

When savage nomads branded their herds of horses with red-hot irons they did not know that the spectrum of radiation of the branding iron obeyed Planck's formula. Bathing their horses in a river they did not imagine that the water consisted of molecules that are isosceles triangles with a vertex angle of 109 degrees and 30 minutes. When they came out of the river, they never gave a thought to the fact that the tan on their skin was due to photons.

Today, after the lapse of thousands of years, everything in nature has remained as before. Each morning the sun rises in the east; the water in the river freezes at 0°C; and red-hot metal cools according to the eternal laws of thermodynamics.

But now we *know* about all these things. Perhaps our knowledge has not made us happier (for it is said, "In much wisdom is much grief"), but this knowledge is irreversible; it is an element of *culture,* which alone distinguishes us from wild shepherds. It is amazing and unexplainable how the reading of good books — a non-material process — can change the whole mentality of a person beyond recognition, and also his speech, his smile, the expression of his face and eyes, and even his gait and gestures. The totality of knowledge we call culture has so radically changed the way of life and the scale of values of civilized peoples that many are inclined to classify them as a different biological species from the one to which our ancestors, from whom we descended, belonged.

Now this huge body of knowledge threatens to crush mankind, which has produced it. One often hears utterances that science has reached a deadlock, that it has bogged down in details and lost its great ideals. All too many people repeat the following words of Thomas Stearns Eliot:

Where is the Life we have lost in living?
Where is the wisdom we have lost in knowledge?
Where is the knowledge we have lost in information?

This argument always impresses people who are weary or disillusioned. But no matter how hopeless things may seem, now and then a spark of talent will suddenly become a blaze, illuminating something very simple and important amidst the chaos of facts and opinions. And then people forget for a time their lamentations and stop bickering over details. They silently share the delight over the new revelation. Like any manifestation of perfect beauty, it is rare, striking, and it disarms one by its heady power.

I would hope that everyone who has taken the trouble to read this book to the end will share with me the joy and wonder that I experienced when I first entered the unusual world of quantum physics.

SUBJECT INDEX

NAME INDEX

Pages underlined contain largely biographical material.
Names marked with an asterisk occur so frequently that only a selection of references is given.*

Princeton Univ. Store, NJ
Sun 11 May 1997 ~ $36.00
+ 2.16 tax
= 38.16